MACROLICHENS OF THE PACIFIC NORTHWEST

Macrolichens

OF
THE

Pacific Northwest

THIRD EDITION

Bruce McCune
Linda Geiser

PHOTOGRAPHS BY

Sylvia and Stephen Sharnoff

Bruce McCune

Martin Hutten

Joe Di Meglio

Richard Droker

Peter Nelson

Jim Riley

DRAWINGS BY

Jason Hollinger

Alexander Mikulin

Oregon State University Press   Corvallis

Library of Congress Cataloging-in-Publication Data
Names: McCune, Bruce, author. | Geiser, Linda, author. | Sharnoff, Sylvia Duran,
    1944– photographer. | Sharnoff, Stephen, 1944– photographer. | McCune, Bruce,
    photographer. | Hutten, Martin, photographer. | Di Meglio, Joseph, photog-
    rapher. | Droker, Richard, photographer. | Nelson, Peter R. (Plant ecologist),
    photographer. | Riley, Jim, photographer. | Mikulin, Alexander G., illustrator.
Title: Macrolichens of the Pacific Northwest / Bruce McCune, Linda Geiser ; photo-
    graphs by Sylvia and Stephen Sharnoff, Bruce McCune, Martin Hutten, Joe Di
    Meglio, Richard Droker, Peter Nelson, Jim Riley ; drawings by Alexander Mikulin.
Description: Third edition. | Corvallis OR : Oregon State University
    Press, 2023. | Includes bibliographical references and index.
Identifiers: LCCN 2023030004 | ISBN 9780870712517
    (paperback) | ISBN 9780870712524 (ebook)
Subjects: LCSH: Lichens—Northwest, Pacific—Identification. | Lichens—
    Northwest, Pacific—Pictorial works. | Field guides.
Classification: LCC QK587.5.N67 M33 2023 | DDC 579.709795—dc23/eng/20230703
LC record available at https://lccn.loc.gov/2023030004

♾ This paper meets the requirements of ANSI/NISO Z39.48-1992
(Permanence of Paper).

 **Oregon State University**
OSU Press

Oregon State University Press
121 The Valley Library
Corvallis OR 97331-4501
541-737-3166 • fax 541-737-3170
www.osupress.oregonstate.edu

Contents

Preface to the Third Edition

Our understanding of the macrolichens of the Pacific Northwest has advanced tremendously in the 25 years since the first edition of this book. The Survey and Manage program of the US Forest Service and Bureau of Land Management stimulated some of these gains. This program came about as part of the fundamental changes in forest management practices by federal agencies, as implemented in the Northwest Forest Plan. It required extensive lichen surveys, resulting in a great increase in local lichenological expertise and information on the distribution and abundance of lichens. Extensive work by the Forest Inventory and Analysis program and Forest Service Air Resources Program produced tens of thousands of specimens and these were accessioned to the Oregon State University herbarium. Combined, these have greatly boosted our knowledge of regional responses to air pollution and climate.

Knowledge of lichens in the Pacific Northwest has also advanced through the efforts of lichenologists worldwide. These advances include refining species concepts; nomenclatural corrections; finding species that were previously known elsewhere, but not in the Pacific Northwest; and simply discovering new species that had not been recognized before. In many cases, these undescribed species lurked beneath rather broad, murky species concepts. This progress was made with traditional studies of morphology, anatomy, and chemistry, as well as DNA sequence data. We added 95 species to the third edition, many of these new to science since 2009. Names changed for many other species, based on taxonomic work. Keys were completely restructured for some genera, especially *Collema* and related genera, *Leptogium* and related genera, and *Stereocaulon*. This edition is also better illustrated, as we increased the total number of illustrations from 417 to 490. For each multispecies genus we added a new section, "ID tips."

Analysis of DNA of lichenized fungi is having a deep impact on lichen taxonomy, in both the organization of genera and recognition of species. For example, at the genus level, consider the split of *Melanelia* into several more natural genera, including *Melanelixia* and *Melanohalea* and, conversely, the lack of support for the narrow genera defined within *Cetraria*, including *Ahtiana*, *Flavocetraria*, *Kaernefeltia*, and *Tuckermannopsis*, resulting in our using a broad but natural genus concept in that case.

The species level is a very active area for molecular systematics—the use of DNA sequences to help resolve the age-old problems of lumping versus splitting species. Expect many changes in our species-level taxonomy to come from molecular tools in the next decade. In many cases, DNA analyses have supported morphological distinctions that we thought *might* be important, for example, the separation of the yellow-sorediate species of *Pseudocyphellaria*.

In other cases, DNA analyses have resulted in a taxonomy that combines species that were once separated by differences that do not reliably distinguish reproductively isolated lineages, for example recognizing that our two species of *Pannaria* actually belong to a single endemic species, *P. oregonensis*. Similarly, molecular methods resulted in lumping North American specimens of *Pseudephebe minuscula* and *P. pubescens* into one species, *P. minuscula*.

Forest Service studies of air quality, climate, and lichens contributed greatly to our understanding of the distribution and abundance of lichens in the Pacific Northwest, as well as foundational taxonomic knowledge. So, the first two editions naturally included a section on lichens as indicators of air quality. To the third edition we have added a new section on lichens and climate change, emphasizing climate indication with lichens in the Pacific Northwest and what it can tell us about trends in climate and lichen communities over the past 25 years.

The longer reference section in the third edition provides concrete evidence of the advances in lichenology in the last decade as they apply to the Pacific Northwest. In addition to the references listed, dozens of recent papers on the ecology of lichens in the Pacific Northwest have been written, including studies of the distribution and abundance of particular species, community responses to forest management practices, landscape-scale analyses of lichen habitats in relation to lichen communities, and the relationships between lichens, air quality, and climate change.

The last decade has also seen advances in lichenology elsewhere in North America, and these have greatly supplemented and strengthened our knowledge of lichens in the Pacific Northwest. We have a continental-scale lichen flora (Brodo et al. 2001) and regional floras for the Southwest (Nash et al. 2002, 2004, 2008) and New England (Hinds and Hinds 2007). The three-volume flora for the Southwest is particularly important for lichenologists in the Pacific Northwest, because it was the first comprehensive lichen flora to include crustose species for any sizable region of North America, apart from the American Arctic (Thomson 1984, 1997). Furthermore, the Southwest shares many species with other parts of western North America. Now crustose lichens in the Pacific Northwest are addressed directly by a comprehensive manual for microlichens (McCune 2017).

Since 2000 we also saw the emergence and growth of Northwest Lichenologists, a group of professional and amateur lichenologists. We hold annual meetings, field trips, and workshops. We invite you, the users of this book, to participate by visiting www.nwlichens.org. The group also sponsors a certification program for lichenologists who wish to establish their credentials for recognition of macrolichens in the Pacific Northwest.

Using This Guide

Coverage

This book can be used to identify macrolichens from Oregon and Washington. All fruticose and foliose lichens known or expected to occur in these two states are keyed, totaling 681 species in 109 genera, plus 2 subspecies and 14 varieties. Photos and detailed descriptions are provided for 273 species, emphasizing lichens prevalent in forested ecosystems, regionally common lichens, and rare species of special concern. Photos are provided for an additional 72 species, along with 45 line drawings. Reasonable coverage for lichens of Idaho and Montana, inland to the Continental Divide, can be expected. Almost all macrolichens known from northern California and southern British Columbia, and most from coastal southeast Alaska, are included as well.

Keys

Keys to the genera begin on page 51. Non-bolded endpoints are not treated further in the book. Keys to species follow, in alphabetical order by genus. Unless otherwise mentioned, the keys assume the following characteristics: upper and lower cortex present, spores hyaline and non-septate, rhizines present. Descriptions, especially for color, refer to dry lichens, unless otherwise stated. Size range indicates average as well as maximum size; for example, "7-12(15) mm" indicates that the usual size is from 7 to 12 mm, inclusive, but some individuals may be as large as 15 mm. A period separates key characters from supplemental information describing individual species. Bold-faced species in the species keys are illustrated and featured with full-page treatment following the species key. We feature not only common and conspicuous species but also those that are rare and thought to be potentially vulnerable to forest management practices. Square brackets [...] around a name indicate species not yet known for this area but known from adjoining areas and likely to be found in Oregon and Washington eventually. Information on range and habitat is provided in the keys for those species not given full-page treatment. References are provided for readers who wish to obtain further information on particular taxa.

Featured species

Featured species are illustrated by full-color photographs, line drawings, or both. Scale bars in the photos are 1 mm, unless otherwise indicated. An expanded description is provided for each featured species. Regionally based air pollution sensitivities, when known, are provided in the Lichens and Air Quality chapter (p. 32).

Descriptions include range within western North America, typical substrates, and habitat requirements in the Pacific Northwest. Lichens whose worldwide distribution is restricted to this region (Pacific Northwest endemics) are noted. Identification tips, differentiation of similar species, and information on ecological, indicator, and economic roles are in the notes section.

Supplemental information

Methods for collecting, handling, and chemical testing of lichens and preparation of collections for herbaria are described under Collecting and Identifying Lichens (p. 18).

Lichens are widely used as air quality indicators. Ratings in this guide have been tailored to air quality conditions in the western United States. The Lichens and Air Quality chapter (p. 32) contains more information on these topics.

Taxonomic authorities, synonyms, and common names for the lichens mentioned in this guide are in the appendix (p. 473).

Abbreviations

States and provinces, compass directions, and spot test reagents and reactions have been abbreviated. Terms describing lichens are not abbreviated and are defined in the glossary.

Alas	Alaska	e	east(ern, ward), east of
Alta	Alberta	e Cas	e of the Cascade Range crest
Appal	Appalachian Mountains		
BC	British Columbia	g	grams
c	central	GL	Great Lakes
C	reagent C (see p. 20)	GNP	Glacier National Park
ca	about	Gr Pl	Great Plains
Cal	California	ha	hectare
Cas	Cascade(s)	I	reagent I, iodine (see p. 21)
CK	reagent C followed by K (see p. 20)	Id	Idaho
cm	centimeters	K	reagent K, potassium hydroxide (see p. 20)
Co	County	KC	reagent K followed by C (see p. 20)
Colo	Colorado	kg	kilograms
Cont Div	Continental Divide	L	Lake
diam	diameter	LM	compound light microscope
DM	dissecting microscope (= stereo microscope), usually 10-40×	lw UV	long-wave ultraviolet

	light (see p. 22)
m	meter(s)
M	molar (moles/liter)
Mex	Mexico
Mont	Montana
Mt(s)	Mountain(s)
n	north(ern, ward)
N	reagent N, nitric acid (see p. 21)
N Am	North America
N Dak	North Dakota
Nev	Nevada
NM	New Mexico
NWT	Northwest Territories
O	orange
OP	Olympic Peninsula
Ore	Oregon
P	reagent P, *p*-phenylene-diamine (see p. 20)
PNW	Pacific Northwest
ppb	parts per billion
ppm	parts per million
Que	Quebec
R	red
RM	Rocky Mountains
s	south(ern, ward)
S Dak	South Dakota

Sask	Saskatchewan
sp	species
spp	species (plural)
subsp	subspecies
sw UV	short-wave ultraviolet light (see p. 22)
TLC	thin-layer chromatography
UV	ultraviolet light (long- or short-wave; see p. 22)
US	United States
w	west(ern, ward), west of
w Cas	west of the Cascade crest
Wash	Washington
Wyo	Wyoming
y	year
Y	yellow
YNP	Yellowstone National Park

Symbols

<	less than
≤	less than or equal to
>	greater than
≥	greater than or equal to
±	more or less
[ ]	species potentially present but not reported
µg	micrograms

Introduction

What are lichens?

Lichens are enormously successful and evolutionarily ancient composite organisms comprising members of two or more biological kingdoms. Specifically, a lichen consists of at least a fungus and its photosynthetic partner (green algae, cyanobacteria, or both) growing together in a mutually controlled, symbiotic relationship. The composite form differs greatly in appearance, physiology, reproduction, and secondary metabolites, compared to free-living fungi, algae, or bacteria.

Lichens are named after their fungal partner (**mycobiont**), usually an ascomycete. (See the glossary for illustrations and definitions of highlighted terms in this and following paragraphs.) The photobiont, as well as other microbial components of the lichen, have their own names. Worldwide, about one-fifth of all known fungi are lichenized (Honegger 1996). In contrast, there are only about 50 genera of photosynthetic partners (Sanders and Masumoto 2021). The most common photosynthetic partners (**photobionts**) are green algae. In the Pacific Northwest, *Trebouxia* and trebouxioid species are the primary photobionts in at least 60% of macrolichen genera. Other green alga photobionts of Pacific Northwest macrolichens are *Asterochloris, Chloroidium, Coccomyxa, Elliptochloris,* and *Symbiochloris* (previously as *Dictyochloropsis*). Because it is seldom possible to determine microscopically the genus of the algal photobiont in lichens, confident identification of the alga requires DNA sequencing.

The primary photobionts of about 15% of Pacific Northwest macrolichen genera are the **cyanobacteria** *Nostoc* and *Rhizonema* (previously *Scytonema*). In this book we often refer to cyanobacteria as "blue green," because under the microscope they are pigmented blue green. Identification of cyanobacterial species also requires DNA sequencing. Lichens with cyanobacterial photobionts can fix atmospheric nitrogen into forms usable by plants and animals. Lichen genera such as *Collema, Fuscopannaria, Koerberia, Lempholemma, Leptochidium, Leptogium, Massalongia, Pannaria, Parmeliella, Peltula, Polychidium, Protopannaria, Pseudocyphellaria, Scytinium,* and *Sticta* all have cyanobacterial photobionts.

A few lichens contain both green algal and cyanobacterial photobionts within the same individuals. This group is disproportionately prominent in the oceanic climate of the Pacific Northwest, including all species of *Solorina, Pilophorus,* and *Stereocaulon* and some species of *Lobaria, Nephroma, Peltigera,* and *Pseudocyphellaria*. In these cases, green algae are the primary photobionts and cyanobacteria are confined to special internal or external nitrogen-fixing structures (**cephalodia**). Cephalodia are surrounded by a dense, cemented layer of fungal tissue that reduces oxygen diffusion. The partially anaerobic environment of the cephalodium increases the density of cyanobacterial nitrogen-fixing cells (heterocysts) compared to the free-living state, boosting nitrogen fixation rates (Hitch and Millbank 1976).

Like most organisms, lichens host a microbiome containing bacteria and nonlichenized fungi, both ascomycetes and basidiomycetes. Although in many cases the microbiome is invisible even under the compound microscope, in some cases it can affect the appearance of the lichen, for example, formation of gall-like structures (Grube and de los Rios 2001, Hawksworth and Grube 2020) and coloration of the cortex as influenced by basidiomycete yeasts (Spribille et al. 2016). Despite the apparent variation in the microbiomes of lichens, perhaps most notable is that a particular lichen as a composite "organism" has predictable characteristics. When you identify a lichen, you can immediately know a lot about its geographic distribution, abundance, and ecological characteristics. In fact, they are just as predictable in those characteristics as plants or animals.

Why are lichens important?

Lichens contribute to biological diversity. The lichenized state is thought to have originated many times in unrelated fungal genera (Tehler 1996). Over subsequent eons a magnificent wealth of genetic, chemical, and ecological diversity emerged. Today we are the inheritors of a planet rich in lichen colors, forms, and species. And many lichens still await discovery. Between 1997 and 2008, about 600 lichen species were reported in North America for the first time (Esslinger 2008). In 2009 we wrote that "over 1,000 different lichen species from the Pacific Northwest are now known." But less than a decade later, McCune (2017) treated 1,721 microlichen species known to occur in the Pacific Northwest. Adding to that the macrolichens treated in this book, and microlichens discovered since 2017, the total lichen flora of the Pacific Northwest is about 2,500 species. Conservation of good air quality and critical habitats will be essential to maintain this diversity.

In addition to their contribution to biological diversity, lichens are ecologically important in the Pacific Northwest as food, shelter, and nesting material for wildlife. Deer, elk, moose, caribou, mountain goats, bighorn sheep, pronghorn antelope, and various species of squirrels, chipmunks, voles, pikas, mice, and bats eat lichens or use them in nest building (Banfield 1974, Bunnel et al. 1999, Conner 1983, Fox and Smith 1988, Gunther et al. 1983, Lehmkuhl et al. 2006, Richardson and Young 1977, Stevenson and Rochelle 1984, Thomas and Rosentreter 1992, Ure and Maser 1982, Ward and Marcum 2005). At least 45 species of North American birds use lichens as nesting material (Richardson and Young 1977). Among invertebrates, bristletails, barklice, katydids, grasshoppers, webspinners, butterflies, moths, lacewing larvae, mites, spiders, snails, slugs, and many beetles live on, mimic, or eat lichens (Gerson and Seaward 1977). Distributions of soil microbes and microflora (Belnap 2001) and aquatic invertebrates (Barley et al. 2006) can be shaped by chemical influences from associated lichen-dominated habitats.

Lichens are not just appetizers or supplementary nesting material, but a critical component of Pacific Northwest forest food webs. For example, flying squirrels rely on the brown beard lichen, *Bryoria*, as their principal winter food source (Maser, C., et al. 1986, Maser, Z., et al. 1985). Flying squirrels, in turn, are one of the two or three primary prey of the northern spotted owl, an endangered species whose protection needs have redefined forest management on public lands in the Pacific Northwest (Spies et al. 2019). In summer, the main staple of flying squirrels is underground-fruiting fungi. As they travel through the forest, the squirrels disperse fungal spores in their droppings. The spores germinate and form mycorrhizal associations with tree roots that are critical to the tree's ability to absorb nutrients (Maser et al. 1986). The trees in turn provide habitat for lichens, flying squirrels, spotted owls, and other organisms.

Lichens play significant roles in mineral and hydrological cycles, notably nitrogen fixation. In certain coastal and western Cascade old-growth forests of the Pacific Northwest, nitrogen-fixing *Lobaria oregana* dominates the lichen biomass. As *Lobaria* leaches during rains or decomposes in litterfall, it contributes significant amounts of nitrogen to the forest ecosystem (Antoine 2004, Holub and Lajtha 2004, Pike 1978). When biomass is high, lichens and bryophytes intercept and hold moisture (Pypker et al. 2005, 2006a, 2006b), moderating humidity and temperature fluctuations within the canopy. They also intercept and release nutrients from atmospheric sources such as rain, dew, fog, dry deposition, and gases—nutrients that might otherwise become lost or unavailable (Knops 1994). Desert crusts of lichens, fungi, cyanobacteria, and moss reduce soil erosion by intercepting surface run-off and facilitating infiltration of water into hardpan soils (Belnap and Lange 2001, Harper and Marble 1988).

Unprecedented human population growth has intensified the conflict between ecological conservation and human needs for more space and resources. It is ever more important to understand the interconnecting web of organisms and the basic requirements necessary to maintain functioning, resilient ecosystems that provide clean air and water, healthy soils, stable climates, and other services vital to our quality of life.

Lichens are also important as indicators. Lichen communities change with plant succession (Lesica et al. 1991, McCune 1993, Neitlich and McCune 1997, Sillett et al. 2000). Land managers can use lichens to indicate forest continuity (Kuusinen 1996, Rose 1976, Tibell 1992), to determine the distribution of specialized microhabitats and microclimates (Pearson and Lawrence 1965), to detect hotspots of biological diversity over the landscape (Karström 1992, Neitlich and McCune 1997, Nitare and Norén 1992, Peterson and McCune 2003), and to assess water quality (Beck and Ramelow 1990), air quality (see p. 31), and climate (see p. 49).

Finally, lichens have economic value. Ornate Pacific Northwest forest lichens, such as *Letharia*, *Usnea longissima,* and *Ramalina menziesii*, are harvested for use in floral displays around the world. Dozens of regional species have culturally significant traditional uses (see p. 28) as dyes, medicines, and textiles. *Bryoria fremontii* has been widely used as food. Hundreds of compounds unique to lichens have potential as natural product pharmaceuticals (see p. 26).

How lichens work

The photosynthetic partner of a lichen produces food for the fungus, converting atmospheric carbon dioxide into sugars and sugar alcohols via photosynthesis. These carbohydrates also provide desiccation and low temperature tolerance to the fungus, similar to carbohydrates in plants. In return, the fungal partner forms a physical structure or "home" for the photosynthetic partner, providing surface area for the capture of water and mineral elements, as well as chemical protection from excessive light, UV radiation, desiccation, and herbivory.

Alternating periods of drying and wetting are necessary for the health of most terrestrial lichens and ensure an adequate distribution of carbohydrates between the partners. In species studied so far, medium to high water content favors carbohydrate transfer to the fungus. At low water content, the alga retains more carbohydrate. Photosynthetically made carbohydrates are transferred from green algae to the fungal partner as sugar alcohols, and from cyanobacteria to the fungal partner as glucose (Spribille et al. 2022).

As composite organisms, lichens grow in habitats where neither partner could subsist well alone. An extraordinary example is the ability of certain desert coastal lichens to thrive on daily dew or fog cycles using a kind of reverse transpiration. Under cool temperatures and high humidity, photosynthesis can be activated in green algal lichens after water vapor uptake alone, partly aided by the high osmotic content of the fungi (Nash 2008). By contrast, activation of photosynthesis in cyanobacterial lichens requires liquid water. Consequently, cyanolichens are typically found in moist microhabitats, such as moss-covered logs or tree bases, soils of shaded roadcuts, and on various substrates in riparian areas.

Growth forms and structural features

Lichens are artificially divided into three growth forms: **crustose** (crust), **foliose** (leafy) and **fruticose** (shrubby). Within each form there is considerable gradation and specialization, including **squamulose** (*Cladonia, Fuscopannaria, Psoroma*) or **umbilicate** (*Umbilicaria, Rhizoplaca, Dermatocarpon*) lichens. The latter have a central point of attachment; the former consist of tiny, leaf-like scales, sometimes producing upright, fruticose stalks called **podetia**. **Macrolichens** include foliose, fruticose, and the larger squamulose forms.

The lichen body is called the **thallus**. Most macrolichen thalli are **heteromerous** (i.e., **stratified** internally). The upper or outermost layer is the **cortex**. The cortex can be thin to thick, pigmented or not, and contains closely packed **isodiametric** or elongated **hyphae**. Many lichens have an outermost epinecral layer of dead, collapsed, and sometimes gelatinized hyphae and photobiont cells. This layer thickens in individuals growing in sunny habitats, making the surface less transparent and therefore less green. Some lichens have surface mineral deposits that give them a powdery white appearance (e.g., most species of *Physconia*). These are often of calcium oxalate and are called **pruina**. In fruticose and foliose lichens, the photobiont layer is found directly internal to, or below, the cortex in the upper **medulla**. The medulla is a layer of loose, cobwebby, white fungal hyphae with many intercellular air spaces. In *Usnea*, the medulla contains a structurally supporting central cord of thick-walled, closely packed hyphae. Most foliose lichens form a lower cortex that is clearly differentiated from the upper cortex by color or by the presence of features such as hair-like, anchoring **rhizines**, or a velvety **tomentum** of short, dense hyphae.

A few macrolichen genera are non-stratified or **homoiomerous**. Non-stratified lichens (e.g., *Collema* and *Leptogium*) generally have a cyanobacterial photobiont dispersed throughout the lichen body and appear dark and gelatinous when wet.

Features commonly used to identify lichens and distinguish species include shape, size, and color of the thallus and **lobes** (branches); the pattern and angle of branching; the degree of adhesion to the substrate; habit of the primary thallus (e.g., spreading, erect, or pendent); and the structure and frequency of reproductive bodies and spores.

Other differentiating features are the presence or absence of surface wrinkles, hair-like **cilia** on the lobe margins, minute bumps called **papillae**, cephalodia, pseudocyphellae, and cyphellae. **Pseudocyphellae** are pores through the upper or lower cortex that are filled with loosely packed medullary hyphae. **Cyphellae** are larger, and the medullary hyphae lining the pores form rounded tips that adhere to each other. They are found only in the genus *Sticta*, and look like smooth, whitish craters dotting the tomentum of the lower surface.

Color is a very important characteristic for lichen identification. It is essential to distinguish taxonomically important color variation from environmentally induced color variation. Strong light can cause individuals to be deeper colored, browned, or blackened (**melanized**). Shade forms, even of yellow or orange lichens, tend to be pale and more greenish. Lichens containing usnic acid are yellowish; but the yellow cast can be hard to detect, depending on how much usnic acid is present. The KC test (see Collecting and Identifying Lichens, p. 20) is useful for detecting usnic acid.

Beginners often struggle with color terms as applied by lichenologists in descriptions and keys. For example, usnic acid gives a distinctive yellowish tinge to lichens, which is recognized by all lichenologists, but initially that yellow tinge can be hard to discern in comparison to the wide range in subtle tones displayed by lichens. A color chart poster (McCune 2019b) can help beginners learn to see colors the way that lichenologists see and refer to them.

Reproduction, growth, and dispersal

Lichens reproduce asexually or sexually or both. A few macrolichens, particularly the pendent epiphytic lichens *Alectoria*, *Bryoria*, *Ramalina*, and *Usnea*, rely largely on fragmentation for dispersal. As an individual grows, pieces break off and become entangled in branches or are carried short distances by wind or animals and may establish new individuals. Although these genera include some of the fastest-growing lichens, fragmentation can lead to a net increase or decrease in population biomass, depending on whether the fragments land in suitable habitats. The fraction that falls to the forest floor provides important winter survival forage for deer and other mammals, particularly in deep snow conditions (Hanley et al. 1989, Rochelle 1980). The largest biomass of these forage lichens occurs in late-successional and old-growth forests (Lesica et al. 1991, McCune 1993).

Overall, lichens grow and disperse slowly compared to vascular plants. Specialized habitat requirements, the need for continuity in the availability of substrate, and sensitivity to climatic conditions and air pollution make many lichen species vulnerable to habitat disturbance or degradation. In the Pacific Northwest, lichen biodiversity is highest in coastal and riparian areas and in high rainfall, low- to mid-elevation forests. Because coastal, riparian, and forested areas receive intensive human use, land management policies will determine the survival chances of some Pacific Northwest lichens over the coming decades (Spies et al. 2019).

Asexual reproduction also occurs via the production and dispersal of tiny propagules containing both the fungus and the primary photobiont. Many types of vegetative propagules have been recognized in lichens. Two kinds, **isidia** and **soredia,** are referred to most often in identification keys. They disperse in air or water and can initiate new individuals. Both types are best observed with a hand lens or dissecting microscope. Isidia are tiny (30 μm to 1 mm high), stratified, usually cylindrical to globular, branched or unbranched structures that are covered by a cortex. The cortex gives isidia a smooth, somewhat shiny surface. Soredia, in contrast, are microscopic (15-50 μm diameter), loose spherical clusters of photobiont cells and fungal filaments. Soredia appear flour-like to granular and occur either diffusely spread over the upper surface or in macroscopic patches, called **soralia**.

Some mystery still shrouds the mechanisms and relative importance of sexual reproduction in lichens. In principle, the fungal partner produces spores that germinate, contact free-living algae, and develop into new individuals. Some lichen photobionts are free living and widespread (*Nostoc*, *Chloroidium*, *Trentepohlia*), but the most common macrolichen photobiont, *Trebouxia*, is rare in nature relative to already lichenized sources of the photobiont, most notably the huge number of lichen soredia, isidia, and small thalli. Many free-living photobionts such as *Nostoc* are not genetically the same as the *Nostoc* in cyanolichens found in the same habitat. A few fungi have been shown capable of obtaining algae from other thalli or vegetative propagules of other species (Friedl 1987, Ott 1987). However it happens, a large number of lichens reproduce sexually and must re-establish the symbiotic state with each reproductive cycle.

Sexual structures are produced by the fungus. Two kinds occur in Pacific Northwest macrolichens: apothecia and perithecia. **Apothecia** are disk-like fruiting structures and are the predominant form. A smooth, fertile layer of **asci** (spore-containing sacs) is exposed on the upper surface of the **disk**. *Acroscyphus, Bunodophoron, Sphaerophorus,* and *Tholurna* have a type of modified apothecium, sometimes called a **mazaedium,** in which the asci disintegrate at maturity, producing a powdery spore mass. **Perithecia** occur in only one macrolichen genus treated here, *Dermatocarpon.* They are flask shaped and immersed in the thallus. All that can be seen with a hand lens is the tiny dark dot indicating the opening of the flask at the surface.

More information

For further reading on lichen biology, see Brodo et al. (2001), Nash (2008), and Purvis (2000). For additional resources on lichen identification, see the following (these attempt to be comprehensive except as noted):

PACIFIC NORTHWEST
Microlichens—McCune (2017)
Common macrolichens, online interactive key—lichens.twinfern-
tech.net/pnw/index.shtml (McCune and Yang 2021)

BRITISH COLUMBIA
Foliose lichens – Goward et al. (1994)
Fruticose lichens – Goward (1999)
Vancouver area, all lichens – Noble (1982, reprinted 2017)

NORTHERN ROCKY MOUNTAINS
Macrolichens, including squamulose lichens – McCune and Goward (1995)

NORTH AMERICA
Selected lichens, all groups – Brodo et al. (2001), Brodo (2016)

GREAT BRITAIN AND IRELAND
Smith et al. (2009)

EUROPE
Wirth et al. (2013)

USEFUL COLOR CHARTS
Lichen Color Chart Poster—wildblueberrymedia.net
Lichen Spot Test Reactions Poster—wildblueberrymedia.net

Collecting and Identifying Lichens

Collection

Lichens can be found in almost any natural habitat in the Pacific Northwest. A few easily obtained tools are needed to collect lichens: a good 10× or 14× hand lens, a sharp pocketknife, and paper collecting sacks or folded packets (damp lichens will mold in plastic bags within a few days). See the Northwest Lichenologists website (nwlichens.org) for many details and practical tips on tools for lichenology. For crustose lichens, we use a 1-cm-wide rock chisel and a 14-ounce rock hammer. Other useful items are a small notebook for recording field notes, and a pencil or waterproof pen. Safety goggles are recommended when using a hammer and chisel to remove lichens from rocks. A spray bottle with distilled water is sometimes useful for hydrating dry, brittle lichens before collection, and garden clippers are handy for snipping fine branches.

When collecting macrolichens, it is important to collect enough material to permit a positive identification but avoid removing entire colonies of lichens or collecting rare lichens. The specimen should be palm sized if possible and include several to many individuals. Most lichens grow very slowly and a palm-sized piece can represent 25 years or more of growth. The material you collect should be in good condition (i.e., alive, minimally discolored, and free of mold). Whenever possible, collect individuals with reproductive structures (**apothecia, perithecia, pycnidia, soredia,** and **isidia**), because these are frequently necessary for identification. Individuals from different colonies are usually best placed in separate packets as chemical variants can occur within species. Removing larger extraneous debris while collecting saves space and helps prevent molding. Leave some substrate attached to the lichen, or at the minimum, clearly record the substrate on the collection package.

Good field notes can greatly improve the value of a collection. Record the location in words and numbers (latitude/longitude or UTM [Universal Transverse Mercator]). Other important notes include substrate, habitat, exposure, abundance on the site, elevation, the date collected, the collector's name, and collection number. These notes can be written on the collection package or in a field notebook.

Pressing

If you wish to make a permanent collection, some additional follow-up preparation will be necessary. Specimens that seem fragile, brittle, or bulky should be lightly pressed. To do this, soak or spray the lichen with distilled water or rainwater until it hydrates. Spruce up the lichen by removing debris such as needles, leaves, other lichens, and mosses, leaving the lichen attached to the substrate. For most foliose lichens it will be important to be able to view both upper and lower surfaces, and removal of some substrate will be necessary.

Place the hydrated lichen on newspaper, recording the packet number. Place newspaper layers between blotter paper separated by corrugated cardboard in a plant press. Close the press and gently tie the straps just snugly enough to hold all the sheets together. Some three dimensionality is preferred to a completely flattened sample. Place the press near a moderate drying source such as a ventilator or electric fan; drying ovens or hot surfaces can alter lichen acids and degrade DNA needed to confirm identifications. Change absorbent layers once a day until the lichens are dry.

Stabilizing lichens on soil

Lichens on soil can be stabilized by dipping the soil part into a shallow dish of diluted standard white glue (polyvinyl acetate) and then allowing it to dry on a piece of wax paper. Avoid getting glue on the lichen itself! The soil will now hold together, and full-strength glue can be applied to the lower surface of the soil to attach it to a 7.5 × 12-cm acid-free card. For more details, see McCune and Rosentreter (2007).

Preparing collections for an herbarium

Once your lichens are cleaned, pressed, and stabilized, they are ready for packaging. Fold a sheet of 100% cotton or other acid-free paper into a packet (fold the bottom third of the paper up, then fold both long edges in about 4 cm, then fold the top third down). Place the lichen on a 7.5 × 12-cm acid-free card and insert in the packet. Specimens that are extremely brittle or fragile or very small can be placed on a piece of tissue, cotton padding, or synthetic cast padding (available from medical suppliers) or tucked inside a small glassine envelope. Crustose lichens and very small macrolichens on rocks or soil should be glued to the card to prevent them from falling to the bottom of the package and crumbling or grinding against each other. Glue on a label or print the label directly on the packet before folding. A sample label is given below:

LICHENS OF OREGON

Platismatia lacunosa

on conifer branch in litterfall,
old forest of *Chamaecyparis lawsoniana* – *Tsuga heterophylla* – *Pseudotsuga*, **trail to Coquille Falls**, Coast Range
Elevation: 430 m   42.7150°N 124.0222°W datum: WGS84
USA: OREGON: COOS COUNTY

Coll. B. McCune   39672
3 October 2022   with Northwest Lichenologists

Identifying lichens using microchemistry

Lichens produce a large number of unique secondary chemical compounds, most of which are not produced by free-living fungi or algae but only by the composite organism. These compounds may reduce herbivory, inhibit decay organisms, and screen damaging ultraviolet light. Some of the compounds are produced in large amounts and have been used in perfumes, over-the-counter cold medicines, sore throat lozenges, and personal care products.

Chemistry is a well-accepted tool in lichen identification and taxonomic research. Today's lichenologists use chromatography, especially thin-layer chromatography (TLC), to verify the presence of specific lichen chemicals. For those who wish to know more about using TLC, the British Lichen Society has published a methods book that provides all of the information needed to get started (White and James 1985). See also the detailed protocol and photos in McCune (2017, Vol. 1). However, many competent identifications can be made using a dissecting microscope and some simple "spot" tests. See the Northwest Lichenologists website (www.nwlichens.org) for sources of reagents.

Beginners often have difficulties interpreting their observations, simply because they have not seen an example of the reaction. A good way to learn is to spot test known lichens. Also, you can compare your results to examples on a poster showing a logical flow chart for spot tests (McCune 2019a). This poster displays the appearance of the most common spot test reactions and shows how the spot tests are used to infer the presence of commonly encountered secondary lichen substances.

The most used chemical spot tests and their reagents are:

K test. Use an aqueous solution of 10% potassium hydroxide.

C test. Use a fresh solution of undiluted commercial bleach or 5.25% sodium hypochlorite. Store C out of direct sunlight and replace every few weeks, or even daily if you work with a clear bottle in a sunny place.

KC test. Moisten the test spot with K, let this soak in, then apply C in the same spot. Look for an immediate reaction. The **CK test** is the seldom-used opposite of the KC test (showing a yellow or orange reaction with barbatic and diffractaic acids).

P test. Use a few crystals of *p*-phenylenediamine (P) dissolved in one drop of 95% ethanol. This light-sensitive solution must be freshly made for each application. The solution can be made on a glass slide or vial or in a well of a ceramic spot plate. As the alcohol evaporates, more can be added to the residue if the resulting solution remains clear to light pink. Note that P is a permanent stain and a skin irritant. Avoid getting P on the skin or clothes. Handle carefully and use forceps to deposit tested pieces of lichen, contaminated paper, and used pipettes carefully in sealed containers. Discard the refuse so that others will not contact it.

Tools and glassware contaminated with P should be wiped with alcohol or washed in water with detergent.

To perform the spot tests, use a small amount of dry lichen under a dissecting microscope. Place the lichen fragment on a small square of white paper on a glass slide to avoid staining the stage of the microscope. If the medulla is to be tested, use a razor blade or carefully tear the material to expose a patch of medulla below the algal or cyanobacterial layer. Place a small amount of reagent on the test surface. Micro-hematocrit capillary tubes of about 1 mm inside diameter are good for delivering the spot test reagents. Finer tips can be made by pulling a tube from each end over an alcohol burner and carefully breaking the center at the desired diameter. An eyedropper will often deliver too much reagent, diluting the reaction color. A toothpick or very fine paint brush can also be used. Mark the dispensers "K," "C," or "P" and store tip down in separate, marked vials as you work. The K, C, and KC test color reactions are usually rapid, often immediate, and may also disappear rapidly, whereas P reactions may take a minute or two to develop fully. In all spot tests, be careful to distinguish between true color reactions and small color changes due to wetting of the algae or transparency of the cortex. If you are unsure, add a drop of water to another small area of the thallus for comparison.

Abbreviations for spot test reactions used in this guide are as follows:

+ = a color change can be observed

- = no change in color will be observed

Y = yellow

O = orange

B = blue

R = red

Negative spot tests are often not listed. So, a lichen cortex that turns red after the application of P but does not react to the K or C tests would be abbreviated as: cortex P+R.

Some less frequently used spot tests are:

N test. Use a 6 molar (6M) solution of nitric acid in water. This is a strong acid requiring extra care. The N test helps to distinguish *Neofuscelia* from *Melanelia, Melanohalea*, and *Melanelixia*. Put the lichen fragment in a depression on a white ceramic spot plate. Wet the fragment with a drop of N. Using a dissecting microscope and strong light, look for a darkening within seconds, with a blue-green tinge, darker and bluer than the green color obtained by soaking in plain water.

I test. Use 0.25 g iodine in 100 ml aqueous 0.5% potassium iodide solution. Iodine solutions react with various kinds of starch. Since most kinds of paper react with the I reagent, it is best to do the test on glass on a white background or white ceramic spot plate. Wet the sample with a drop of reagent and look for a color change to blue, violet, or black.

UV test. Use short-wave (254 nm) and long-wave (366 nm) ultraviolet light to detect key lichen substances. This test is fast and effective, often giving clear results. Colors can be orange, gray, gold, brown, red, white, or bluish white and vary in intensity according to amount and type of lichen substance present. Fluorescence is produced by substances in the cortex (e.g., xanthones) or in the medulla (e.g., alectoronic acid, alpha-collatolic, divaricatic, evernic, homosekikaic, and squamatic acids). Expose the medulla with a razor blade, because some UV-absorbing substances in the cortex, such as usnic acid, can hide fluorescent substances in tissues below. While some substances fluoresce in both short-wave and long-wave UV, others are more strongly fluorescent in one or the other, so it is often helpful to note this with each wavelength.

Portable UV lamps with fluorescent bulbs can be obtained from most scientific and geological supply companies. Avoid inexpensive LED-based UV flashlights; these include some visible light and give misleading results or include only long-wave UV.

To perform a UV test, first darken the room and choose the desired wavelength. In many cases it is instructive to try both long- and short-wave UV. Some lamps illuminate both wavelengths simultaneously. Hold the lamp close over the specimen on your working surface. If the specimen fluoresces, record the color and intensity you observe. Warning: UV light can cause irreparable eye damage. Always wear UV-filtering glasses or goggles when working with a UV light source and **never** look directly at a UV lamp to see if it is on. UV lamps should be used only for short periods of time. Short-wave UV light is far more injurious than long-wave commercial "black lights."

Sectioning lichens

Although most macrolichens can be identified with a hand lens or dissecting scope, viewing thin sections of thalli or apothecia under a compound light microscope (LM) can sometimes be a big help. Sectioning lichens is easy and a few pointers will help get you started:

Choose a new, sharp razor blade, either a single-edged safety razor or a double-edged blade broken lengthwise into two single-edged blades. The latter are much sharper than safety razors, but be very careful when breaking them. Using very sharp blades gives best results.

Make the sections under a dissecting scope. If you have youthful eyes or head-mounted magnifiers you might be able to get by without the dissecting scope, though we strongly recommend these for lichen identification in general.

Section the dry lichen with a slicing motion, making the section almost invisibly thin, supporting the side of the razor blade with your forefinger on the opposing hand.

For challenging situations, embed a small lichen piece in water-soluble wax. We use PEG 3350, sold as an osmotic laxative powder under various brand names, such as MiraLax. To use this, place the piece to be sectioned on a glass slide. Make a tiny pile of PEG on and around the piece (it is easiest to do this if you have transferred the PEG to a narrow-mouthed vial). Gently heat the bottom of the slide with a lighter until the PEG melts. Let cool until it hardens (about 1 minute; hairline cracks will form). Trim excess wax with blade. Slice with sharp razor blade and transfer dry sections to a drop of water on a second glass slide. The PEG will dissolve instantly. Place a cover slip and view.

Lichen Chemical Compounds: Production and Roles in Lichens

Primary compounds

Primary compounds are molecules needed for basic functions within living cells. These include proteins, amino acids, polyols, carotenoids, chlorophylls, polysaccharides, nucleic acids, and vitamins. Primary compounds also help to maintain the lichen symbiosis. For example, sugars made by lichen algae via photosynthesis are converted to sugar alcohols by the fungus, which may then be used as food for the fungus and to create osmotic potential to slow desiccation.

Lichen primary compounds may also play wider ecological roles in promoting wildlife health. For example, serine protease, an enzyme occurring in lichen leachates, can decompose prions in soils and in animal digestive systems that cause neurodegenerative diseases, such as the famous "mad cow disease." Water extracts of serine protease from Pacific Northwest forest epiphytes (e.g., *Parmelia sulcata*, *Lobaria pulmonaria*, *Usnea cavernosa*, and *U. rubicunda*), widespread matt-forming lichens in tundra/ alpine and woodland forest habitats (*Cladonia rangiferina*), and a rock-dwelling lichen (*Xanthoparmelia coloradoënsis*), all degrade prions from prion-infected hamsters, mice, and deer (Johnson et al. 2011, Rodriguez et al. 2012). Deer and other ungulates and rodents readily consume lichens and may thus gain some protection against these diseases.

Secondary compounds of the primary mycobiont

Lichens are slow growing, long-lived composite organisms that are subject to extreme temperatures, drought, nutrient deficiencies, and exposure to pathogens and predators. Secondary compounds are extracellular molecules produced to support this way of life. They have specialized roles in the composite organism, including regulation of internal water relations by creating hydrophobic surfaces, regulation of photobiont metabolism, mineralization of essential elements, light screens, antimicrobials (against pathogenic viruses, bacteria, fungi, and protozoans), allelopathic agents (chemical inhibition of bryophytes, vascular plant growth), defense from predators (e.g., herbivory by invertebrates, rodents, ungulates), and as a general response to stress (e.g., changing climates or air pollution) (Lawrey 2009). Over 1,050 secondary compounds are produced by the primary lichen mycobionts, and most are unique to lichens. These include polycyclic aromatic compounds, such as depsides and depsidones, dibenzofurans, xanthones, chromones, anthraquinones, naphthoquinones, fatty acids, terpenoids, steroids, and pulvinic acid derivatives (**Table 1**).

UV protection. The lichen cortex must provide adequate light exposure while also moderating water loss and protecting the photobiont from harmful ultraviolet radiation. Lacking the protective, waxy cuticle of plants, lichens instead assemble a surface layer of densely packed cortical fungal hyphae containing water-absorbent polysaccharides (primary compounds) and then produce secondary compounds such as usnic acid and atranorin that screen light, especially in the UV range, protecting the alga's photosynthetic pigments (**Table 1**). These secondary compounds are commonly deposited as water-insoluble crystals on the surface of the hyphae. When wet, the outer layer is translucent, allowing photosynthesis, but when it dries it becomes protectively opaque.

Defense against microbes and herbivory. Despite being mainly fungus, living lichens resist microbial decay. The wide variety of secondary compounds with antimicrobial activity includes many that also have anti-herbivory activity, for example usnic acid, stictic acid, and diffractaic acid. These compounds defend against a wide spectrum of organisms (Lawrey 2009; **Table 1**). On the other hand, lichens are well known as forage for some animals, presumably with the ability to tolerate or avoid lichens rich in defensive compounds.

Secondary compounds of endolichenic fungi and bacteria

Modern genetic and molecular biology techniques have enabled the discovery of an array of fungi, bacteria, and cyanobacteria associated with lichens. Some

Table 1 (*facing page*). Bioactivities of common lichen secondary compounds. Adapted from Ranković (2019). Light screens can function by absorbing UV (A, absorbing and reradiating as heat) or by fluorescence (F), a simplification because many substances absorb at some UV wavelengths and fluoresce at others or are weakly fluorescent (f). Some are pigments (P), meaning that they differentially absorb or reflect at visible wavelengths. Compound names that end in "ic" are acids (e.g., evernic acid); we have omitted the word acid for compactness.

Table 1. Bioactivities of common lichen secondary compounds.

Chemical class	Compound	light screen	antioxidant	antimicrobial	allelopathic	cytotoxic	anti-herbivory	anti-cancer	anti-inflammatory	anti-diabetes	neuroprotective	analgesic/ anti-pyretic	Additional compounds
Fatty acids	lichesterinic			X	X				X				rangiformic
	protolichesterinic		X	X	X	X		X	X				
para- and meta-depsides, tridepsides, and benzyl esters	atranorin	A	X	X				X	X	X	X		squamatic thamnolic
	barbatolic			X									
	diffractaic	A	X	X			X	X	X			X	
	divaricatic		X	X				X		X			
	evernic	F	X	X	X			X	X				
	gyrophoric	f	X	X		X		X		X			
	perlatolic	F							X		X		
	sekikaic	F	X	X						X			
	sphaerophorin	F	X			X		X					
	tenuiorin		X										
depsides, depsidones, and diphenyl ethers	alectoronic			X									
	fumarprotoce-traric	A	X	X		X		X			X		
	lobaric	F	X	X				X		X	X		
	norstictic		X	X				X					
	physodic	f	X	X				X			X		
	protocetraric		X	X		X		X					
	psoromic			X		X		X			X		
	salazinic	A	X	X				X		X	X		
	stictic		X	X			X	X			X		
dibenzofurans, usnic acids	usnic	A, P	X	X	X	X	X	X	X	X	X	X	didymic pannaric
anthraquinones	emodin	F, P		X		X		X					teloschistin xanthorin
	parietin	F, P				X		X					
melanins	fungal melanins	A, P											
xanthones				X									arthothelin lichexanthone thiophaninic
terpenoids	zeorin		X	X							X		nephrin
pulvinic acid derivatives	vulpinic	F, P		X			X	X					calycin pulvinic rhizocarpic

172 diverse secondary compounds have been isolated from endolichenic fungi (fungi living inside lichens) alone, including peptides, alkaloids, steroids, coumarins, xanthones, anthraquinones, terpenes, and terpenoids (Singh et al. 2017). These compounds have activities as potential anti-cancer/anti-tumor drugs (44%), antimicrobials (37%), antioxidants (2%), UV light screens (1%), and anti-inflammatory agents (1%) (Agrawal et al. 2020). Additional primary and secondary compounds have been isolated from the bacterial communities associated with lichen cortices, and their properties are also an active area of research (Cardinale et al. 2012; Grimm et al 2021).

Secondary compounds of lichen photobionts

Lichen photobionts also produce secondary compounds. Microcystins, for example, are peptides consisting of seven amino acids linked together in a ring. These potent liver toxins can poison humans and animals and are primarily associated with aquatic cyanobacterial blooms. The production of microcystins in lichens is now known to occur globally across many different climates and cyanolichen taxa. Kaasalainen et al. (2012) screened 803 *Nostoc*-containing cyanolichen specimens from five continents, finding either the genes for making microcystins or the toxin itself in 12% of the specimens. Microcystin genes were found in 21% of specimens collected in Oregon, specifically in *Lobaria oregana*, *Nephroma tropicum*, *Peltigera collina*, *P. membranacea*, *P. neopolydactyla*, and two other *Peltigera* species. Of these, only *P. neopolydactyla* contained more than 10 µg/g of microcystins, the detection limit.

Microcystins likely contribute to the chemical defense against herbivores. Microcystin-producing cyanolichens are a potential health risk for a variety of lichen-consuming herbivores and humans. Many mollusks and arthropods, but also mammals such as voles, squirrels, and ruminants, are known to feed on lichens, which might also introduce toxins into food chains. For example, the grazing trails of mollusks on *Peltigera membranacea* in the Pacific Northwest show that they preferentially consume the nutrient-dense cyanobacterial layer over the fungal medulla below without apparent harm. Lichens are important winter feed for reindeer and caribou, which graze heavily on *Cladonia* species with green algal symbionts (Danell et al. 1994). Even though the relatively high nitrogen content of cyanolichens suggests a better nutritional value (Storeheier 2022), reindeer distinctly avoid eating cyanolichens, even during starvation.

Lichens as Natural Products

The unique, bioactive secondary compounds produced by lichens, principally those of the primary mycobiont, are of great interest as natural products. Natural products are chemical compounds sourced from living organisms (bacteria, archaea, fungi, plants, animals), frequently used in cosmetics, dietary supplements, agricultural chemicals, and in drug discovery. About 50% of all modern prescription drugs are of natural product origin. Many lichen compounds not only have well-demonstrated antibiotic, anti-viral, anti-fungal, and antioxidant properties but also show promising activity against cancer and autoimmune conditions such as neurodegenerative diseases (dementia, Alzheimer's), rheumatoid and osteoarthritis, and diabetes (Molnár and Farkas 2010; Ranković 2019; Elkhateeb et al. 2020).

The first step in developing natural products is establishing "activity," or the ability of the substance to produce a desired effect. These include laboratory tests such as microbial inhibition, enzyme assays, and inhibition of normal and cancerous animal and human cell lines. Some studies have included invertebrate and small animal testing. Such tests can be conducted on lichen crude extracts and/or specific compounds extracted in small amounts directly from lichens. Because some secondary metabolites can be cytotoxic, high activity and low toxicity are the key desired attributes for a natural product. That is, natural products should be at least as effective as current drugs and preferably have lower toxicity.

Most of the common lichen secondary compounds have multiple types of activity (**Table 1**). For example, vulpinic acid has potential applications in sunscreens, antibiotics, and anti-cancer drugs. Usnic acid has demonstrated antimicrobial, antioxidant, analgesic (pain relief), anti-pyretic (fever reducing), anti-inflammatory, anti-cancer, and anti-diabetes activity. Fumarprotocetraric acid has antimicrobial, antioxidant, anti-neurodegenerative, and anti-cancer activity, and can act as a muscle relaxant. In high doses, all three substances are also cytotoxic.

Although many lichen compounds have been isolated and have known chemical structures and demonstrated in vitro activity, commercial applications of lichen substances in prescription medicines are rare. In general, the limited availability and slow regrowth of lichens in nature and the absence of synthetic or biosynthetic techniques are bottlenecks for the large-scale production of secondary compounds needed for commercial drug testing, approval, and manufacturing. A few over-the-counter products such as toothpastes, deodorants, tinctures, and shaving cream made from extracts of wild-harvested *Usnea*, a sore throat herbal tea including *Lobaria pulmonaria*, and throat-coating lozenges using *Cetraria islandica* can be found in the United States and Europe.

Recent advances in biochemistry and biotechnology might help widen the production bottleneck. Some promising examples include synthetic laboratory production of novel usnic acids (Kartsev et al. 2018); continued progress in the culture of lichen mycobionts; turning on biosynthesis of lichen compounds using environmental stressors such as pH, heat, and light (Stocker Wörgötter 2013); and biosynthesis of lichen compounds in bacteria by inserting and turning on the relevant polyketide genes (Kim et al. 2021, Pyrczak-Felczykowska et al. 2019).

Traditional Uses of Lichens

With most of us living in urban areas today, it is easy to forget that throughout human history, provisioning food, medicines, shelter, and textiles required an intimate knowledge of the properties and uses of local plants, animals, fungi, and microbes. In North America's Pacific Northwest, about 50 different indigenous cultures have traditional uses for lichens as medicines, dyes, textiles, and food. We know about them because of the contributions of traditional knowledge holders, past and present. This knowledge remains a cornerstone of cultural identity while also contributing to modern knowledge of lichens. The following synopsis gleans from Crawford's (2007) excellent treatise on regional lichen uses and compendium (2019) of traditional lichen medicinal uses globally. These references provide detailed, culture-specific information about names, harvest and preparation methods, and uses of lichens.

Medicines

The pharmaceutical treasure trove provided by the exceptionally diverse secondary chemistry of lichens has been long recognized and utilized by the region's indigenous peoples. A place-based knowledge continues to guide traditional medicine practitioners, who may select seasons, substrates, and/or geographical locations for the harvest of specific lichens, thereby optimizing the probability of correct identification and the medicinal potency of the material collected.

The genera most frequently used in regional traditional medicines are *Alectoria, Bryoria, Cetraria (Vulpicida), Cladonia, Letharia, Lobaria, Parmelia, Peltigera, Stereocaulon,* and *Usnea.* All have uses as antimicrobials to treat or prevent infections in the form of salves and ointments, teas, extracts, and/or bandaging that are applied externally to rashes, sores, and wounds or taken internally to treat eye infections, colds, ulcers, and pulmonary ailments. Some are also used to prepare analgesics and anti-inflammatories to reduce swelling; treat arthritis, burns and scalds, bee stings, headache, and stomach pain; assist labor; and sooth teething babies. Although modern pharmaceuticals are also in use today, we use the present tense in examples below to highlight the continuing significance of traditional medicines to the cultures identified here.

Specific remedies include the use of *Alectoria* and *Usnea*, and especially *U. longissima*, beard lichens containing usnic acid, for wound dressing (Ditidaht), compresses for open sores and eyes (Cayuse, Umatilla), staunching bleeding (Chugach), and prevention of infections in newborns. Eye conditions and slow-healing sores can also be treated with preparations of *Cladonia bellidiflora* (Tlingit, Haida) and *C. chlorophaea* (Okanagan-Colville), respectively. *Bryoria fremontii* can be used in poultices to reducing swelling (Atsugewi) or as a tonic for upset stomach, indigestion, or diarrhea (Nimiipuu). Weak decoctions or infusions of *Letharia* can be used to treat stomach disorders, ulcers, and

hemorrhaging while stronger decoctions or poultices can be applied externally to sores and wounds (Syilx), bruises, swellings, arthritis, or eye conditions
(Cayuse-Umatilla). In an unusual remedy for rashes, eczema, and wart sores,
the Niitsitapi rub fire-blackened *Letharia* on the affected skin. Aqueous infusions of *Lobaria* can be used as a purifying tonic or in a bath to treat arthritis
(Gitksan). The Nuxalk peoples collect *Lobaria* only from dogwood or crabap-
ple and boil it to make an eyewash or a hot drink to treat stomach pains, or
pulverize it and apply it to the skin. The Gitga'at consider *Lobaria oregana* best
for treating sore throats when it is harvested from subalpine fir and boiled
with juniper. The Saanich may have used *Lobaria pulmonaria* and *Parmelia
sulcata* interchangeably for birth control purposes. *Parmelia sulcata* can be
rubbed on gums of teething babies to relieve discomfort (Métis). *Peltigera* species are used alone or as an ingredient in medicinal mixtures to dress wounds
(Oweekeno), to induce urination (Ditidaht), or treat tuberculosis (Dena'ina).
Cetraria canadensis can be used as a tea to treat coughs and colds or chewed
fresh to help the lungs (Ulkatcho).

Dyes

At least 40 of the region's lichen species can be used as natural dyes, including
species of beard lichens (*Alectoria, Bryoria, Usnea*), *Cladonia*, cyanolichens
(*Lobaria, Nephroma, Peltigera, Pseudocyphellaria, Solorina*), parmelioids
(*Evernia, Hypogymnia, Platismatia, Xanthoparmelia*), blackish and yellow ce-
trarioids, and orange lichens (*Xanthomendoza, Xanthoria*) (Brodo et al. 2001).
The value of lichens as dyes relates to their varied secondary chemistry and
especially to their light-absorbing pigments (e.g., usnic, fumarprotocetraric,
gyrophoric, vulpinic, lobaric, evernic, and diffractaic acids; atranorin, mela-
nins, parietin, sphaerophorin, and cyanobacterial mycosporines). The resulting
colors can differ markedly among species within a genus and the method of
extraction. The two most frequently used methods for extracting pigments
are boiling in water and fermentation in ammonia (traditionally from urine).
Boiling often extracts bright colors, especially yellows and oranges and shades
of browns, even though most secondary metabolites are not particularly water
soluble. Ammonia fermentation tends to yield rich browns and tans, with some
species yielding shades of pink and even purple.

The iconic ceremonial robe of the Tlingit, Haida, Tsimshian, and other coastal
peoples of British Columbia and Southeast Alaska has been central to these
cultures for many generations. Weaving the curvilinear and circular forms
following the pattern board is considered one of the most complex weaving
techniques in the world and it takes a full-time weaver about a year to create a
single robe. Whereas today's weavers may use commercial dyes, Emmons and
Boas (1907) noted after interviewing expert Chilkat weavers that "yellow was
obtained from the lichen *Evernia [Letharia] vulpina*, known as 'saxoti.' It does
not occur on the coast, but is most abundant back of the mountain range, and

was procured in trade from the interior. The color is extracted by boiling the lichen in the fresh urine of children until the desired shade is developed. Into this decoction the yarn is dropped, boiled for a very short time, and allowed to steep; it is then removed, rinsed, and hung in the sun to dry." Crawford (2007) mentioned that the Tlingit traded for *Letharia* from the Nuxalk, who obtained it from the Ulkatcho Dakelh in interior British Columbia.

Fiber

Beard lichens have additional traditional uses as textiles for bedding, clothing, cooking, and hygiene. These include stuffing mattresses and pillows with *Usnea*, notably *U. longissima* and *U. cavernosa* (Haida, Hanaksiala, Makah, Yuki, Pomo, Yoba), and lining steam pits and wiping the slime from salmon with *Usnea*, *Lobaria*, and *Alectoria* (Haida, Kwakwaka'wakw, Oweekeno). Baby diapers, sanitary napkins, and toilet wipes can be made from *Usnea* species, *Ramalina menziesii*, and *Alectoria sarmentosa* (Ditidaht, shíshálh, and Pomo). The traditional ponchos and footwear of the St'át'imc include *Alectoria sarmentosa* and *Usnea* species interwoven with silverberry bark fibers.

Food

Bryoria fremontii is an important traditional food throughout the inland forests of the Pacific Northwest, where it is widespread and abundant. Most cultures have preparations of it ranging from a delicacy to famine food (Crawford 2007; **Fig. 1**). The lichen is exceptionally abundant in the traditional territory of the Secwépemc, who call it "wila." *Bryoria fremontii* forms varying amounts of vulpinic acid, a bright yellow, toxic secondary metabolite. Gatherers choose low-vulpinic acid individuals using clues such as dark color and microhabitat preferences such as location and tree species. The lichen is always thoroughly washed to remove unwanted detritus and bitterness, and usually pit-cooked for at least half a day, typically with root vegetables such as camas bulbs and wild onions. The cooled lichen is gelatinous and eaten fresh or dried and stored. In modern times it may be served sweetened or with fruit, with meat in a soup, or made into a pudding with flour and butter (Brodo et al. 2001).

Except for *Lobaria pulmonaria*, which may have been eaten by the Coast Salish of Vancouver Island, *Bryoria fremontii* appears to be the only lichen widely consumed in the Pacific Northwest. People of more northerly arctic and boreal regions use reindeer lichens (*Cetraria* and *Cladonia* species), the epiphyte *Hypogymnia physodes*, cyanolichens (*Lobaria scrobiculata*, *Nephroma arcticum*, *Peltigera canina*), and the saxicolous *Umbilicaria* in traditional dishes. In general, these lichens are fermented or processed in some way and often boiled with other foods or added to soups to make them more palatable.

With the extensive effort required to collect and prepare *Bryoria fremontii*, one might presume that it is a valuable source of calories. Surprisingly, this

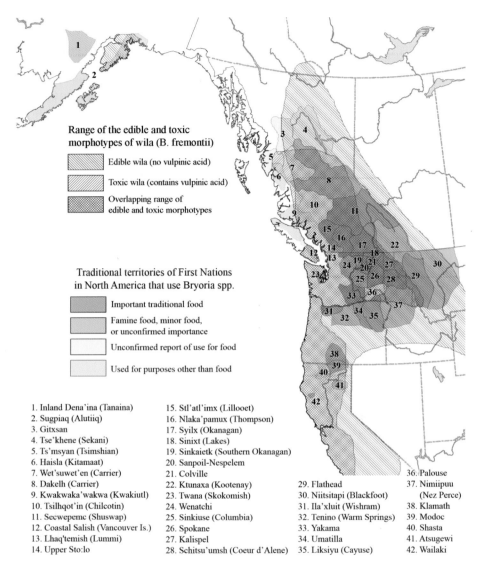

Range of the edible and toxic
morphotypes of wila (B. fremontii)

Edible wila (no vulpinic acid)

Toxic wila (contains vulpinic acid)

Overlapping range of
edible and toxic morphotypes

Traditional territories of First Nations
in North America that use Bryoria spp.

Important traditional food

Famine food, minor food,
or unconfirmed importance

Unconfirmed report of use for food

Used for purposes other than food

1. Inland Dena'ina (Tanaina)
2. Sugpiaq (Alutiiq)
3. Gitxsan
4. Tse'khene (Sekani)
5. Ts'msyan (Tsimshian)
6. Haisla (Kitamaat)
7. Wet'suwet'en (Carrier)
8. Dakelh (Carrier)
9. Kwakwaka'wakwa (Kwakiutl)
10. Tsilhqot'in (Chilcotin)
11. Secwepemc (Shuswap)
12. Coastal Salish (Vancouver Is.)
13. Lhaq'temish (Lummi)
14. Upper Sto:lo

15. Stl'atl'imx (Lillooet)
16. Nlaka'pamux (Thompson)
17. Syilx (Okanagan)
18. Sinixt (Lakes)
19. Sinkaietk (Southern Okanagan)
20. Sanpoil-Nespelem
21. Colville
22. Ktunaxa (Kootenay)
23. Twana (Skokomish)
24. Wenatchi
25. Sinkiuse (Columbia)
26. Spokane
27. Kalispel
28. Schitsu'umsh (Coeur d'Alene)

29. Flathead
30. Niitsitapi (Blackfoot)
31. Ila'xluit (Wishram)
32. Tenino (Warm Springs)
33. Yakama
34. Umatilla
35. Liksiyu (Cayuse)

36. Palouse
37. Nimiipuu
    (Nez Perce)
38. Klamath
39. Modoc
40. Shasta
41. Atsugewi
42. Wailaki

Figure 1. Map of the traditional territories of Pacific Northwest cultures that use
Bryoria as food and the distribution of toxic and edible morphotypes of *Bryoria
fremontii*. Figure courtesy of Stuart Crawford.

assumption remains unproven (Zhao et al. 2021). Lichen carbohydrates are
primarily medium-size polysaccharides made from glucose, called lichenins
(β-D-glucans) and isolichenins (α-D-glucans), and long-chain hemicelluloses
such as galactomannans. Human cells do not produce enzymes that can di-
gest these types of compounds. Historic human feeding trials with lichenin
found that the full caloric value passed through the digestive systems of

the participants (Crawford 2007). In a personal replication of the extensive Secwépemc traditional cooking process for wila, Crawford reported that none of the lichen carbohydrates were converted to component sugars (glucose, galactose, mannose). On the other hand, he did find that camas bulbs cooked with (compared to without) wila retained nearly double the calorie content in the food mixture. This was due to absorption of the nutrient-rich juices released by the camas bulbs in the fire pit by the mucilaginous lichen carbohydrates. Even if lichen carbohydrates do mostly transit human digestive systems as fiber, similar non-lichen β-glucans and galactomannans are known prebiotics. That is, some of the lichen carbohydrate might serve as food for beneficial bacteria in the lower intestine. Such bacteria manufacture essential short-chain fatty acids that have various beneficial physiological roles. Lichens can also be a source of proteins, vitamins, and minerals (Zhao et al. 2021). Globally, people utilize dozens of lichen species for food, typically specially processed and cooked with other foods, as in North America. In India, lichens are also used as spices and as preservatives, the latter presumably related to their activity as antioxidants, antimicrobials, and/or insecticides.

Lichens and Air Quality

Lichens, air quality, and a sustainable future

We rely on natural ecosystems for so many essential things: clean air, clean and abundant water, fertile soils, pollinators, food, wood and fiber, fish, wildlife, decomposition, carbon sequestration, and more. Natural ecosystems are in turn sustained by biodiversity, the intricate web of soil microbes, fungi, invertebrates, plants, and animals, all interacting interdependently. In an era with unprecedented human population and accompanying pressures on resources, achieving sustainability is ever more meaningful and necessary in all that we do. Reducing air pollution contributes to a sustainable future by helping to protect human health, biodiversity, the natural ecosystems that support human well-being, and by slowing climate change. Lichens can help us assess how well we are doing.

Why lichens are good indicators of air quality

Scientists use indicator species to inform us about ecosystem health. For example, we could define a person's health by a combination of their internal temperature, blood pressure, and cholesterol levels. These are indicators of health, not a comprehensive examination. Nor do the indicators in themselves usually allow the diagnosis of a specific problem. Rather, these indicators tell us when we need to examine a problem more closely. Lichens can inform us on ecosystem health with data that are easy to obtain yet tell us a lot (McCune 2000).

Physical and physiological characteristics of lichens underlie their indicator value for individual air pollutants. Unlike most vascular plants, which largely obtain nutrients and contaminants from the soil, lichens, especially epiphytes, are primarily comprised of nutrients and contaminants obtained from the atmosphere. Further, lichens can contain relatively high levels of pollutants because they lack the waxy cuticle, guard cells, and other specialized tissues that help vascular plants exclude pollutants. When wetted, pollutants deposited to their surfaces as vapors or fine particles dissolve and are absorbed, along with the wash from particulates deposited to canopy leaves and bark. These pollutants are concentrated when the lichen dries. Precipitation can leach pollutants again, creating a dynamic equilibrium between accumulation and leaching. Equilibration is fast—weeks to months for mobile elements needed for nutrition, like nitrogen or sulfur, and years for some metals that bind tightly to cell walls. Analyzing levels of pollutants in lichens can thus indicate or summarize their levels in the air.

In addition to accumulating pollutants, lichen algal and cyanobacterial partners are especially vulnerable to sulfur dioxide, ammonia, and sulfuric and nitric acids, and reactive gases, such as fluorides. These harm essential processes like photosynthesis and respiration by altering enzyme activity, oxidizing photosynthetic pigments, and physically compromising cellular ultrastructure or membrane integrity (Carter et al. 2017). Different species have different sensitivities to these pollutants, and thus the species present at a site or their condition can also indicate acidic deposition (acid rain) or fluoride emissions. Similarly, species differ in their requirements for nutrients such as nitrogen, so community composition can also be used to detect enhanced levels of fertilizing pollutants such as nitrates and ammonium.

Lichen indication methods

Various methods have been devised for using lichens as indicators. Selecting an approach depends on one's resources, pollutants, and questions of interest.

Community analyses yield the most comprehensive ecological assessments of air pollution effects and can provide evidence of harm, especially if influences like climate, forest type, and disturbance history are considered. Tracking indicator species, indices of species groups, or critical load exceedances, are the simplest community indication methods. See further discussion under nitrogen and sulfur pollutants and greenhouse gases below.

Changes in physiology, growth rates, or appearance of lichens can be used to document harm from air pollution. For example, carbon dioxide uptake, chlorophyll ratios, and enzyme activity rates are used to assess photosynthesis (Calatayud et al. 2000, Gries et al. 1995, Riddell et al. 2008). Weighed transplants or permanent photo points can be used to detect changes in growth rates and cover. Ultrastructural changes can be detected by microscopy (Ra et al. 2004). Bleaching, reddening, or necrosis of lobes; smaller, thicker thalli; and hyper-production of reproductive propagules are outward signs of air-pollution damage to lichens. If the nitrogen deposition is high, nitrogen-loving lichens (eutrophs) and green algae may even grow over previously established lichens. Observe these responses by comparing the same species across similar habitats (e.g., open-grown mature oaks) in your study area.

Chemical analysis of lichen thalli is most often used to track individual contaminants. Elemental analysis can indicate the influence of specific industries like aluminum manufacturing (fluorine), coal-fired power plants (mercury, sulfur, and $\delta^{34}S$), mining (ore metals), or agriculture (nitrogen, sulfur, phosphate); vehicle exhaust (chromium, nickel, vanadium, titanium); fallout from nuclear explosions and accidents (^{106}Ru, ^{137}Cs, ^{144}Ce); acid rain (calcium, sulfur, nitrogen); sea-salt (sodium, sulfur, magnesium); or dust (aluminum, iron, calcium, silicon). Element concentrations in lichens or mosses can be used to map urban pollution hotspots (Donovan et al. 2016) or, if calibrated to instrumented measures of pollutants, to estimate deposition (Root et al. 2013). Chemical analysis of semi-volatile organic pollutants in Pacific Northwest lichens indicated deposition patterns of pesticides, polychlorinated biphenyls, and polycyclic aromatic hydrocarbons (Landers et al. 2008). In general, epiphytes are preferred for study, being less affected by local soil dust, soil water, or seasonal snow burial than species like *Xanthoparmelia, Rhizocarpon, Umbilicaria,* and *Dermatocarpon,* which are often used in non-forested areas. Good target epiphytic lichens in the region are *Platismatia glauca, Letharia vulpina, L. columbiana,* common *Hypogymnia* species, *Parmelia sulcata,* and *Evernia prunastri. Alectoria sarmentosa* and *Bryoria fremontii* are widespread, but they are not as good accumulators and tend not to occur at low elevations or in areas with acidic or enhanced nutrient nitrogen deposition (e.g., urban areas). *Usnea* species can be hard to tell apart in the field. Collections should be single species. Multiple thalli from many trees should be collected to obtain a sample that is representative of the population at the test site. Ideally the dry weight of the dried sample will be about 10 g and will be free of other forest debris, dead material, bark, or other lichens and mosses.

Lichen indication of regional air pollutants, sources, and effects

The Clean Air Act charges the US Environmental Protection Agency with regulating air pollutants in the United States; the individual states implement those regulations. Regulated pollutants include four greenhouse gases, six criteria air pollutants, and 188 hazardous air pollutants. According to the 2017 National Emissions Inventory, the biggest sources of these pollutants in Oregon and Washington are mobile sources, fires, industrial processes, and fuel combustion. Lichens can be used to assess status, trends, effects, and sources of many of these regional pollutants (**Table 2**).

GREENHOUSE GASES (GHGs)

In sheer tonnage, GHGs are the predominant regional pollutants, comprising about 95% of emissions (US EPA 2020). Greenhouse gases are the chief cause of climate change, a very serious problem. The most recent Intergovernmental Panel on Climate Change report (IPCC 2022) calls for significant reductions by 2030 to stave off a catastrophic future. Of the four GHGs tracked in 2017, carbon dioxide contributed ~95% of the hundred-year climate warming potential in Oregon and Washington.

Climate strongly influences lichen community composition. A few lichens tolerate large fluctuations in climate, but most have adapted to more specific regimes (Smith et al. 2017). Even a 1 °C change in mean annual temperature can drastically affect the probability of finding some rare coastal lichens (Glavich et al. 2005b). Climate change–related shifts in the geographic distribution of these species and overall biodiversity can be tracked by monitoring lichen communities (Jovan 2008, Will-Wolf 2010). Lichen community vulnerability indices (Smith et al. 2020) and species turnover (Robert J. Smith, unpublished analyses) can help focus adaptive management actions.

Air quality, climate change, and the health of people and nature are intertwined. Warming climate will likely increase future ground-level ozone and ammonia emissions. And, although wildfires are a natural and beneficial part of Pacific Northwest ecosystems, warming is also projected to continue increasing fire frequency and intensity, generating smoke, GHGs, criteria pollutants, and hazardous air pollutants that will challenge cleanup efforts. Conversely, continued reductions in other sources of air pollution will improve health outcomes for people and climate resiliency of ecosystems. For example, reducing nitrogen deposition by just 2 kg/ha/y may offset the effects of several degrees C of warming on lichen communities (Geiser et al. 2021).

CRITERIA AIR POLLUTANTS

The Clean Air Act criteria pollutants are fine particulate matter < 10 and < 2.5 microns in diameter (PM_{10}, $PM_{2.5}$), nitrogen oxides (NO_x), sulfur dioxide (SO_2), ozone, carbon monoxide, and lead. Fine particulates are associated with respiratory and cardiovascular harm, a leading cause of early human mortality and acute and long-term respiratory and cardiovascular conditions. They also contribute to smog, which reduces visibility. Nitrogen and sulfur oxides are harmful as gases and lead to formation of other nitrogen and sulfur compounds that contribute to particulate matter and enhanced deposition of fertilizing pollutants—especially nutrient nitrogen and acid deposition.

Nitrogen oxides and ammonia. The primary nitrogen-containing air pollutants emitted to the atmosphere are nitrogen oxides (NO and NO_2) and ammonia (NH_3). Dominant regional sources are mobile sources, fires, agriculture, biogenics, and fuel combustion. An essential component of proteins and nucleic acids, N can have beneficial fertilizing effects (Ra et al. 2005), but too much can adversely affect entire ecosystems (Fenn et al. 2003). After climate change gases, excess deposition of N-containing pollutants is the greatest air-quality concern in the Pacific Northwest—not to be confused with N_2, the largely inert, dominant gas of the earth's atmosphere.

Nitrogen (N) deposition. Nitrogen is primarily accumulated by lichens as fine particulates or solutes of ammonium (NH_4) and nitrates (NO_3) in precipitation and canopy throughfall, from dry deposition of ammonia, and in small quantities as nitric acid (HNO_3) vapor. Pacific Northwest lichens are adapted to different levels of N deposition and availability. *Oligotrophs* grow only in nutrient N-poor environments, *mesotrophs* have an intermediate nutrient N requirement, and *eutrophs* thrive in nutrient N-enhanced environments (**Fig. 2**). A few lichens have broad ecological amplitudes and tolerate a large range of nitrogen availability. In a general way, assignment to oligotroph, mesotroph, or eutroph classes corresponds to the more traditional classification used by lichenologists: *acidophytes, neutrophytes* and *nitrophytes*, because lichens of low-fertility environments frequently grow in high-rainfall, leached environments on acidic conifer bark, whereas moderate to high-fertility environments often have neutral hardwood or even alkalinized substrates. However, our grouping, commensurate with available information, is based solely on response to total N deposition. As anthropogenic N is added to background N levels, the excess N favors eutrophs (*Candelaria, Xanthoria, Xanthomendoza, Physcia, Physconia,* and *Melanelixia*) at the expense of endemic and ecologically valuable oligotrophs and mesotrophs, including forage (*Alectoria, Bryoria, Usnea*) and N_2-fixing (*Lobaria, Peltigera, Pseudocyphellaria, Nephroma*) lichens (Geiser and Neitlich 2007). The distribution of common eutrophs such as *Candelaria pacifica, Physcia adscendens,* and *Xanthoria polycarpa* indicate hotspots of nutrient nitrogen deposition.

Sulfur dioxide. Lichens were originally used as air quality monitors of sulfur dioxide (SO_2) emissions related to coal combustion in Europe (Nash 2002). Some lichens are extremely sensitive to SO_2. Sources of regional SO_2 emissions are fires (60%); combustion of coal, tar, diesel, heating oil, and marine bunker fuels; petroleum processing; and ore smelting. Pollution controls on industries required by the Clean Air Act have brought about continuous emission declines. SO_2 may still affect the most sensitive lichens very close to point sources, for example, northern Washington's Puget Sound (Johnson 1979; Taylor and Bell 1983) but acid deposition is a greater concern.

Acid deposition. Acid deposition occurs when sulfuric and nitric acids formed from atmospheric conversions of SO_2 and NO_x are deposited to vegetation, soils, and surface waters. It is not a widespread problem in the Pacific Northwest: rainwater pH monitored by the National Atmospheric Deposition program is largely within background ranges (> 5.2). However, it can be problematic in major urban centers and areas downwind like the Columbia River Gorge (Fenn et al. 2007). Acid deposition is exacerbated by fog, which greatly concentrates pollutants (Muir 1991). Total sulfur (S) deposition is a good metric for assessing acid deposition, and modeled data are available for the entire region. Most lichens have a preferred substrate pH range; therefore, community composition can be used to pinpoint sites along regional gradients of acid deposition (Geiser et al. 2021), quantified in units of S deposition. Cyanolichens and large foliose and pendent species tend be most sensitive.

N and S critical loads. A critical load is the depositional loading below which no known harm occurs to a specific ecosystem component from a given atmospheric pollutant according to current knowledge. Units are typically in kilograms of total atmospheric deposition per hectare per year (kg/ha/y). Preventing exceedance of lichen critical loads provides broad protection for other sensitive biota and can help prevent declines in the biodiversity, growth, health, and ecological integrity of forests from fertilizing and acidifying pollutants. Nutrient nitrogen (N) and acid deposition (S) critical loads have been established for epiphytic macrolichen community composition (Geiser et al. 2021); species richness (total number of species); sensitive species richness; diversity and abundance of ecological functional groups, such as forage and cyanolichens (Geiser et al. 2019); and for trees, herbs, mycorrhizal fungi, surface waters, freshwater diatoms, and fish (Clark et al. 2019, Horn et al. 2018, Pardo et al. 2011). Thanks to the Clean Air Act and the hard work of the state and federal air quality agencies, most criteria pollutants are decreasing in the region, but elevated N deposition is still observed in many urban and rural areas (**Fig. 3**).

Regional air pollution sensitivity ratings. Table 3 provides N and S air pollution sensitivity ratings for epiphytic macrolichens in this book. You can use these in several ways. The presence of species that are sensitive to N and S deposition (especially if there are many, abundant, and in good health) indicates good air quality and low deposition of N- and S-containing air pollutants. Conversely, sites dominated by N- and S-tolerant species are likely enhanced in nutrient nitrogen or acidified.

Another way to use the ratings is to calculate a community "airscore." Conduct a survey of epiphytic macrolichens on a 35-m-radius circular plot containing at least 10 trees (preferably > 40 years old) and rate each macrolichen species' abundance by the number of individuals you observed (1 = 1-3 individuals, 2 = 4-10 individuals, 3 = 10+ individuals on up to half of the trees, 4 = found on more than half the trees). Search for at least 30 minutes but not more than two hours. Look up the deposition rating of each species, then multiply it by the abundance rating you gave it. Sum the abundance-adjusted ratings for all the species, then divide by the sum of the abundance ratings. This is your N or S airscore, expressed in units of kg of N or S deposition per hectare annually. Refer to **Figure 4** to place your site in a regional context. Scores under 4 indicate good air quality, 4-6 are marginal, and 7 or more indicate enhanced N or S deposition. Note that very warm or dry sites (e.g., southern Oregon valleys, dry forests of eastern Oregon and Washington) will have slightly higher airscores for N deposition because hot, dry climates and high nutrient deposition cause somewhat similar stresses on community composition. Sulfur airscores are not affected by climate. There are many other ways to use lichens to assess spatial and temporal trends in N- and S-containing pollutants, including the use of radioisotopes in source apportionment, and differentiating reduced versus oxidized forms of nitrogen (Anderson et al. 2017, Jackson et al. 1996, Jovan et al. 2012).

Ozone. Ground-level ozone is a secondary pollutant produced by heat and sunlight in the presence of oxygen, nitrogen oxides, and volatile organic compounds. Certain vascular plants, notably ponderosa pine, are more sensitive than lichens and are better indicators of this pollutant. Lichens are most likely to be affected where ozone episodes frequently exceed 120 ppb, especially if the episodes occur when lichens are hydrated and metabolically active. Such episodes only rarely occur in the Pacific Northwest, and regional lichen sensitivities are not established. Also, ozone responses can be difficult to separate from nitric acid responses (Riddell et al. 2008).

HAZARDOUS AIR POLLUTANTS (HAPs)
The 188 hazardous air pollutants vary in lifetimes and toxicities, but all HAPs are carcinogens and can be associated with other serious health risks, such as reproductive harm. They include volatile and semi-volatile organic compounds and metals. The US Environmental Protection Agency's EJScreen maps neighborhoods with greater than average cancer risk from air toxics. Lichen and moss biomonitors have been used to indicate levels of various HAPs in remote parks and wilderness areas of the United States (Landers et al. 2008) and across Europe (Frontasyeva et al. 2020). They are providing an accessible and inexpensive air quality monitoring approach for research and community scientists in urban neighborhoods of Portland (Donovan et al. 2016) and Seattle (Derrien et al. 2020). Analysis of lichens for metals can complement instrumented monitors by providing high spatial resolution.

MICROPLASTICS
Microplastics, fragments of any type of plastic under 5 mm in length, are now nearly ubiquitous in the environment. All of us inhale and consume microplastics daily and have measurable blood levels. Microplastics even occur in precipitation at remote locations of the western United States (Brahney et al. 2020). The health and environmental impacts are still being assessed, but some key sources in the western United States are now known to be roads, oceans, and agricultural dust (Brahney et al. 2021). Lichens (Loppi et al. 2021) and mosses (Roblin and Aherne 2020) can be used to assess types of plastics, size ranges, and amounts of plastics at a given location.

MORE INFORMATION

Learn more about community science and biomonitoring with these videos and webpage:

US Forest Service. "Critical Loads Video Series: Lichens" (video, 9:15). https://www.youtube.com/ playlist?list=PLNsZX2SBTlVl21CSjtvzZYp8Pigp3vKvm.

Nichols, Jesse. "What moss tells us about air pollution" (video, 8:59). *Grist.* https://www.youtube.com/watch?v=TtFC2cB5h2E.

U. S. Forest Service, Pacific Northwest Research Station. "Duwamish Valley communities collaborate on urban air quality biomonitoring projects." https://www.fs.usda.gov/pnw/projects/duwamish-valley-communities-collaborate-urban-air-quality-biomonitoring-projects.

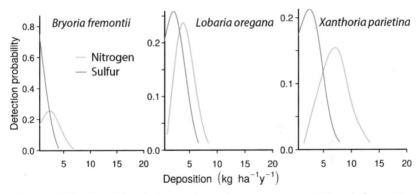

Figure 2. Sulfur (purple) and nitrogen (green) response curves of *Bryoria fremontii, Lobaria oregana,* and *Xanthoria parietina* across their western United States distributions. The probability of detecting each species increases with deposition to an optimum (peak), then decreases. The depositions at the optima serve as a quantitative pollution sensitivity rating (Table 3). *Bryoria* is most likely to be detected where N deposition is < 4.2 kg/ha/y (oligotroph); *Lobaria,* from 4.2 to 8.0 kg/ha/y (mesotroph); and *Xanthoria,* > 8.0 kg/ha/y (eutroph). Their tolerances for acidic deposition (S optima) follow the same increasing order, though all are S-sensitive (S optima < 4.2 kg/ha/y).

Figure 3. Lichen community indication can be used to map estimated regional nitrogen deposition and to document effects, such as shifts from ecologically valuable, pollution-sensitive species to pollution tolerant species. In this map of lichen community-based airscores from surveys conducted between 1993 and 2018 in the northwestern United States, high deposition is indicated by N-tolerant, eutroph-dominated lichen communities and high airscores (reds and purples). Low deposition is indicated by N-sensitive, oligotroph-dominated lichen communities and low air scores (yellows and oranges). Surveys conducted by the US Forest Service (Geiser et al. 2021); all scores climate adjusted.

Figure 4. Once a critical load has been established, modeled deposition alone can be used to map and quantify predicted harm to that component of the ecosystem. Here we use a map of modeled total N deposition for 2015-2017 (CMAQ v5.1) to detect areas exceeding the critical load (3.2 kg/ha/y) for epiphytic lichen species richness. Species richness is the total number found at a site and is a measure of biodiversity. The critical load was selected to allow for no greater than a 20% reduction in species richness, a value that accounts for measurement errors. N deposition in most areas does not exceed the critical load; however, species richness declines of 40% to 60% are predicted for Portland and Seattle. National Forest boundaries are shown by red lines.

Table 2. Regional air pollutants and lichen indication methods. Main rows provide the pollutant category, substances in that category regulated under the Clean Air Act, and major regional emissions sources. The sum of Oregon and Washington emissions (US EPA 2020) is given in million short tons for each category, except microplastics. Carbon dioxide contributes ~95% of the 100-year warming potential (CO_2eq) of the GHGs. Mobile emissions are dominated by gasoline-powered cars and diesel trucks. Wildfires contribute about 93% to 97% of fire emissions.

Pollutants	What to measure	What it can indicate	References
Greenhouse Gases (4) carbon dioxide, methane, nitrous oxide, sulfur hexafluoride 157 million short tons from mobile sources, fires, industrial processes	Community vulnerability indices, species turnover rates	Places on landscape to anticipate rapid ecological changes associated with climate change; focus areas for ecological forestry; temporal and spatial trends in lichens.	Smith et al. 2020, Smith et al. 2017
Criteria Air Pollutants (6) sulfur dioxide, nitrogen oxides, (ammonia), lead 8 million short tons from mobile sources, fires, fuel combustion, biogenics, industrial processes	Community air quality indices; indicator species; elemental N, S, Pb and N, S, Pb isotope ratios; physiological or morphological condition; critical loads exceedances	Temporal and spatial trends in deposition of acidic and fertilizing compounds and quantification of ecological harm. Estimation of N or S deposition. For radioisotope analysis: source attribution of nitrogen-, sulfur-, and lead-containing pollutants, and differentiation of reduced vs. oxidized forms of nitrogen.	Anderson 2017, Geiser et al. 2019 and 2021, Hoffman et al. 2019, Jackson et al. 1996, Jovan et al. 2012; McMurray et al. 2013, Root et al. 2014
Hazardous Air Pollutants (188) other metals, semivolatile organic compounds, persistent organic pollutants 0.6 million short tons from fires, biogenics, mobile sources	Individual compounds	Bioaccumulation, spatial and temporal deposition patterns, source attribution, urban hotspots.	Derrien et al. 2020, Donovan et al. 2016, Frontasyeva et al. 2020, Landers et al. 2010
Microplastics from roads, ocean aerosols, agricultural dust	Types of plastics, size ranges, quantities	Types and amounts of microplastics deposited from the air.	Loppi et al. 2021, Roblin and Aherne 2020, Brahney et al. 2021

Table 3. Air quality ratings for Pacific Northwest lichens for sulfur (S) and nitrogen (N) pollutants. Fxl gp = ecological functional group. Cya = cyanolichen, For = forage lichen, Mtx = matrix lichens (other green algal macrolichens); p – pendent, sh = shrubby, s = small, m= medium, l = large. S class = Acid sensitivity class. S = sensitive; I = intermediate, T = tolerant. N class = Nutrient nitrogen class. O = oligotroph; M = mesotroph, E = eutroph. Class breaks are at < 4.2, 4.2-8, and > 8 kg/ha/y total S or total N deposition. S deposition is a measure of acidic deposition (acid rain); N deposition is a measure of nutrient nitrogen or fertilizing sources of nitrogen in deposition. S optm, N optm = total nitrogen (N) or sulfur (S) deposition in kg/ha/y corresponding to the lichen's optimum, or peak, detection frequency across its western United States distribution. Empty cells have insufficient information to assign values. Deposition modeled by CMAQ model version 5.0.2, grid cell diameter 12 km (US EPA 2019).

Lichen	Fxl gp	S class	S optm	N class	N optm
Alectoria imshaugii	For p	S	0.4	O	2.0
Alectoria lata	For p	S	1.8		
Alectoria sarmentosa	For p	S	0.4	O	0.8
Alectoria vancouverensis	For p	S	1.6	O	2.8
Bryoria americana	For p			O	0.8
Bryoria bicolor	For p	I	5.2	O	0.8
Bryoria chalybeiformis	For p	S	0.4	O	2.5
Bryoria fremontii	For p	S	0.4	O	1.8
Bryoria furcellata	For p	S	0.4	O	2.4
Bryoria fuscescens	For p	S	0.4	O	2.5
Bryoria glabra	For p			O	0.8
Bryoria kockiana	For p	I	5.3	O	0.8
Bryoria nadvornikiana	For p	S	1.2	O	2.8
Bryoria pseudofuscescens	For p	S	0.4	O	1.8
Bryoria pseudofuscescens var. friabilis	For p	S	1.4	O	3.4
Bryoria pseudofuscescens var. pikei	For p			O	2.7
Bryoria simplicior	For p	S	0.4	O	1.5
Bryoria tenuis	For p	S	1.3	O	0.8
Bunodophoron melanocarpum	For sh	S	1.3	O	0.8
Candelaria pacifica	Mtx s	T	8.8	E	12.3
Cetraria californica	Mtx s	I	5.9	O	0.8
Cetraria canadensis	Cya s-m	S	0.4	O	1.7
Cetraria chlorophylla	Mtx m-l			O	2.5
Cetraria coralligera	Mtx m-l	S	1.7	M	5.3
Cetraria fendleri	Mtx m-l	S	1.5	M	4.5

Lichen	Fxl gp	S class	S optm	N class	N optm
Cetraria juniperina	Cya s-m	S	0.4	O	1.7
Cetraria merrillii	Mtx s	S	0.4	O	1.6
Cetraria orbata	Mtx m-l	S	3.0	M	6.8
Cetraria pallidula	Mtx m-l	S	0.4	O	2.4
Cetraria pinastri	Cya s-m	S	1.4	M	4.6
Cetraria platyphylla	Mtx m-l	S	0.4	O	2.3
Cetraria sepincola	Mtx m-l	S	1.2	O	0.8
Cetraria sphaerosporella	Mtx m-l	S	0.4	O	1.8
Cetraria subalpina	Mtx m-l	S	2.1	M	4.4
Cetrelia cetrarioides	Mtx m-l	T	8.8	E	9.3
Cladonia albonigra	Mtx s	S	1.2	O	0.8
Cladonia bacillaris + macilenta	Mtx s	S	1.5	M	4.3
Cladonia bellidiflora	Mtx s	S	1.6	O	0.8
Cladonia cariosa	Mtx s	S	1.5	M	5.5
Cladonia carneola	Mtx s			O	0.8
Cladonia cenotea	Mtx s	T	8.8	O	0.8
Cladonia coccifera	Mtx s	S	1.4	O	0.8
Cladonia coniocraea	Mtx s	T	8.8		
Cladonia cornuta	Mtx s	I	5.2	O	0.8
Cladonia deformis	Mtx s	S	1.2	O	0.8
Cladonia fimbriata	Mtx s	T	8.8		
Cladonia grayi	Mtx s			O	0.8
Cladonia multiformis	Mtx s	S	1.6	M	7.2
Cladonia norvegica	Mtx s	S	1.6	M	4.3
Cladonia pleurota	Mtx s	S	1.2	O	0.8
Cladonia pyxidata	Mtx s	S	1.8	M	5.4
Cladonia scabriuscula	Mtx s	I	5.3	O	0.8
Cladonia squamosa	Mtx s	S	2.5		
Cladonia straminea	Mtx s	S	1.3	O	0.8
Cladonia sulphurina	Mtx s	S	2.2	O	0.8
Cladonia transcendens	Mtx s	I	5.5	E	9.5
Cladonia umbricola	Mtx s			O	0.8
Cladonia verruculosa	Mtx s	S	2.6	M	7.2
Collema furfuraceum	Cya s-m	S	3.3	E	8.8
Collema nigrescens	Cya s-m	I	5.5	E	10.0
Collema subflaccidum	Cya s-m	S	0.8	M	4.6
Erioderma sorediatum	Cya s-m	S	1.5	O	0.8
Esslingeriana idahoensis	Mtx m-l	S	0.4	O	2.5

Lichen	Fxl gp	S class	S optm	N class	N optm
Evernia divaricata	For sh	S	1.9	M	5.2
Evernia mesomorpha	For sh	S	1.1	O	3.0
Evernia prunastri	For sh	T	8.8	E	8.7
Flavoparmelia caperata	Mtx m-l	I	5.1	E	12.3
Flavopunctelia flaventior	Mtx m-l	I	5.8	E	9.9
Flavopunctelia soredica	Mtx m-l	S	1.7	M	4.9
Fuscopannaria ahlneri	Cya s-m	I	5.3	O	0.8
Fuscopannaria leucostictoides	Cya s-m	I	5.3	O	0.8
Fuscopannaria mediterranea	Cya s-m	I	5.2	M	6.6
Fuscopannaria pacifica	Cya s-m	I	5.2	O	2.7
Fuscopannaria praetermissa	Cya s-m	S	1.6	O	0.8
Gabura insignis	Cya s-m	S	1.6	O	0.8
Heterodermia leucomela	Mtx m-l	I	5.2	E	8.2
Hypogymnia apinnata	Mtx m-l	I	4.5	O	0.8
Hypogymnia austerodes	Mtx m-l			O	3.4
Hypogymnia bitteri	Mtx m-l	S	2.1	O	0.8
Hypogymnia duplicata	Mtx m-l	S	1.6	O	0.8
Hypogymnia enteromorpha	Mtx m-l	S	2.4	M	7.2
Hypogymnia heterophylla	Mtx m-l	S	1.7	O	2.4
Hypogymnia hultenii	Mtx s	I	4.9	O	0.8
Hypogymnia imshaugii	Mtx m-l	S	0.4	O	2.3
Hypogymnia inactiva	Mtx m-l	S	2.6	M	7.6
Hypogymnia lophyrea	Mtx s			O	0.8
Hypogymnia occidentalis	Mtx m-l	S	0.4	O	2.3
Hypogymnia oceanica	Mtx m-l	I	4.6	O	0.8
Hypogymnia physodes	Mtx m-l	T	8.8	E	8.5
Hypogymnia rugosa	Mtx m-l			O	3.2
Hypogymnia tubulosa	Mtx m-l			M	7.8
Hypogymnia vittata	Mtx m-l			O	0.8
Hypogymnia wilfiana	Mtx m-l	S	0.4	O	2.4
Hypotrachyna afrorevoluta	Mtx m-l			E	9.2
Hypotrachyna sinuosa	Mtx m-l	S	2.9	E	8.7
Imshaugia aleurites	Mtx m-l	T	8.8		
Imshaugia placorodia	Mtx m-l	S	1.9	M	5.2
Lathagrium fuscovirens	Cya s-m	S	1.3	M	6.3
Leioderma sorediatum	Cya s-m	S	1.6	O	0.8
Leptochidium albociliatum	Cya s-m	I	5.3	M	5.5
Leptogidium contortum	Cya s-m	S	1.4	O	0.8
Leptogium cyanescens	Cya s-m	S	2.4	M	6.9

Lichen	Fxl gp	S class	S optm	N class	N optm
Leptogium pseudofurfuraceum	Cya s-m	S	1.9	M	6.0
Leptogium saturninum	Cya s-m	S	2.5	M	7.9
Letharia columbiana	For sh	S	0.4	O	1.7
Letharia vulpina	For sh	S	0.4	O	2.0
Lobaria anomala	Cya l	S	1.5	O	0.8
Lobaria anthraspis	Cya l	S	1.1	O	2.9
Lobaria hallii	Cya l	I	5.3	O	0.8
Lobaria linita	Cya l			O	0.8
Lobaria oregana	Cya l	S	1.7	O	0.8
Lobaria pulmonaria	Cya l	S	1.8	O	3.0
Lobaria scrobiculata	Cya l			O	0.8
Loxosporopsis corallifera	Mtx s				
Melanelixia californica	Mtx m-l	T	8.8	E	12.3
Melanelixia glabratula	Mtx m-l	T	8.8	E	9.5
Melanelixia subargentifera	Mtx m-l			E	12.3
Melanelixia subaurifera	Mtx m-l	S	3.5	E	8.4
Melanohalea elegantula	Mtx m-l	S	0.4	O	4.0
Melanohalea exasperatula	Mtx m-l	S	0.4	M	4.5
Melanohalea subolivacea	Mtx m-l	S	0.4		
Menegazzia subsimilis + terebrata	Mtx m-l	S	2.6	E	9.0
Montanelia disjuncta	Mtx m-l	I	5.2	E	9.2
Montanelia panniformis	Mtx m-l	S	0.4		
Montanelia sorediata	Mtx m-l	S	0.4	O	2.5
Montanelia tominii	Mtx m-l	I	5.1	M	6.0
Nephroma arcticum	Cya l			O	0.8
Nephroma bellum	Cya l			O	0.8
Nephroma isidiosum	Cya l	I	5.3	O	0.8
Nephroma laevigatum	Cya l	S	1.6	O	3.9
Nephroma occultum	Cya l	I	5.3	O	2.9
Nephroma parile	Cya l			O	0.8
Nephroma resupinatum	Cya l	S	1.7	M	6.4
Nephroma tropicum	Cya l	S	1.4	O	0.8
Niebla cephalota	Mtx m-l	I	5.2	E	12.3
Nodobryoria abbreviata	For p	S	0.4	O	1.9
Nodobryoria oregana	For p	S	0.4	O	2.7
Parmelia pseudosulcata	Mtx m-l	T	8.8	O	0.8
Parmelia saxatilis	Mtx m-l	S	3.2		

Lichen	Fxl gp	S class	S optm	N class	N optm
Parmelia squarrosa	Mtx m-l			O	0.8
Parmelia sulcata	Mtx m-l	T	8.8	E	8.5
Parmeliella parvula	Mtx s	S	1.3	O	0.8
Parmeliella triptophylla	Mtx s	I	5.3	O	0.8
Parmelina coleae	Mtx m-l	S	3.0	E	12.3
Parmeliopsis ambigua	Mtx s	S	0.4	O	2.4
Parmeliopsis hyperopta	Mtx s	S	0.4	O	2.6
Parmotrema arnoldii	Mtx m-l	S	3.1	E	8.9
Parmotrema perlatum	Mtx m-l	S	2.8	E	9.1
Parmotrema crinitum	Mtx m-l	S	1.6	O	3.5
Peltigera collina	Cya l	I	2.7	M	8.0
Phaeophyscia ciliata	Mtx s	I	4.9		
Phaeophyscia decolor	Mtx s	S	1.4	M	4.4
Phaeophyscia hirsuta	Mtx s	S	1.5	M	4.3
Phaeophyscia hirtella	Mtx s	S	1.4	O	3.6
Phaeophyscia kairamoi	Mtx s	S	0.8	O	4.1
Phaeophyscia nigricans	Mtx s	S	1.5	M	5.8
Phaeophyscia orbicularis	Mtx s	I	5.1	E	12.3
Phaeophyscia sciastra	Mtx s	S	1.3	M	6.3
Physcia adscendens	Mtx m-l	T	8.8	E	12.3
Physcia alnophila	Mtx m-l	S	3.0	E	8.6
Physcia biziana	Mtx m-l	S	1.4	E	12.3
Physcia caesia	Mtx m-l	S	1.3	M	6.0
Physcia dimidiata	Mtx m-l	I	5.6	E	12.3
Physcia dubia	Mtx m-l	S	1.5	M	5.6
Physcia occidentalis	Mtx m-l	S	0.4	M	4.9
Physcia stellaris	Mtx m-l	S	1.8		
Physcia tenella	Mtx m-l	T	8.8	E	12.3
Physciella chloantha	Mtx s	S	1.3	M	5.2
Physciella melanchra	Mtx s	S	2.7	E	8.8
Physconia americana	Mtx m-l	I	4.6	E	12.3
Physconia enteroxantha	Mtx m-l	I	5.1	E	12.3
Physconia fallax	Mtx m-l	T	8.8	E	12.3
Physconia muscigena	Mtx m-l	S	0.7	M	4.7
Physconia perisidiosa	Mtx m-l	I	5.1	E	12.3
Platismatia glauca	Mtx m-l	T	8.8	M	8.0
Platismatia herrei	Mtx m-l	S	2.2	M	6.7
Platismatia lacunosa	Mtx m-l	S	1.5	O	0.8

Lichen	Fxl gp	S class	S optm	N class	N optm
Platismatia norvegica	Mtx m-l			O	0.8
Platismatia stenophylla	Mtx m-l	S	2.3	M	7.0
Polychidium muscicola	Mtx s	S	0.4	O	0.8
Protopannaria pezizoides	Cya s-m	S	1.3	O	0.8
Pseudocyphellaria citrina	Cya l	I	5.1	O	0.8
Pseudocyphellaria mallota	Cya l	S	1.2	O	0.8
Pseudocyphellaria rainierensis	Cya l	S	1.4	O	0.8
Punctelia jeckeri	Mtx m-l	S	3.4	E	9.1
Ramalina dilacerata	For sh	S	2.4	M	7.4
Ramalina farinacea	For sh	T	8.8	E	8.4
Ramalina labiosorediata	For sh	I	5.7	M	7.0
Ramalina menziesii	For p	I	4.9	M	5.9
Ramalina obtusata	For sh	I	5.3		
Ramalina roesleri	For sh	I	4.4	O	0.8
Ramalina sinensis	For sh	S	1.8	M	5.2
Ramalina subleptocarpha	For sh	T	8.8	E	12.3
Ramalina thrausta	For p	S	1.4	O	0.8
Ricasolia amplissima	Cya l	S	1.3	O	0.8
Scytinium cellulosum	Cya s-m	S	1.6	M	7.6
Scytinium palmatum	Cya s-m	S	2.4	M	8.0
Scytinium polycarpum	Cya s-m	T	8.8	E	8.8
Scytinium teretiusculum	Cya s-m	S	1.1	E	8.2
Sphaerophorus tuckermanii + venerabilis	For sh	S	2.2		
Steineropsis laceratula	Cya s-m	S	1.3	O	0.8
Sticta fuliginosa	Cya l	S	2.7	M	8.0
Sticta limbata	Cya l	S	2.6	E	8.2
Sticta rhizinata	Cya l	S	1.5	O	0.8
Sulcaria spiralifera	For p	S	2.4	O	2.3
Usnea cavernosa	For p	S	1.8	M	4.7
Usnea cornuta	For sh			E	12.3
Usnea dasopoga	For p	S	2.4	M	6.3
Usnea flavocardia	For sh	S	3.5	E	8.9
Usnea fragilescens	For sh	S	2.1	O	4.0
Usnea glabrata	For sh	S	1.6	O	4.1
Usnea hirta	For sh	S	1.7	M	4.8
Usnea intermedia	For sh	S	2.1	M	5.8
Usnea lapponica	For sh	S	1.6	O	4.1
Usnea longissima	For p	S	2.6	O	0.8

Lichen	Fxl gp	S class	S optm	N class	N optm
Usnea pacificana	For sh	S	2.4	M	7.6
Usnea rubicunda	For sh	S	1.6	O	2.6
Usnea scabrata	For p	S	2.5	O	2.6
Usnea silesiaca	For sh	T	8.8	O	0.8
Usnea subfloridana	For sh	S	3.4	M	6.0
Usnea subgracilis	For sh	S	1.6	O	3.6
Usnea wasmuthii	For sh			E	12.3
Xanthomendoza fallax	Mtx s			E	12.3
Xanthoria candelaria	Mtx s	T	8.8		
Xanthoria parietina	Mtx m-l	S	3.0	E	10.0
Xanthoria polycarpa	Mtx s	T	8.8	E	12.3
Xanthoria sorediata	Mtx s	S	1.3	M	5.0
Xanthoria tenax	Mtx s	I	6.1	E	12.3

Lichens and Climate Change

Lichen communities: sentinels of changing climate

Epiphytic macrolichens are especially useful climate indicators due to their ubiquity, year-round presence, exposed position, and species-specific climate tolerances. Community composition can help pinpoint a site on a climate gradient, assess the biological and ecological impacts of climate change, and focus management responses. For example, Smith et al. (2020) provided climate niche limits for 444 lichen species encompassing 24 climate variables. From a list of lichen species present at a site, one can use these to calculate climate vulnerability indices, such as the percentage of species exceeding upper climatic limits, the mean community vulnerability, and the climate safety margins.

Regional climate has largely warmed and slightly moistened, though not uniformly

Regional climates have been warming since the first edition of this book. Temperatures increased at 91% of US Forest Service lichen air quality and climate monitoring sites in Oregon and Washington by the 20-year revisits conducted from 2013 to 2018. Cooling occurred at some coastal sites (-0.2 °C), and other low-elevation sites experienced no or little warming (0 to +0.2 °C). Warming increased with elevation, peaking at the Cascade crest and eastwards (+0.4 °C). Despite the warming, the average site was slightly moister (climatic moisture deficit [CMD] -2.5 mm); high levels of drying occurred only in southwestern Oregon (CMD up to -51 mm) (Wang et al. 2016).

Lichens have responded and there are some ecological implications

Lichen species distributions have responded to this warming. A few species have emerged as warming "winners" and are now more commonly encountered in the Cascade crest vicinity. These winners are all common green algal lichens, including generalist beard lichens (*Alectoria sarmentosa, Nodobryoria oregana*), foliose tube lichens (*Hypogymnia apinnata, H. enteromorpha, H. imshaugii, H. physodes*), and parmelioids (*Parmeliopsis hyperopta, Platismatia glauca,* and *Pl. herrei*). However, more species are decreasing, especially those requiring cool, moist conditions year-round. Warming "losers" include the beard lichens (*Bryoria americana, B. glabra, Usnea flavocardia, U. glabrata*), medium to large cyanolichens (*Leptogium saturninum, Nephroma tropicum, N. laevigatum, N. parile, Sticta fuliginosa, S. limbata*), and some green algal lichens (*Cetraria californica, C. subalpina, Cetrelia cetrarioides, Hypogymnia duplicata*). Declines in these and other ecologically valuable forage and nitrogen-fixing cyanolichens are expected to continue with future warming. In comparison, despite high levels of warming, species turnover attributable to warming was relatively low in the dry, continental forests of the eastern and southern Cascades.

Forest management practices may help to mitigate effects of climate change on the lichen biota

Forest management practices can offset macroclimatic changes; for example, by reducing hazardous fuels and changing vegetation density. Both of these can alter microclimates and disturbance regimes. So if warming and drying are happening at the macro level, it could make sense to retain vegetation. Stand density prescriptions such as variable-retention forestry in moist, West Cascades forests can retain a diversity of microclimates for non-timber vegetation, reduce soil and understory moisture stress, and protect soil carbon. These and other ecological forestry practices can be consistent with sustaining lichen diversity, reducing hazardous fuels, promoting timber production, and sustaining soil resilience even in drier, East Cascades forests with low average canopy cover. In general, forest management actions that r.educe the risk of severe wildfires, promote species migration by providing refugia, and reduce other stressors such as invasives and pests and diseases, will help mitigate the early effects of climate warming.

More information

See Smith et al. (2017, 2020) for recent research on lichens and climate change.

Key to the Macrolichen Genera in the Pacific Northwest

Names in square brackets are potentially present but have not yet been found in Oregon and Washington. Non-bolded endpoints are not treated further in the book.

Introductory Key

1a Fungus a basidiomycete; fruiting structures mushroom-like
or club-shaped.. **Key A**, page 2

1b Fungus an ascomycete; fruiting structures various but not
mushroom-like

  2a Thallus gelatinous, without obvious internal layers, black to
brown or gray; photobiont blue green except in two
intertidal spp...**Key B**, page 3

  2b Thallus stratified (or too finely filamentous to tell), color
various; photobiont blue green and/or green

    3a Thallus yellow, orange, red, or fluorescent yellow green
.... **Key C**, page 4

    3b Thallus another color (including white, green, dull
yellow green, gray, or brown)

      4a Thallus foliose, with dorsiventral lobes

        5a Thallus umbilicate (having a central holdfast),
on rock .... **Key D**, page 4

        5b Thallus not umbilicate, more broadly attached,
or if centrally attached then not on rock

          6a Primary photobiont blue green (stratified
cyanolichens); upper surface usually some
shade of dark gray, brown, or black .... **Key E**, page 5

          6b Primary photobiont green (look for grass green
algal layer); upper surface variously colored .... **Key F**, page 7

      4b Thallus fruticose, rarely with dorsiventral lobes

        7a Thallus blackish filamentous, prostrate or erect,
minute or large...................................... **Key G**, page 12

        7b Thallus not of black filaments; color various

          8a Thallus of hollow stalks (podetia) that are ±
round and pointed or tipped with apothecia
or cups..**Key H**, page 13

          8b Thallus not of hollow stalks (either not
stalked or not hollow).......................**Key I**, page 14

Key A: Basidiomycetes

1a Fruiting body with typical mushroom form, tan or brown, with
   decurrent gills; lichenized thallus greenish and granular to small
   squamulose; on rotten wood and peaty soil; widespread, common
   .... *Lichenomphalia* (not treated further)
1b Fruiting body short, club shaped, pinkish brown, yellowish, tan,
   or whitish; lichenized thallus greenish, granular or membranous;
   on soil; widespread, occasional........... *Multiclavula* (not treated further)

Lichenomphalia

Multiclavula

Key B: Gelatinous, Blackish

*Thallus gelatinous, non-stratified; non-filamentous; black to brown
or dark gray, photobiont mostly blue green*

1a Habitat aquatic, submerged most of the year
  2a Habitat freshwater; photobiont blue green; ascocarps apothecia
    3a Lobes 2-15 mm broad, loosely attached; lower surface veined
    ....*Peltigera hydrothyria*
    3b Lobes < 2 mm broad, coalescing and tightly attached in dark, thin
      rosettes that are difficult to remove from rock; lower surface difficult
      to view..................................... ***Scytinium rivale*** (see *Leptogium* key)
  2b Habitat rocky intertidal; photobiont green; ascocarps perithecia
    4a Spores elongate ellipsoidal, 11-18 x 3-5 µm; gelatinous matrix
      enclosing asci in center of ascocarp; thallus with short, thin lobes
      to 1 cm wide, shiny greenish gray; algal partner *Prasiola borealis*,
      algal cells often in regular groups, the groups separated by fungal
      mycelium; Alas to s BC .... [*Kohlmeyera complicatula*]
    4b Spores ellipsoidal, 10-13 x 3.5-7 µm; gelatinous matrix filling only the
      ostiole; thallus of tubular to flattened lobes; algal partner *Blidingia*
      minima var. *vexata*; algal cells irregularly arranged in surface view;
      fungal hyphae layered between algae; BC to Cal
      ....*Turgidosculum ulvae*
1b Habitat mostly not aquatic but occasionally in seasonally wet habitats or
  seepage
  5a Thallus glossy black, on rock, in section with a thick white central layer.
    Widespread, rare in PNW, common in SW.............*Lichinella nigritella*
  5b Thallus brown, gray, or dull black, substrate various, in section
    completely dark or with a whitish paper-thin central layer
    6a Thallus generally dull, black or olive-black, lacking a cortex, the upper
      surface of interwoven hyphae, lacking isodiametric cells in surface
      view under LM
      7a Spores nonseptate; thallus tiny; upper surface of interwoven
        hyphae; rare spp
        8a Apothecia lacking thalloid margin; thallus granulose to
          subfruticose ....*Leciophysma* (not treated here)
        8b Apothecia with thalloid margin; thallus either of flat, interlaced
          lobes with fruiting warts or of small knobby clusters
          .... *Lempholemma*
      7b Spores septate; thallus tiny to large; upper surface various; including
        common species .... *Collema* (and related genera)
    6b Thallus with a cellular cortex, generally of ± isodiametric cells; thallus
      often faintly shiny, brown, gray or black
      9a Hyaline hairs present on and near the lobe margin; thallus weakly
        stratified, a broken lobe showing a pale layer sandwiched between
        dark algal layers; on moss and soil or soil over rock
        .... *Leptochidium albociliatum*

9b Hyaline hairs lacking on upper surface or if present then thallus homogeneous at 10-40X; substrate various .... *Leptogium* and *Scytinium*

Key C: Yellow, Orange, Red

1a Thallus foliose
   2a Thallus cortex K+ purple .... *Xanthoria* and *Xanthomendoza*
   2b Thallus cortex K-
      3a Lobes < 0.5 mm wide and < 2 mm long................. *Candelaria pacifica*
      3b Lobes > 1 mm wide .... *Cetraria* (*Vulpicida* group)
1b Thallus fruticose
   4a Thallus medium to large (generally > 2 cm diameter)
      5a Thallus bright yellow to fluorescent green; widespread .... *Letharia*
      5b Thallus orange; usually near the coast
         6a Main branches with a tough central cord ....*Usnea rubicunda*
         6b Main branches without a central cord .... *Teloschistes flavicans*
   4b Thallus minute (generally < 2 cm long), yellow to orange
      7a Soredia present
         8a Thallus dwarf erect fruticose, generally < 2 mm high; on rock, occasionally trees... *Xanthoria candelaria*
         8b Thallus small, prostrate to suberect fruticose, generally > 2 mm high; sheltered crevices in calcareous rock in steppe and dry forests....................................*Teloschistes contortuplicatus*
      7b Soredia lacking
         9a On seacoast rocks...*Caloplaca coralloides*
         9b On other substrates
            10a Thallus knobby, short-branched, yellow to orange, typically forming compact globular clusters on twigs; apothecia generally present .... *Xanthoria*
            10b Thallus of short, branched, erect, orange filaments, the filaments forming a mat generally < 5 mm high; apothecia absent; common on tree trunks in many habitats, generally where moist .... *Trentepohlia* (a free-living alga)

Key D: Umbilicate, Foliose

1a Photobiont blue green; thallus often < 1 cm wide .... *Peltula*
1b Photobiont green; thallus generally > 1 cm wide
   2a Ascocarps forming perithecia (visible as black dots on the surface) or fruiting structures lacking; isidia and soredia lacking
      3a Perithecia present.. *Dermatocarpon*
      3b Perithecia lacking

4a Habitat usually semi-aquatic (streamside rocks, trickle lines on
 cliffs, seepy cracks in outcrops, etc.) ....*Dermatocarpon*
4b Habitat various but seldom semi-aquatic...................... *Umbilicaria*
2b Ascocarps forming apothecia; isidia or soredia occasionally present
 5a Thallus white to gray, pruinose, appearing areolate, KC+R;
  apothecia brown, sunken in the thallus; spores minute, spherical,
  2-4 μm diameter, many per ascus; on calcareous rock in
  continental climates .... [*Glypholecia scabra*]
 5b Thallus and apothecia otherwise; spores larger, 8/ascus
  6a Thallus yellowish to greenish, occasionally whitish; apothecia
   lecanorine ....*Rhizoplaca*
  6b Thallus brown, black, gray, whitish gray or red; apothecia
   lecideine .... *Umbilicaria*

Key E: Stratified Foliose Cyanolichens

1a Thallus with veins on the lower surface (occasionally the veins ± obscure);
 lower cortex absent (appearing dull and fibrous under lens)
 2a Medulla white; lower surface never orange...................................*Peltigera*
 2b Medulla orange; lower surface orange with brown veins*Solorina crocea*
1b Thallus without veins on the lower surface; lower cortex present or not
 3a Thallus minutely foliose or subfoliose; lobes < 2 mm wide
  4a On bark or wood
   5a Cortex or medulla P+O; thallus with definite lobes...........*Pannaria*
   5b Cortex and medulla P-; thallus with definite lobes or nearly
    squamulose .... *Fuscopannaria*
    (*Parmeliella* and *Steineropsis* are also keyed under *Fuscopannaria*.
    Koerberia biformis, a minute blackish olive species on trees in Cal,
    would also key here.)
  4b On rock, moss, humus, or soil, or moss over bark or wood
   6a On mosses or humus over rock or bark, sometimes directly on rock;
    thallus brown or bluish brown; lobes never longitudinally ridged
   7a Thallus brown or dark brown; lobes isidiate to lobulate; spores 2-4
    celled.. *Massalongia*
   7b Thallus gray or brownish gray; lobes isidiate becoming sorediose;
    spores simple ....*Fuscopannaria* (including *Moelleropsis*)
   6b On rock; thallus blackish, gray green, or olive colored; lobes often
    longitudinally ridged or striate
   8a Lower surface dark ....*Placynthium* (not treated)
   8b Lower surface pale.. *Tingiopsidium*
 3b Thallus medium to large foliose; lobes > 2 mm wide
  9a Lower surface with discrete white or yellow spots or pores (cyphellae
   or pseudocyphellae) < 2 mm diameter
  10a Lower surface with broad (generally 0.5-2 mm) whitish craters
   (cyphellae)... *Sticta*

8b Upper cortex K- or cortex too dark to see the K reaction;
    upper and lower surfaces various .... **Group 3** (Lead 26)
7b Lobes generally broader, generally > 1.5 mm wide; thallus
    medium to large, often > 3 cm .... **Group 4** (Lead 40)

Group 1

With faint yellowish tint (usnic acid)

9a Soredia present
  10a Upper surface with whitish pores (pseudocyphellae); soredia laminal
    and marginal .... *Flavopunctelia*
    (*Cetrelia* can have yellowish tints but has only marginal soralia and
    lacks usnic acid)
  10b Upper surface lacking pores; sorediate or not
    11a On rock
      12a Medulla UV+ white, K- .... *Arctoparmelia*
      12b Medulla UV-, K+Y to R .... *Xanthoparmelia mougeotii*
    11b On bark and wood
      13a Soredia marginal and/or terminal
        14a Lobes narrow, < 2(3) mm broad; lower surface black; rhizines
          branched and abundant; medulla C-, KC-; w Cas
          .... *Hypotrachyna sinuosa*
        14b Lobes medium, > 2 mm broad; lower surface brown; rhizines
          generally simple, sparse; medulla C+R, KC+R; substrate various,
          mainly e Cont Div .... *[Flavopunctelia soredica]*
      13b Soredia laminal
        15a Lobes < 2 mm broad, tightly appressed .... *Parmeliopsis ambigua*
        15b Lobes > 2 mm broad, loosely appressed. Medulla K-, P+R; Cal
          and e N Am, rare in w Ore .... *Flavoparmelia caperata*
9b Soredia lacking
  16a On bark or wood
    17a Thallus suberect, pale greenish; pycnidia mainly marginal and
      submarginal, black; at low elevations .... *Cetraria pallidula*
    17b Thallus closely appressed, pale greenish in shade to nearly black
      where exposed; pycnidia laminal, black; mainly subalpine
      .... *Cetraria sphaerosporella*
  16b On soil or rock
    18a On rock, moss over rock, or occasionally soil; thallus erect or
      appressed; upper surface yellow green or greenish to blackish, lower
      surface pale tan, brown, or black but differing in color from the
      upper surface .... *Xanthoparmelia*
    18b On soil or alpine sod; thallus erect or suberect; upper and lower
      surfaces of lobes similar in color .... *Cetraria nivalis* and *C. cucullata*

Group 2

Narrow-lobed foliose, pale above, cortex K+Y, usnic acid lacking

19a On soil in semiarid sites; apothecial margin lecideine; thallus closely appressed to calcareous soil, upper surface K+ dirty Y. Uncommon e Cas .... *Buellia elegans*

19b Not on soil; apothecial margin lecanorine; thallus and substrate various; upper surface K+Y

  20a Isidia present; thallus closely appressed to bark ..... *Imshaugia aleurites*

  20b Isidia lacking; thallus and substrate various

    21a Lower surface densely pock-marked with tiny pits (use lens); mainly w Cas.......................................*Hypogymnia hultenii* and *H. lophyrea*

    21b Lower surface not pitted or thallus too closely appressed to easily tell

      22a Lower cortex lacking; thallus loosely attached; marginal cilia present..*Heterodermia leucomela*

      22b Lower cortex present or too closely appressed to tell; marginal cilia present or not

        23a Soredia present

          24a Soralia orbicular and laminal; thallus tightly appressed to bark; spores hyaline, nonseptate; medulla UV+ white .... *Parmeliopsis hyperopta*

          24b Soralia otherwise or, if orbicular and laminal, then growing on rock; spores brown, septate; medulla UV-.............. *Physcia*

        23b Soredia absent

          25a Disk pale brown to medium brown; spores hyaline, nonseptate; medulla P+ deep Y; s RM, n to e Mont .... *[Imshaugia placorodia]*

          25b Disk dark brown to black (or white pruinose); spores brown, septate; medulla P- or P+ pale Y .... *Physcia*

Group 3

Narrow-lobed foliose; gray (or whitish pruinose),
brown or black; cortex K-

26a Thallus brown or brownish black *and* on rock

  27a Pycnidia marginal, often protruding isidia-like; pseudocyphellae, if present, marginal; lobes often ascending; mostly subalpine to alpine ....*Melanelia hepatizon* and *Cetraria commixta*

  27b Pycnidia immersed, mainly laminal if present; pseudocyphellae, if present, laminal; lobes and habitat various

    28a Lower surface lacking rhizines; lobes narrow and thick, often nodulose; pseudocyphellae lacking; mostly alpine..*Allantoparmelia*

    28b Lower surface rhizinate; lobes thin to moderately thick; pseudocyphellae present or not; widespread

      29a Upper surface gray to brown, often with pale angular markings .... *Parmelia*

(Dark forms of normally pale gray spp, e.g., *P. omphalodes* and *P. saxatilis*, will key here.)

29b Upper surface a rich brown to almost black, lacking angular markings or angular markings very subtle

30a Cortex N+ dark blue green; thallus with globular isidia that often burst open to become soredia-like; pseudocyphellae lacking .... *Neofuscelia*

30b Cortex N- (turning greener just from being wet); isidia present or not; pseudocyphellae present or not

.... *Melanelia* and related genera

26b Thallus another color *or* on another substrate

31a Lower surface white to pale brown

32a Soredia lacking

33a Thallus with few or no marginal rhizines; on calcareous rock and moss over rock.. *Anaptychia roemerioides*

33b Thallus with abundant long, pale marginal rhizines or cilia; on bark or rock

34a Lower cortex lacking; on bark .... *Anaptychia crinalis*

34b Lower cortex present; on rock or moss over rock

.... *Phaeophyscia constipata*

32b Soredia present

35a Lobes very narrow and linear, generally < 0.3 mm broad, with sparse granular soredia or isidia on some lobe tips and margins; occasional (generally overlooked) on bark and rock

.... *Phaeophyscia nigricans*

35b Lobes broader (mostly > 0.3 mm), with discrete laminal, marginal, or terminal soralia

36a On calcareous rock; upper surface pruinose

.... *Anaptychia elbursiana*

36b On trees and shrubs; upper surface scarcely or not at all pruinose

37a Lobe tips lacking pale cortical hairs; lower surface always pale below; lower cortex a compact mass of elongate fungal cells

.... *Physciella*

37b Lobe tips with pale cortical hairs (inspect protected areas of the thallus — the hairs are fragile and often rubbed off); lower surface pale or dark below; lower cortex is a compact mass of ± isodiametric cells .... *Phaeophyscia hirsuta*

31b Lower surface brown to black or thallus too closely appressed to tell

38a Lobes suberect or erect.. *Cetraria*

38b Lobes ± appressed

39a Rhizines squarrosely branched; upper cortex frequently heavily pruinose.. *Physconia*

39b Rhizines simple or sparsely dichotomously branched upper cortex occasionally pruinose .... *Phaeophyscia*

Group 4

Medium- to broad-lobed foliose; not yellowish (usnic acid lacking)

40a Thallus brown, greenish brown, olive-brown, or brown-black
  41a Thallus erect or suberect, loosely or partially attached or ± free from substrate; rhizines not abundant.. *Cetraria*
  41b Thallus broadly attached to substrate; rhizines generally abundant
    42a Cortex N+ dark blue green; isidia present, globular, often sorediose; pseudocyphellae lacking; on rock...................................... *Neofuscelia*
    42b Cortex N-; isidia present or not; pseudocyphellae present or not; substrate various
      43a Lower surface lacking rhizines; lobes narrow and ± thick; on rock; pseudocyphellae lacking... *Allantoparmelia*
      43b Lower surface with rhizines; lobes and substrate various; pseudocyphellae present or not........ *Melanelia* and related genera
40b Thallus gray, greenish gray, or bluish gray, sometimes with browned edges and lobe tips
  44a Thallus erect or suberect, loosely or partially attached or ± free from substrate; rhizines absent to moderately abundant
    45a Upper or lower surface with pseudocyphellae *or* lobe margins with black cilia; primarily w Cas
      46a Upper and/or lower surfaces lacking pseudocyphellae; marginal cilia generally present; lobes medium to broadly rounded .... *Parmotrema*
      46b Upper and/or lower surfaces with small white pseudocyphellae; marginal cilia lacking; lobes broadly rounded
        47a Lobes with laminal isidia and a network of sharp ridges .... *Platismatia norvegica*
        47b Lobes with mainly marginal soredia, isidia, or lobules and smooth to weakly ridged
          48a Lobes isidiate and/or lobulate; lower surface pale .... *Pseudocyphellaria rainierensis*
          48b Lobes with marginal soralia; lower surface black with brown edges.. *Cetrelia cetrarioides*
    45b Upper surface lacking pseudocyphellae *and* margins lacking cilia; common inland and on the coast
      49a Lower surface black (edges may be brown); marginal pycnidia abundant; lobes < 4 mm broad; apothecia commonly present; isidia and soredia lacking.......................... *Esslingeriana idahoensis*
      49b Lower surface black, brown, or white, often mottled with those colors; marginal pycnidia apparent or not; lobes often > 4 mm broad; apothecia present or not; isidia or soredia present or not .... *Platismatia*
  44b Thallus broadly attached to substrate; rhizines generally abundant
    50a Lower surface pale brown or tan, rarely black; upper surface with pores; medulla C+R, KC+R .... *Punctelia*

50b Lower surface black (generally brown edged); upper surface with or without pores; medulla spot tests various
  51a Upper surface with whitish angular markings (either conspicuous or subtle); medulla C- .... *Parmelia*
  51b Upper surface sometimes cracked but without whitish angular markings; medulla C- or C+R
    52a Rhizines simple or sparsely branched; soredia absent
    ....*Parmelina coleae*
    52b Rhizines dichotomously branched; soredia present
    ....*Hypotrachyna*

Key G: Fruticose, Blackish Filamentous

1a Filaments over 2 cm long and over 0.1 mm in diameter; photobiont a single-celled green alga
  2a Thallus forming black, richly branched, prostrate flat mats on siliceous rock ....*Pseudephebe*
  2b Thallus color various, sparsely branched, prostrate or erect; substrate various but most common on trees .... *Bryoria*
1b Filaments < 2 cm long or < 0.1 mm in diameter; photobiont blue green, often filamentous (LM needed to proceed with confidence from this point)s
  3a Filaments very fine, main branches < 20 μm (= 0.02 mm) diameter
    4a Fungal hyphae on the surface of the filaments very irregular and bent (LM) .... *Cystocoleus ebeneus*
    4b Fungal hyphae parallel and straight to somewhat knobby, coalescing (LM) .... *Racodium rupestre*
  3b Filaments more coarse, main branches generally > 70 μm diameter
    5a On rock, mosses over rock, soil, or moss over soil
      6a Thallus mat forming; on rock ....*Ephebe* (not treated here)
      6b Thallus cushion forming; substrate various
        7a Spores brown, 2- to 4-celled; on soil and mosses. Cephalodia fruticose; thallus reduced; arctic-alpine to montane; rare
        ....*Solorina spongiosa*
        7b Spores hyaline, spores multicellular to muriform; on all substrates
          8a On mosses over rock; spores 2-celled; thallus densely cushion forming, richly branched, with glossy round branches; apothecia usually present ....*Polychidium muscicola*
          8b On various substrates including moss over rock; spores multicellular to muriform; thallus minutely fruticose or more often with fruticose projections from a dorsiventral thallus
          ....*Scytinium* (see *Leptogium* key)
    5b On bark
      9a Hyaline hairs present on surface (LM); on bark w Cas...... *Ricasolia*
      9b Hyaline hairs lacking; substrate various

10a Cortical cells in jigsaw pattern (LM); photobiont *Scytonema*
  (cylindrical filamentous)
    ....*Leptogidium contortum* (see *Polychidium* key)
10b Cortical cells forming a mosaic but not jigsaw patterned
  11a Photobiont *Nostoc* (beaded-filamentous)
    ....*Scytinium* (see *Leptogium* key)
  11b Photobiont *Scytonema* (cylindrical-filamentous)
    .... [*Leptogidium dendriscum*] (see *Polychidium* key)

Key H: Fruticose, Hollow Stalks

1a Stalks tipped with bright red apothecia...*Cladonia*
1b Stalks tipped with brown apothecia or apothecia lacking
  2a Stalks richly branched (generally branched more than 5 times over the
    length of a mature individual), the branch tips pointed.........*Cladonia*
  2b Stalks simple to few-branched, pointed, blunt, or tipped with cups
    3a Thallus wholly of erect to prostrate stalks, lacking squamules; habitat
      mainly subalpine to alpine
      4a Thallus white to cream; stalks pointed.............................*Thamnolia*
      4b Thallus yellowish brown to brownish, or yellowish green,
        occasionally purplish brown; stalks blunt
        5a Thallus yellowish green; podetia branched
          ....*Cetraria madreporiformis*
        5b Thallus yellowish to brownish; podetia branched or not
          ....*Dactylina*
    3b Thallus differentiated into a basal, crustose to squamulose primary
      thallus and an erect secondary thallus (podetia); habitat various
      6a Stalks < 1 mm diam and to 2(5) mm tall, tipped with a black
        powdery apothecium (mazaedium).................. *Tholurna dissimilis*
      6b Stalks either > 1 mm diam or more than 4 mm tall, never tipped
        with a black mazaedium
        7a Podetia unbranched, short (generally < 1.5 cm), mostly naked
          and ending in a brown to pinkish apothecium; primary thallus
          continuous, warty-squamulose.................................... *Baeomyces*
        7b Podetia unbranched or branched, short or tall, often bearing
          squamules or soredia; primary thallus squamulose, generally
          persistent but sometimes disappearing......................... *Cladonia*

Key I: Fruticose, not Hollow-Stalked

1a Thallus free-living (loose) on soil......................................**Group 1** (Lead 5)
1b Thallus attached, substrate various
  2a Thallus brown to black, olive black, pale brown, or grayish brown
    ....**Group 2** (Lead 9)
  2b Thallus some shade of gray, green, yellow green, or whitish
    3a Cortex lacking; thallus whitish, of minute, delicate pseudopodetia
      forming a short (generally < 1 cm), leprose-appearing turf; common
      in many open habitats ....*Leprocaulon* (not treated here)

3b Cortex present; thallus various
    4a Thallus white, gray, olive gray, or brownish white (usnic acid
        lacking) ....**Group 3** (Lead 18)
    4b Thallus yellowish green to pale green (containing usnic acid)
        ....**Group 4** (Lead 25)

Group 1

Thallus free-living (loose) on soil

5a Thallus gray, olive, or brown above (usnic acid lacking)
    6a Branches dorsiventral but tightly rolled, gray or brown above, brown to
        black below .... *Dermatocarpon*
    6b Branches round, not dorsiventral, olive, brown, greenish, or greenish
        gray..*Aspicilia*
5b Thallus pale greenish to yellow green (usnic acid present)
    7a Lobes in part round in cross section ....*[Rhizoplaca haydenii]*
    7b Lobes flat or rolled
        8a Medulla K-; thallus of broad lobes that are somewhat divided and
            rolled under but not repeatedly branched ....*Rhizoplaca*
        8b Medulla K+Y, O, or R, rarely K-; thallus of elongate lobes that are
            repeatedly branched...*Xanthoparmelia*

Group 2

Thallus brown to black, olive black, pale brown, or grayish brown

9a On bark or wood; at all elevations
    10a Thallus dark reddish brown; cortical cells knobby and jigsaw-like in
        surface view (LM); soredia absent ....*Nodobryoria*
    10b Thallus dark brown, olive, tan, occasionally grayish; cortical cells
        smooth and elongate in surface view (LM); soredia often present
        11a On conifers on the immediate coast; thallus tufted, small (< 2(3) cm
            long).. *Cetraria californica*
        11b Characters various but not both small-tufted and on the coast
            12a Branches twisted, distinctly furrowed, dull brown to yellowish
                brown; isidia and soredia lacking............ *Sulcaria* (key in *Bryoria*)
            12b Branches otherwise; isidia and soredia present or not ....*Bryoria*
9b On soil, rock, or alpine sod; mainly at mid to high elevations
    13a Thallus of erect, ± flattened lobes (though sometimes channeled or
        tubular); mainly at mid to high elevations......*Cetraria islandica* group
    13b Thallus of branches that are ± round or angular in cross section
        14a Thallus reddish brown
            15a Pseudocyphellae absent; medulla C-, KC-; thallus prostrate
                .... *Nodobryoria subdivergens*
            15b Pseudocyphellae present; medulla C+R or C-, KC+R or KC-;
                thallus erect, often forming ± compact turfs
                ....*Cetraria* (*aculeata* and *muricata*)
        14b Thallus olive, brown, tan, or black

16a Thallus forming compact tufts of stiff, thin or thick branches, dark brown to black or olive black; attached to subalpine or alpine rocks..*Cornicularia normoerica*

16b Thallus of slender branches, color various; on soil, alpine sod, or rock

   17a Thallus generally black, pale brown or straw-colored at the base; cortex and medulla KC+R, P+ yellowish.......*Alectoria nigricans*

   17b Thallus color various; spot tests otherwise ....*Bryoria*

Group 3

Thallus white, gray, olive gray, or brownish white

18a Thallus with unbranched to sparingly branched stalks, generally < 2 cm long

  19a Ascocarp a mazaedium; mazaedia on tips of short stalks; on bark or wood, less often on rock

    20a Thallus white or pale yellowish white; stalks often > 1mm diameter; medulla deep yellow to orange...........*Acroscyphus sphaerophoroides*

    20b Thallus pale gray to greenish gray; stalks < 1 mm diam; medulla white... *Tholurna dissimilis*

  19b Ascocarp not forming a mazaedium or ascocarp not found; substrate various

    21a Stalks (actually long isidia on a crustose thallus) < 2 mm high and < 0.2 mm thick, white; on bark..... *Loxosporopsis corallifera*

    21b Stalks generally > 5 mm high or > 0.5 mm thick, grayish to greenish gray, sometimes brownish in exposed habitats; on rock, wood, or soil

Loxosporopsis corallifera

      22a On rock (rarely wood); stalks commonly tipped with black apothecia

      .... *Pilophorus*

      22b On soil in coastal bogs; stalks never with apothecia

      ....*Siphula ceratites*

18b Thallus with repeatedly branched stalks, mostly > 2 cm long

  23a Stalks lumpy and/or tomentose, with minute cauliflower-like outgrowths; spores produced in apothecia; mostly on rock or moss over rock...*Stereocaulon*

  23b Stalks smooth and glossy; spores produced in a globose, irregularly bursting ascocarp with a loose spore mass (mazaedium) inside; mostly on trees

    24a Stalks roundish to irregular in cross section; common

    ....*Sphaerophorus*

    24b Stalks distinctly flattened, consistently broadly elliptical in cross section; rare.. *Bunodophoron melanocarpum*

Group 4

Thallus yellowish green to pale green (containing usnic acid)

25a Main branches with a tough central cord (stretch branches with fingers or slice with razor blade); most spp with numerous perpendicular branchlets .... *Usnea*

25b Main branches without a central cord; most spp with few or no perpendicular branchlets

  26a On alpine sod

    27a Branches soft, the cortex often cracked and eroding; whitish elongate pseudocyphellae lacking .... *Evernia divaricata*

    27b Branches cartilaginous, tough, the cortex not cracked and eroding; whitish elongate pseudocyphellae present.......... *Alectoria vexillifera*

  26b On bark or wood, rarely on rock

    28a Branches greenish above, whitish below, flat, dorsiventral .... *Evernia prunastri*

    28b Branches not differentiated into an upper and lower surface, mostly pale greenish above and below

      29a Branches soft, irregular in cross section; rare

        30a Branches often becoming black-spotted; on trees and rocks on the immediate coast.. *Niebla cephalota*

        30b Branches not black-spotted; not on the coast .... *[Evernia mesomorpha]* and E. divaricata

      29b Branches cartilaginous, roundish or irregular in cross section

        31a Thallus flat to irregular in cross section........................ *Ramalina*

        31b Thallus roundish in cross section

          32a Branch tips tightly curled into a tiny fiddlehead with minute terminal soralia; main branches generally < 1 mm, without prominent elongate pseudocyphellae ....*Ramalina thrausta*

          32b Branch tips curved or straight but not curled into fiddleheads; main branches becoming thick (often > 1 mm), with elongate often spiraling pseudocyphellae .... *Alectoria*

KEYS AND DESCRIPTIONS OF SPECIES

Acroscyphus

Acroscyphus sphaerophoroides

Description: Thallus fruticose, grayish white to pale yellowish white, a knobby cluster or forming more elongate branches that are 0.5-5.0 mm diam and to 20(40) mm long, tough, solid, tipped with apothecia; fertile branches cylindrical to slightly swollen around the apothecia; medulla light orange to deep brownish orange; apothecia common, to 3 mm diam, forming a black powdery spore mass (mazaedium) within a cup-shaped thalline margin that is pale yellowish or light orangish brown at the edge; spores 1-septate, dark; cortex and medulla K+Y or R, C + rose or R, KC+R, P+ deep Y (atranorin, gyrophoric acid, usnic acid and five other substances).

Range: Coastal BC to Wash Cas (Snohomish County).

Substrate: On wood and rock.

Habitat: Low elevation to subalpine, open rocky areas or forests.

Notes: One of the rarest macrolichen in the PNW, this species is known from two sites in Wash (1300 m, *Abies amabilis – Tsuga* forest) and several sites in BC. Although widespread, occurring at both low and high latitudes worldwide, including Bhutan, China, Japan, Mex, S Am, and S Africa, it is widely scattered and globally rare. Dispersed by animals, rather than wind, the black powdery spores are easily transferred by touch.

Sources include: Joneson and Glew (2003), Tibell (1996).

Alectoria

Thallus fruticose, erect and tufted, pendulous, or prostrate, yellow-green, pale green, or sometimes blackening, in one sp typically blackish; branches generally terete, occasionally flattened or angular; soredia absent (ours) but isidioid spinules arising from the pseudocyphellae in one sp; pseudocyphellae present, ± conspicuous, elongate, whitish; apothecia lateral, disk tan, dark brown, or blackish; pycnidia rare; usnic acid (cortex KC+Y) present in all but one sp; medulla K+Y or K-, P+Y or P-, KC+R or KC-; on bark, wood, soil, rock, or arctic-alpine sod; conifer forests and tundra.

Sources include: Brodo and Hawksworth (1977); Halonen et al. (2009).

ID tips: A basic lichen trick to learn in the PNW, easily mastered by small children, is to distinguish *Alectoria* from pendulous *Usnea*. Both genera are abundant in our forests, so you really need to know how to do this. Gradually pull a strand by the ends: *Alectoria* pops apart with a clean break, while *Usnea* shows a stretchy central cord like a slender rubber band. Although normally pale greenish, all *Alectoria* species will partially blacken in exposed environments. But only one species, *A. nigricans*, is predominantly and consistently blackish.

1a Thallus tufted
  2a Isidioid spinules present in tufts along the branches or arising from the pseudocyphellae; on bark and wood ....***A. imshaugii***
  2b Isidioid spinules absent; substrate various
    3a Thallus mottled grayish brown and black or entirely black; cortex and medulla K+Y, C+R, KC+R, P+ yellowish (alectorialic acid). Arctic-alpine s to Wash and NM; rare in Wash ....*A. nigricans*
    3b Thallus yellowish green; cortex and medulla P-
      4a Thallus often blackening from the apices downward; branches terete; apothecia occasional, lateral; medulla KC- and CK+Y or O (diffractaic chemotype) or rarely KC+R and CK- (alectoronic chemotype), other spot tests negative; on alpine sod, rock crevices, and soil, rarely on bark; arctic-alpine, s to BC and Alta ....[*A. ochroleuca*]
      4b Thallus not blackening, or if so, then blackening toward the base; branches usually somewhat angular; apothecia abundant to absent, lateral; medulla KC+R or -, other spot tests negative; on conifer branches, rarely on soil and rocks; usually at high altitudes; infrequent in Wash and Ore Cascade, Coast, and Siskiyou ranges .... *A. lata*

1b Thallus pendent or prostrate

   5a Apices of branches sorediate or granular, generally hooked or curled into a tiny "fiddlehead"....................................(See ***Ramalina thrausta***)

   5b Apices of branches not sorediate or granular, not hooked

      6a Thallus soft, flaccid; branches dull, irregularly pitted and uneven; cortex frequently broken in rings, exposing the medulla; on bark, wood or alpine sod....................................... (See *Evernia divaricata*)

      6b Thallus firm; branches shiny, ± smooth; cortex unbroken except for elongate whitish pseudocyphellae

         7a On alpine sod; main branches often becoming broad, flattened, and dorsiventral, generally > 3 mm diam. Thallus ± rigid; arctic-alpine s to Wash and Mont; uncommon .... *A. vexillifera*

         7b On bark or wood, rarely on rock or mosses over rock, low elevations to subalpine; main branches terete to angular or occasionally ± flattened, generally < 2.5 mm diam

            8a Medulla C-; low to high elevations.......................... ***A. sarmentosa***

            8b Medulla C+R; low to mid elevations (usually < 700 m) ....***A. vancouverensis***

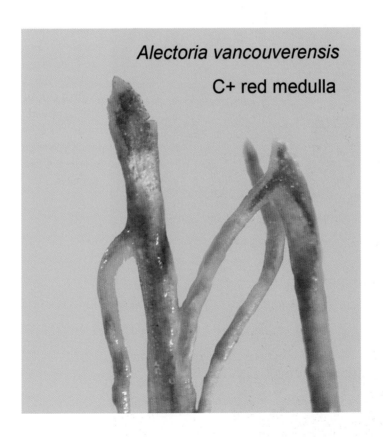

Alectoria vancouverensis

C+ red medulla

Alectoria imshaugii

Description: Thallus tufted fruticose, erect or somewhat drooping, to 8(12) cm, pale greenish (or tinged blue-black when dying or in exposed sites), occasionally with an almost translucent appearance; branches with clusters of isidia-like spinules; pseudocyphellae elongate, slightly paler than the surrounding cortex; apothecia rare; medulla K+Y, P+Y or O, UV- (thamnolic acid) or K-, P-, UV+ blue white (squamatic acid), KC- or rarely KC+R (alectoronic acid).

1 mm

Range: S BC to Cal, e to w Mont; common in low- to mid-elevation forests w Cont Div.

Substrate: Bark and wood of both conifers and hardwoods.

Habitat: Common in moist low- to mid- elevation forests e Cas, but only occasional w Cas.

Notes: Often mistaken as an *Usnea* at first glance, but lacking the central cord of *Usnea*. Easily distinguished from other local species of *Alectoria* by its clusters of isidia-like spinules.

Alectoria sarmentosa

Description: Thallus pendent fruticose, to 40(80) cm long, pale greenish or yellowish green, occasionally blackening; main branches round in cross section or somewhat angular, occasionally ± flattened, generally < 2.5 mm in diameter; pseudocyphellae raised, usually elongate, whitish, to about 1 mm long; apothecia fairly common, the disk pale tan or blackish; medulla KC+R (alectoronic acid) or KC-, usually P-, rarely P+Y (thamnolic acid).

Range: Alas to Cal, e to Alta and Mont; mainly w Cont Div, but mostly avoiding the immediate coast.

Substrate: On bark or wood, rarely on rock or mosses over rock.

Habitat: Common in low- to mid-elevation conifer forests, associated with older forests; toward the eastern edge of its range increasingly restricted to moist, old-growth forest.

Notes: Another species, *A. vancouverensis* is essentially identical in form, reacts C+ bright red in the medulla, and differs in range and habitat. A very rare sorediate form of *A. sarmentosa* is known from coastal states and provinces.

Alectoria vancouverensis

Description: Thallus pendent fruticose, to 40(55) cm long, pale greenish or yellowish green; main branches round in cross section or somewhat angular, occasionally ± flattened, generally < 2.5 mm in diameter; pseudocyphellae raised, usually elongate, whitish, to about 1 mm long; apothecia uncommon; medulla C+R, KC+R (olivetoric and alectoronic acids).

Range: Coastal BC s to Cal, in and w Cas.

Substrate: Bark and wood.

Habitat: Most common in the transition between valley forests and mountain forests (generally below about 700 m); often occurring in the same stands as *A. sarmentosa*, but usually dropping out below the elevation of highest dominance by *Alectoria*; more common on the immediate coast than *A. sarmentosa*.

Notes: Essentially identical in form to *A. sarmentosa* but differentiated by the C+R medulla. At this point we do not have a reliable morphological character that can be used to predict the C reaction (i.e., *A. vancouverensis* vs. *A. sarmentosa*).

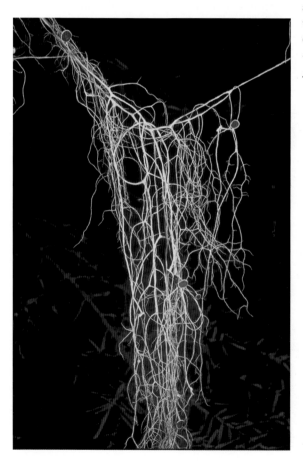

Allantoparmelia

Allantoparmelia alpicola

Description: Thallus foliose, dark brown, loosely to tightly adnate, occasionally cushion-like, to 7(12) cm diam; lobes to 1.5 mm broad, convex, and often nodulose; upper surface lacking pseudocyphellae, soredia, and isidia, but often lobulate; lower surface generally folded or irregular; attached by irregular holdfasts; medulla K+Y, P+Y, C-, KC+O (alectorialic and barbatolic acids).

Range: Circumpolar s to Wash and Mont; fairly common in its restricted habitat.

Substrate: On rock.

Habitat: Generally subalpine to alpine.

Notes: *Allantoparmelia alpicola* can be distinguished from several superficially similar small, dark Melanelia species (especially *M. stygia*) by the absence of pseudocyphellae and rhizines. *Allantoparmelia almquistii*, a similar but more northern species, differs from *A. alpicola* in chemistry (K-, P-, C+R; olivetoric acid). *Allantoparmelia almquistii* occurs in subalpine and arctic-alpine habitats in oceanic climates in the north Pacific, south to BC.

Sources include: Esslinger (1977, 1978a).

Anaptychia

Thallus foliose, small, adnate to ascending, rhizinate, light colored; lobes < 3 mm broad; lower surface white to tan or light brownish; upper cortex prosoplectenchymatous; lower cortex present or not, prosoplectenchymatous if present; apothecia with thalline margin and dark disk; spores 1-septate, dark at maturity; spot tests negative; substrate various.

Sources include: Esslinger (2007), Hollinger et al. (2022), Kurokawa (1962, 1973).

ID tips: Easily confused with *Physcia, Physconia,* and related genera; see **Table 8**, p. 314.

1a Thallus subfruticose, with elongate narrow lobes .... ***A. crinalis***
1b Thallus foliose, with appressed to ascending lobes; on calcareous rock, moss over rock, or alpine sod
  2a Soredia present. Lobe tips occasionally with cortical hairs, but mostly these are small and hard to find, or absent; soralia marginal and terminal on lateral lobes, becoming lip shaped, granular, often darkening to nearly black; lower surface white to tan; lower cortex poorly developed, fibrous, in section very thin and similar in structure to the medulla; widespread in w N Am .... *A. elbursiana*
  2b Soredia absent
    3a Lobe tips (and often the lobes) pruinose, also with minute hyaline hairs. Lobes 0.8-2.0 mm broad; algae in clumps, giving the thallus a spotted appearance; lower cortex distinct, prosoplectenchymatous; apothecia unknown; dry forests to grasslands at mid to high elevations, mainly e Cont Div

    ....*A. roemerioides*
    3b Lobe tips and lobes epruinose and lacking minute hyaline hairs. Thallus often turf-like, with numerous ascending lobules; on moss over rock or soil; arctic-alpine, Alas, Nunavut, BC, and Colo

    ....*A. bryorum*

Anaptychia crinalis

Description: Thallus subfruticose, forming tangled masses of linear lobes, whitish to pale brown or pale gray, lobes elongate, slender, with long marginal cilia; cortex lacking hyaline hairs; lobes mostly < 1(1.5) mm broad, flat to channeled, the edges and upper surface corticate but lacking a cortex on the lower surface; apothecia uncommon, ciliate margined, the disk brown to black but often whitish pruinose; cortex and medulla K-, KC-, P- (no lichen substances).

Range: In the PNW restricted to the immediate coast; rare; more common northward.

Substrate: On bark or wood; elsewhere also on rock.

Habitat: Rare in coastal forests and scrub.

Notes: Heterodermia leucomela can be very similar in form, having elongate, narrow lobes with marginal cilia. *H. leucomela* differs in its K+Y cortex (atranorin) and presence of soredia on the lower surface.

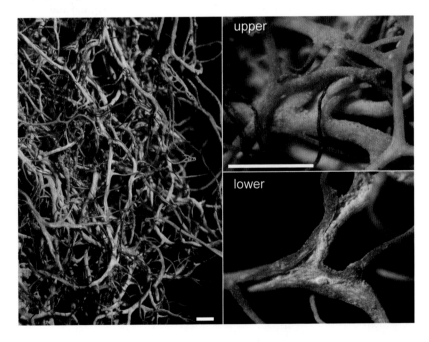

Arctoparmelia

Thallus foliose, yellowish green; superficially similar to *Xanthoparmelia* but differing primarily in chemistry and a lower surface that is often velvety, ivory white to gray, black, or purplish; lobes often narrower than in *Xanthoparmelia*; cortex with atranorin and usnic acid (KC+Y, UV-); medulla UV+ white, KC+R (alectoronic acid); on rock.

Sources include: Hale (1986).

ID tips: Arctoparmelia is mostly restricted to boreal and arctic regions, where it is common and abundant. It becomes very rare south of the Canadian border in w N Am, where it is found in cold, rocky sites, for example the cold air drainage at the base of north-facing coarse talus slopes. In the field it is likely to be overlooked as a *Xanthoparmelia*.

1a Soredia present
    2a Soredia diffuse on the upper surface. Arctic s to w Mont and Colo,
        expected in n Id and ne Wash ....*A. subcentrifuga*
    2b Soredia in discrete roundish soralia. Lower surface dark; mainly arctic
        with rare disjuncts s to s BC, Wash Cas .... *A. incurva*
1b Soredia lacking
    3a Lower surface gray to black or purplish black. Arctic and boreal s to n
        BC and Sask .... [*A. separata*]
    3b Lower surface pale. Mainly arctic and boreal s to central BC
        .... [*A. centrifuga*]

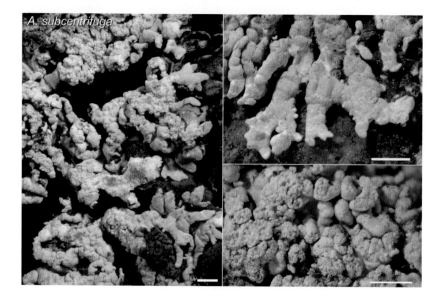

A. subcentrifuga

Aspicilia

Thallus crustose (not treated here) or prostrate to erect fruticose, the fruticose forms to 2 cm broad, olive green to brownish or gray; branches solid, nearly circular, flattened, or somewhat irregular in cross section; cortex sometimes with white pores (pseudocyphellae); apothecia rare, usually with a thalline margin, the disk blackish; medulla K- or K+R (often developing slowly); on soil or plant detritus, less often on rock, mostly in arid or semiarid habitats; transitions from crustose forms on rock to fruticose forms can be found.

Sources include: McCune and Rosentreter (2007), McCune and Di Meglio (2021).

ID tips: While the many crustose *Aspicilia* species are poorly understood and difficult to identify, the fruticose species keyed here are more tractable. The genus has been split into *Circinaria* and others, but these are not well defined morphologically or phylogenetically. We therefore take a broad view here and include all in *Aspicilia*.

1a Thallus of minute ± appressed mats; branches ± flattened
  2a Thallus nearly squamulose, of irregular short, broad lobes; shortgrass prairie to e slope Cas..*A. reptans* group
    (Six species are keyed and illustrated in McCune and Di Meglio [2021].)
  2b Thallus of narrow, elongate, ± flattened lobes, often with distinctly forked tips like snake tongues; occasional in steppe; w Mont to e BC, s to Cal and Great Basin... *A. filiformis*
1b Thallus of erect hemispherical to spherical cushions or knobby balls; branches terete or nearly so
  3a Branching dichotomous, very compact; branches short, thick, blunt, tipped with conspicuous pseudocyphellae; on calcareous soil in dry areas, occasionally with the next sp; uncommon in e Mont to e Ore s to Colo, Utah, Nev ....*A. rogeri*
  3b Branching irregular to dichotomous, ± compact; branches short to elongate, tapered, often delicate, not tipped with conspicuous pseudocyphellae (which are instead scattered); on calcareous soil in dry steppe; e BC to Sask, s to Ariz and NM........................... *A. hispida*

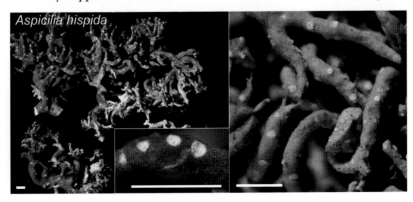
Aspicilia hispida

Baeomyces

Baeomyces rufus

Description: Thallus of two parts, a greenish crustose or partly squamulose primary thallus and an erect, unbranched, hollow, fissured stalk to 6 mm tall, bearing a terminal apothecium; disk usually convex, pale brown, cream, or pink; thallus K+Y, P+O (stictic acid group).

Range: Circumpolar s to Cal and NM; fairly common in and w Cas, rarer inland.

Substrate: Soil or rock, often overgrowing thin bryophyte mats, rarely on decaying wood or bark.

Habitat: Often on damp or seepy roadcuts or stream cutbanks, boulders, and outcrops, usually sloping and shaded.

Notes: This distinctive but often inconspicuous species can form greenish patches covering extensive areas of steep, damp roadcuts. The stalks can, however, be very short or nearly absent. Once you learn to recognize it with stalks, you will start seeing it more frequently as a sterile crust. As an added bonus, look for the *Baeomyces*-specific parasitic crustose lichen *Epilichen scabrosus*, which forms yellowish patches with black apothecia.

Sources include: Thomson (1967a).

Brodoa

Brodoa oroarctica

Description: Thallus foliose, loosely or tightly appressed, of narrow to medium wide (< 2 mm) lobes that are thickened and puffy but solid; upper surface light gray to brown or black, depending on exposure, with pale angular markings toward the lobe tips; lower surface light brown to ± black; rhizines lacking; cortex K+Y, medulla K-, C-, P-, KC+R.

Range: Arctic s to Ore and NM.

Substrate: Noncalcareous rock, occasionally overgrowing adjacent mosses and alpine sod.

Habitat: Arctic-alpine tundra to subalpine rocky ridges, exposed or somewhat sheltered by surrounding rock.

Notes: Some small dark forms might be confused with *Allantoparmelia alpicola* or *Melanelia stygia*. *Allantoparmelia alpicola* has a K+Y and P+Y medulla and usually has more tightly appressed, narrower, less inflated appearing lobes. *Melanelia stygia* has distinct (but often dark) pseudocyphellae that are readily visible with a hand lens.

Sources include: Goward (1986).

Bryoria, with Sulcaria

Thallus fruticose, erect, tufted, prostrate, or pendent, generally some shade of brown, occasionally gray green or black; branches roundish to ± angular, slender; isidia absent, although isidioid spinules occasionally arise from the soralia; soredia present or not; pseudocyphellae absent to abundant, when present often inconspicuous, generally elongate fissures or splits in the cortex, white, yellow, or brownish; apothecia lateral, the disk brown; spot tests various (see below); mostly on trees and shrubs at all elevations, but a few spp occur on rock and soil, mainly in the subalpine to alpine.

Sources include: Boluda et al. (2019), Brodo and Hawksworth (1977), Glavich (2003), McCune and Stone (2022), Myllys et al. (2006, 2011, 2014, 2015, 2016a, 2016b), Spribille et al. (2016), Velmala et al. (2009, 2014).

ID tips: *Bryoria* is a prominent, ecologically and culturally important genus in the PNW. Furthermore, the genus presents challenging species identifications, so we devote some extra space here to that problem.

When determining fissural vs. tuberculate soralia, observe the shape of the emerging soralia. Fissural soralia emerge with pointed ends, while tuberculate soralia emerge as circular structures. Tuberculate soralia can, however, develop pointed ends as they become large.

Spot tests are important for identifying *Bryoria* species and varieties. The K, C, and KC spot tests are best made on acetone extracts as follows: place a wad of *Bryoria* fibers in a depression in a white porcelain spot plate or on a piece of glass on a white background. Add just enough acetone (2-5 drops) to immerse the sample. Allow the acetone to evaporate completely. Pluck the dry fibers out of the depressions, leaving only the acetone extract. Add 2-3 drops of K, observing any color change (to yellow or red). Run the drops down the sides of the depression using a different part of the depression for each chemical; much of the extract is deposited on the sides. After about two minutes to complete the K reaction, add a drop of C, noting any instant color change. Note also the KC reaction where the two solutions contact. Beware of mixed collections, which will give seemingly contradictory results. For example, the common mixture of *B. pseudofuscescens* var. *pikei* and var. *friabilis* is C+R and K+ strong Y.

For P tests in this genus, place a few to many strands of thallus on a small square (about 1 × 1 cm) of white paper. Add a drop of P reagent to the lichen with a capillary. Look for red or yellow colors that may diffuse from the lichen into the paper or may appear on the thallus and soredia themselves. Most spp of *Bryoria* that are P+R contain fumarprotocetraric acid. Species with alectorialic and barbatolic acid are K+ deep Y and P+Y.

Studies of DNA sequences in *Bryoria* have led to surprising conclusions. The best known of these is that two previously accepted species are genetically identical, the apparent difference arising from the presence vs. absence of a basidiomycete yeast in the cortex (Spribille et al. 2016). This somehow results in a color difference from the production of the yellow vulpinic acid; see more under *B. fremontii* and comparison photos of cortex under *Nodobryoria* (p. 266).

Furthermore, some of our traditional species separations have been upended (Velmala et al. 2014). For example, differences in secondary substances, ecology, and geographic distributions among *B. capillaris*, *B. friabilis*, and *B. pseudofuscescens* were not supported by phylogenetic analyses, yet European and N Am versions of the same chemotypes often do tend to fall in distant branches of the evolutionary tree. Constructing a practical taxonomy from these results has been challenging, and we have taken a somewhat different approach, recognizing chemically distinguished varieties within *B. pseudofuscescens* as described by McCune and Stone (2022).

Species of *Sulcaria* are keyed below with *Bryoria*.

Introductory Key

1a Thallus prostrate to decumbent and on rock, soil, or alpine sod
....**Group 1**, Lead 4
1b Thallus erect or pendent, usually on bark or wood
  2a Soredia present...**Group 2**, Lead 10
  2b Soredia absent
    3a Thallus bicolored, black at base with pale brown tips; on coastal trees, rocks, and mossy turfs...**Group 3**, Lead 18
    3b Thallus variously colored but not strongly bicolored; habitat various.
    ....**Group 4**, Lead 19

Group 1

Prostrate on rock, soil, or alpine sod

4a Soredia present
  5a Soredia bearing tufts of isidioid spinules; thallus P+R; soredia P+R
  .... *B. furcellata* (Lead 14a)
  5b Soredia lacking isidioid spinules; thallus P-; soredia P+R or P-
  6a Soredia P-.. *B. simplicior* (Lead 14b)
  6b Soredia P+R. On rocks, soil, and tundra sod; arctic-alpine and subalpine s to Ariz; frequent but generally not plentiful
  .... *B. chalybeiformis*
4b Soredia lacking
  7a Thallus reddish brown, dull; P- throughout
  ....*Nodobryoria subdivergens*
  7b Thallus brown to dark brown or olivaceous black, generally shiny
    8a Thallus K+Y, P+Y (alectorialic acid)
    ....very dark forms of *Alectoria nigricans*

8b Thallus K-, P- or P+O or R

    9a Medulla P+O or R; soredia lacking; spinules (short, sharp lateral branchlets) frequent. Arctic to subarctic with rare disjuncts south in Can RM, OP, and s BC ....*B. nitidula*

    9b Medulla P- but soralia P+O or R, if present; spinules few or none.. ....*B. chalybeiformis* (Lead 6b)

Group 2

Erect or pendent; sorediate

10a Soredia yellow...***B. fremontii***

10b Soredia not yellow, mostly white to greenish

    11a Acetone extract K+Y, C- or C+ pink, KC+R, P+O or R (alectorialic, barbatolic, and fumarprotocetraric acids); on bark and wood; e boreal forests, rare w to Wash and BC ....*B. nadvornikiana* (The very rare sorediate form of *Bryoria pseudofuscescens* var. *pikei* will key here. It has a P+Y thallus.)

    11b Acetone extract K-

        12a Thallus or acetone extract P+Y; pseudocyphellae present ....*B. kockiana* (Lead 29a)

        12b Thallus or acetone extract P- or P+R; pseudocyphellae present or not

            13a Thallus tufted to prostrate

                14a Soralia white, bearing tufts of isidioid spinules, P+R; medulla P+R. On bark and wood, occasionally on rock or soil; boreal forest and on the coast .... *B. furcellata*

                14b Soralia greenish black, lacking isidioid spinules, P-; medulla P-. On conifer twigs and small branches in open sites; Alas to Newf s to Colo; a more northern, mainly boreal sp, occasional e Cas to n Cal...*B. simplicior*

            13b Thallus pendent

                15a Thallus very pale brown to whitish; medulla and soralia P+R

                    16a Cortex P+R; soralia usually abundant ....***B. fuscescens*** (rare chemotype)

                    16b Cortex P-; soralia sparse. Coastal Alas to OP ....*B. trichodes*

                15b Thallus usually brown or dark brown; medulla P-, soralia P+R

                    17a Angles between the main branches U-shaped, broadly rounded; branches even in diam; thallus olive brown or olive black, shiny; soralia fissural, rarely tuberculate .... ***B. glabra***

                    17b Angles between the main branches ± V-shaped, acute, generally not broadly rounded; branches uneven in diam, at least in part; thallus pale to dark brown or blackish, shiny or not; soralia fissural or tuberculate, generally broader than the branches on which they occur ....***B. fuscescens***

Group 3

Bicolored

18a Thallus with third-order perpendicular branches, tufted, erect. Inner cortex and medulla P+R; Alas s to n Ore near the coast .... *B. bicolor*
18b Thallus usually lacking third-order perpendicular branches, erect to decumbent. Inner cortex and medulla P+R; Alas s to n Ore near the coast
.... *B. tenuis*

Group 4

Erect or pendent; soredia lacking; not bicolored

19a Acetone extract K+Y or R
  20a Acetone extract K+Y to R, C-, KC-; thallus P+Y (norstictic acid)
    21a Thallus dull red brown or chestnut brown; perpendicular side branches frequent; pseudocyphellae to 4 mm long, abundant, conspicuous, white, spiraling around the branches; n Cal to Coos Co., Ore; rare .... *Sulcaria spiralifera* var. *spiralifera*
    21b Thallus dark olive brown to brownish black; perpendicular side branches absent or sparse; pseudocyphellae to 1 mm long, not prominent; common in PNW
    .... ***B. pseudofuscescens* var. *pseudofuscescens***
  20b Acetone extract K+Y, C+ pink or C-, KC+R or pinkish or KC-
    22a Thallus with deep longitudinal fissures, P- or P+ pale Y (atranorin)
    .... ***Sulcaria badia***
    22b Thallus smooth, not deeply fissured, but often with elongate pseudocyphellae that may be slightly depressed, P+Y or P+R
      23a Thallus pale to dark, gray green to brown or olive brown, P+R (fumarprotocetraric and alectorialic acids)
      .... *B. nadvornikiana* (Lead 11a)
      23b Thallus pale brown to gray or greenish gray, occasionally dark brown, P+Y (alectorialic acid)
        24a On exposed coastal trees; pseudocyphellae long, linear whitish; perpendicular short side branches frequent. Immediate coast, Ore to Cal...................... *Sulcaria spiralifera* var. *pseudocapillaris*
        24b Habitat otherwise; pseudocyphellae present but inconspicuous; perpendicular short side branches lacking or nearly so
        .... ***B. pseudofuscescens* var. *pikei***
19b Acetone extract K-
  25a Thallus tufted
    26a Branches angular, reddish brown; apothecia common; soredia lacking.. ***Nodobryoria abbreviata***
    26b Branches medium brown to dark brown or almost black; apothecia unknown; soredia generally present............ *B. simplicior* (Lead 14b)

25b Thallus pendent

   27a Acetone extract KC+ pinkish orange or red (gyrophoric acid)
   ....*B. pseudofuscescens* **var.** *friabilis*

   27b Acetone extract KC-

      28a Cortex or medulla P+Y or P+R; thallus dark brown, pale brown, or whitish

         29a Cortex and medulla P+Y; boreal. On conifer bark and wood; Alas to e N Am s to n Id and nw Mont; rare .... *B. kockiana*

         29b Cortex P-, medulla P+R; mainly w Cas .... ***B. americana***

      28b Cortex and medulla P- or pale yellowish; thallus mid to dark brown or yellowish brown

         30a Thallus forest green to olive green, shiny; lichen substances lacking; pseudocyphellae present; occasional in mountain conifer forests, mainly on conifers, occasionally hardwoods; w Cas .... *B. pseudofuscescens* var. *inactiva* (TLC with heavy extract can reveal a trace of norstictic acid (var. *pseudofuscescens*) or gyrophoric acid (var. *friabilis*). All of these varieties differ from esorediate forms of *B. glabra* in having pseudocyphellae.)

         30b Thallus brown, reddish brown, or yellowish brown

            31a Thallus reddish brown (hold against dark background); cortical cells jigsaw shaped in surface view (photo p. 266) .... ***Nodobryoria oregana***

            31b Thallus medium to dark brown, rarely yellowish brown; cortical cells elongate and smooth or with minute granules .... ***B. fremontii***

Bryoria americana

Description: Thallus pendent, filamentous, mostly 5 15 cm long, reddish brown to dark brown; main branches 0.1-0.2(0.4) mm diam; soredia rare; pseudocyphellae abundant or sparse, oval or elongate; apothecia occasional; medulla, soralia, and acetone extract K-, P+R (fumarprotocetraric acid; occasionally P-), C-, KC-.

Range: Alas to Santa Cruz, Cal, mainly in the Coast Range, w Cas in Ore and Wash.

Substrate: Bark or wood of conifers.

Habitat: Wet coastal forests at low to high elevations.

Notes: A related species, *B. trichodes*, is similar but often relatively pale (often almost white or pale greenish white), with whitish oval pseudocyphellae and anisotomic dichotomous branching. In contrast, *B. americana* is dark brown, often with a reddish brown tint, with sparse elongate pseudocyphellae and dichotomous branching. Both *B. americana* and *B. trichodes* are usually seen very near the coast, especially in old stands of *Pinus contorta*. Reliable identification of *B. americana* requires a P test.

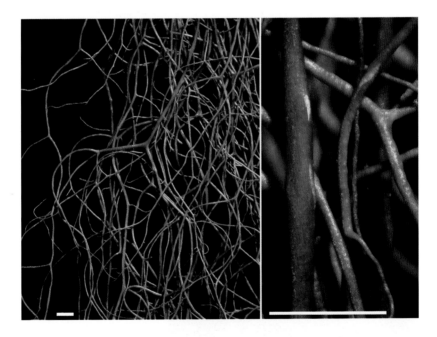

Bryoria fremontii

Description: Thallus pendent, filamentous, often very long (to 50 cm), medium to dark brown, generally shiny; branches uneven in diam, becoming twisted and foveolate, generally 0.4-1.5 mm in diam; soredia yellow, sparse or frequently absent; pseudocyphellae absent or sparse, white to pale brown; apothecia absent or sparse; all spot tests negative.

Range: BC to Alta s to Colo and Cal; common e Cas but uncommon to rare w Cas.

Substrate: On bark or wood, mainly on conifers, occasionally on hardwoods.

Habitat: Achieves greatest biomass in dry, open *Pinus* and *Pseudotsuga* forests and savannas, often forming long, thick beards, but occurring in small to moderate amounts in open and semi-open habitats to near timberline. Also in the treetops in mesic low-elevation forests.

Notes: B. fremontii is characterized by its medium to dark brown color, relatively coarse branches, large size of the thallus, and sometimes sparse yellow soralia. No other *Bryoria* in our region that achieves such high biomass has all negative spot tests. The usual brown form intergrades with a yellowish form,

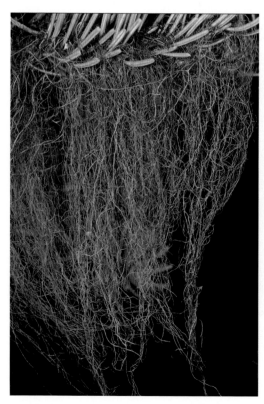

previously separated as *B. tortuosa,* having pale yellow pseudocyphellae and sometimes a yellowish cortex. The yellow pigment can be seen in surface view under LM as yellowish granules that glow in polarized light (comparison photos under *Nodobryoria*, p. 266). These yellow forms are, however, genetically the same as *B. fremontii*, the yellow pigment (vulpinic acid) somehow induced by a cortical lichenicolous fungus, a basidiomycete yeast. The yellowish morph occurs from BC to Cal, inland to c BC and w Mont; occasionally in and w Cas, rarer inland.

Bryoria fuscescens

Description: Thallus pendent fruticose, to 15(30) cm, pale brown to dark brown or blackish, dull or shiny; branches 0.3-0.4(0.6) mm diam, somewhat uneven in diameter; angles between the main branches ± V-shaped, acute, generally not broadly rounded; soralia fissural or tuberculate, generally broader than the branches on which they occur; apothecia rare; soredia P+R (fumarprotocetraric acid), otherwise P-; rarely with cortex and medulla P+R throughout.

Range: Widespread, throughout Ore and Wash.

Substrate: Conifer bark or wood, occasionally on rocks or hardwoods.

Habitat: Ubiquitous in forests e Cas, occasional w Cas.

Notes: This is the most common sorediate *Bryoria* species in the PNW, but seldom a dominant species and never forming long, thick beards. Distinguishing features are its sprinkling of soralia, V-shaped branching, and short to medium length. *Bryoria fuscescens* intergrades in appearance with *B. glabra*, which is usually dark, uniformly olive colored, shiny, and has U-shaped branch axils. *B. fuscescens* also intergrades with a finely branched morph with the appearance of steel wool, what used to be separated as *B. lanestris*. That name is now synonymized with *B. fuscescens*. Another morph previously segregated as a distinct species, *B. subcana*, is very pale, with a P+R cortex and medulla. The variant has also been shown to fall within *B. fuscescens*.

Bryoria glabra

Description: Thallus pendent fruticose, to 15 cm, olive, olive brown, or olive blackish, shiny; branches 0.2-0.4 mm diam, somewhat uneven in diameter; angles between the main branches U-shaped, broadly rounded; pseudocyphellae lacking; soralia fissural; apothecia rare; soredia P+R (fumarprotocetraric acid), otherwise P-.

Range: Alas to Cal, e to Alta and Mont; most common in northern coastal areas, uncommon inland to w Mont.

Substrate: Conifer bark or wood.

Habitat: Coastal and mountain conifer forests, mainly low to mid elevations.

Notes: Distinguished by dark olive, glossy thallus, springy texture, presence of soredia, and U-shaped branching, *B. glabra* intergrades in appearance with *B. fuscescens* (which has more V-shaped angles between the branches). Specimens morphologically intermediate between these species are common.

Bryoria pseudofuscescens var. *friabilis*

Description: Thallus pendent, finely filamentous, to 15 cm long, pale to medium brown, becoming darker in the herbarium; branches often uneven, contorted, wrinkled and foveolate; pseudocyphellae abundant, brownish to white, elongate, twisting around branches; apothecia rare; acetone extract K-, C+R, KC+R (gyrophoric acid).

Range: BC to n Cal e to Mont; rare in n RM, more common w Cas.

Substrate: On bark of hardwoods and conifers, rarely on rocks.

Habitat: W Cas, widely but thinly distributed from valley bottoms to mountain forests, occasionally in old orchards and riparian hardwood forests (where it is the most common *Bryoria* sp); e Cas, an old-growth associate, most often found in moist older forests at lower elevations.

Notes: Very difficult to identify in the field with certainty (requires verification with a spot test), but many individuals have a characteristic faintly milky olive color when dry. The K-, C+R, KC+R spot tests clinch the identification. Variety *pseudofuscescens* is similar but is K+R, C-. Variety *pikei* is K+ strong Y, C- or C+O, KC+O.

Bryoria pseudofuscescens var. *pikei*

Description: Thallus finely filamentous, pale brown to gray or greenish gray, occasionally dark brown, to 15(30) cm long, 0.1-0.3 mm in diameter; pseudocyphellae generally present, inconspicuous, elongate, white, to 0.25 mm long; apothecia rare; common in low- to mid-elevation moist conifer forests; thallus K+Y, KC+ orange, P+Y, C+ O or C- (alectorialic and barbatolic acids).

Range: Alas to n Cal, e to Mont, Newf, with a disjunct in Colo; most common in moist low- to mid-elevation forests.

Substrate: On bark or wood of both conifers and hardwoods, but typically on conifers.

Habitat: Most common in understory of conifers.

Notes: This is the palest of our common *Bryoria*, typically ranging in color from pale brown to pale grayish. Although chemical tests are needed for verification, very pale specimens with fine branches collected in full or partial shade are likely to be var. *pikei*. Pale forms are also occasionally found in other species, so these need to be carefully distinguished by spot tests. Near the coast and in the Coast Range, very pale forms of several species may be found: *Sulcaria spiralifera* var. *pseudocapillaris*, pale forms of *B. fuscescens*, and *B. trichodes* (see key for differences). A very rare sorediate form of var. *pikei* is known from Ore.

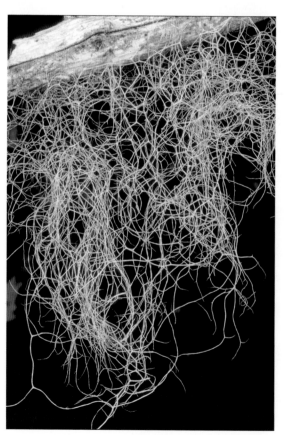

Bryoria pseudofuscescens var. *pseudofuscescens*

Description: Thallus pendent, filamentous, to 15(30) cm long, pale to dark brown or blackish, often olivaceous; branches even to uneven in diam, straight or twisted, mainly 0.15-0.3 mm diam; pseudocyphellae generally abundant, white, short to elongate fusiform, depressed, often forming partial spirals, to 1 mm long; apothecia very rare; acetone extract K+Y to R or dull brownish red (norstictic acid), C-, KC-; thallus P+Y.

Range: Alas to n Cal, e to sw Alta, w Mont, and nw Wyo; fairly common, but rare e Cont Div in Mont.

Substrate: Conifers, rarely on hardwoods.

Habitat: Moist valleys to mountains; e Cas mostly in *Abies grandis* – *Thuja* habitats to subalpine (*Abies lasiocarpa*) forests; in a broader range of habitats w Cas.

Notes: This variety can form long, thick, very dark beards and is often the dominant *Bryoria* in cool, moist mountain forests. It resembles var. *friabilis* and var. *pikei*, but they have a K- or K+Y reaction, rather than the K+R reaction of an acetone extract. A direct K test on the thallus on filter paper will usually show a clear yellow for var. *pikei*, but will usually be a muddy ambiguous reaction in var. *pseudofuscescens*. For that reason we recommend testing an acetone extract, which yields a clear red reaction for var. *pseudofuscescens*.

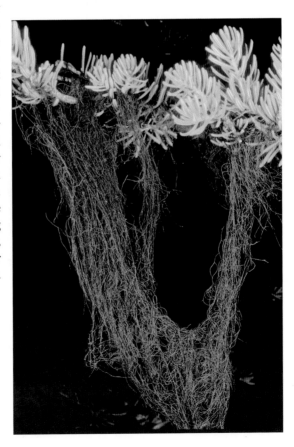

Bunodophoron

Bunodophoron melanocarpum

Description: Thallus fruticose, forming loose suberect or drooping tufts to 5 cm long; color white, light gray, cream, or orangish tan; branches slightly to distinctly flattened in cross section, solid, mostly 1-2 mm diam; cortex shiny; soredia and isidia absent, though slender branchlets may function as propagules; apothecia terminal, roundish to flattened cup shaped, bursting open irregularly exposing a black spore mass (mazaedium); spores globose, dark, 1-celled; medulla white, K+O or K-, UV+ blue white (sphaerophorin and stictic acid complex).

Range: s Alas to OP, near the coast, rare in Wash.

Substrate: On bark and wood of conifers.

Habitat: Forests in cool northern oceanic climates.

Notes: Formerly included in *Sphaerophorus*, this species is separated by its flattened branches. Branches in *Sphaerophorus* are sometimes weakly flat-

tened or irregular in cross section, but the branches are distinctly flattened in *Bunodophoron*. This common genus of the s hemisphere is represented by only one sp in our area.

Sources include: Ohlsson (1973b), Tibell (1975), and Wedin (1993a, 1993b; 1995).

Caloplaca

Thallus crustose, fruticose, or subfoliose, small, but sometimes expanding and coalescing to cover large areas; lobes narrow (generally < 1(2) mm broad); upper surface some shade of yellow or orange, occasionally greenish or even beige or tan in shaded sites; in lobate species lower surface lacking a cortex; soredia present or not; apothecia lecanorine, spores 1-septate, polarilocular; upper cortex K+ purple or R (anthraquinones, which are yellow or orange pigments); on a wide range of substrates.

Sources include: Arup (1995); McCune (2017).

ID tips: *Caloplaca* in the broad sense is a huge genus, very common and speciose throughout the PNW, occurring on almost all substrates and in most habitats. These are keyed in McCune (2017). Here we include only a few of the most conspicuous species. The traditional segregation of the genera in the Teloschistaceae by growth form is artificial. Many rearrangements and segregate genera have been defined. Here the family Teloschistaceae is represented by *Caloplaca* (broad sense), *Seirophora*, *Teloschistes*, *Xanthomendoza*, and *Xanthoria* (broad sense).

1a Thallus crustose ....not treated here
1b Thallus fruticose or subfoliose
  2a Thallus fruticose; on rock, often on the immediate coast
    3a Thallus sorediate..***Xanthoria candelaria***
    3b Thallus lacking soredia... ***C. coralloides***
  2b Thallus subfoliose
    4a On bark or wood .... See ***Xanthoria***
    4b On rock
      5a Lower cortex present .... ***Xanthoria elegans***
      5b Lower cortex absent
        6a Upper surface rough and granular; always on calcareous rock, especially bird perches. Widespread but infrequent, in hot dry habitats e Cas..*C. trachyphylla*
        (See photo with the next species.)
        6b Upper surface smooth or finely pruinose; on both calcareous and noncalcareous rock ....***C. biatorina*** (and *C. saxicola*)

Caloplaca biatorina

Description: Thallus lobate crustose to subfoliose, orange or dull red orange, closely appressed, usually < 2 cm wide; marginal lobes ± flat to convex, mostly < 0.5 mm wide; upper surface smooth or finely pruinose; lower surface lacking rhizines and cortex; apothecia common, both disk and margin colored like the thallus; soredia and isidia lacking; cortex K+ purple.

Range: Widespread and abundant, circumpolar, arctic to temperate.

Substrate: Both calcareous and noncalcareous rock.

Habitat: Exposed to somewhat sheltered rock, often where nitrogen-enriched by bird or small mammal manure, often on steep or vertical surfaces.

Notes: Often covering large areas of rock faces, this species is commonly mistaken for *Xanthoria elegans*, which differs in having a lower cortex. *C. trachyphylla* is similar to *C. biatorina* but has a rough, warty surface, occurs exclusively on calcareous rock in continental climates, and is somewhat larger.

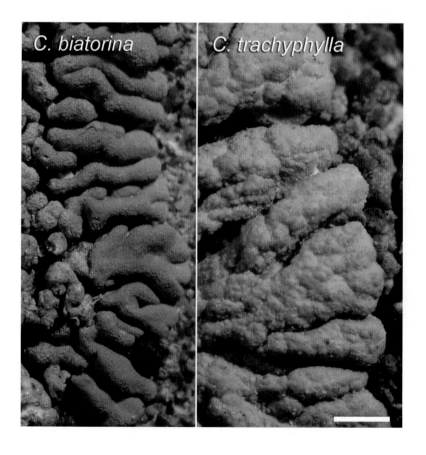

Caloplaca coralloides

Description: Tiny erect or drooping fruticose, the thallus rising only 2-3 mm above the substrate and mostly < 1 cm long, yellow to yellow-orange (occasionally greenish where shaded); branches roundish, < 1 mm diam, dichotomously branched; apothecia usually present, lateral or terminal; soredia and isidia lacking; cortex K+ purple.

1 cm

Range: S Cal, n to c Ore, on the immediate coast.

Substrate: Noncalcareous rock.

Habitat: Steep hard rock faces on coastal cliffs and sea stacks, just above normal high tide.

Notes: This is a distinctive species difficult to confuse with any other. In similar coastal rock habitats one will also find the yellow-orange fruticose *Xanthoria candelaria*, which is readily distinguished by its soredia and smaller size.

Candelaria

Thallus minutely foliose, to 1 cm wide or coalescing to cover large areas, sorediate, the soredia often coalesced into a granular crust, or in some cases the lobes poorly developed and the thallus mainly granular; upper surface egg yolk yellow to pale lemon yellow or yellowish green; lobes minute, mostly 0.1-0.3 mm wide; soredia marginal and terminal or on the lower surface of the lobes; lower surface white to yellow, at least partly corticate, rhizinate; apothecia small, yellowish, uncommon; spores 8-64 per ascus; spot tests negative except cortex UV+ deep reddish.

Sources include: Westberg and Nash (2002), Westberg and Arup (2011).

ID tips: Small *Xanthoria* species can be distinguished from *Candelaria* by their K+ purple reaction and an often more orangish color. *Candelariella* is a crustose genus that lacks the distinct elongate lobes of *Candelaria* when viewed with a lens. All three of these genera frequently grow together.

The two *Candelaria* species likely to occur here are best separated by spore number, so for critical identifications it is worth taking a little extra time to search for apothecia when making collections. While both species are usually sterile, a little extra effort to find fertile material can be rewarding.

Both *C. pacifica* and *C. concolor* are quite plastic in their degree of development, ranging from masses of soredia to clearly foliose with branched lobes. Both can be profusely sorediate on the margins, often coalescing and obscuring the corticate parts.

1a Spores 8 per ascus; soredia on lower surface and margins of lobes; lower
    surface partly lacking a cortex ....*C. pacifica*
1b Spores 16 or more per ascus; soredia on margins of lobes; lower surface
    corticate except at the very edge where soredia are formed; widespread
    but infrequent in our area .... *C. concolor*

Candelaria pacifica

Description: Thallus minutely foliose, sorediate, often coalesced into a granular crust, yolk yellow to pale lemon yellow or greenish yellow; lobes minute, 0.1-0.3 mm wide; soredia marginal and terminal and on the lower surface of the lobes; lower surface yellow, partly corticate and rhizinate; apothecia small, yellowish, uncommon; spores 8 per ascus.

Range: Widespread in w N Am, common throughout the PNW inland to at least e Mont.

Substrate: On nutrient-rich bark or wood (mainly hardwoods), roofing; rarely on rock.

Habitat: Valleys and foothills, often in urban, suburban, and agricultural areas; occasional in the mountains.

Notes: Although fertile *C. pacifica* is easily distinguished by its 8-spored asci from the widespread *C. concolor*, *C. pacifica* is usually sterile. In that case, look for the partly sorediate lower surface of *C. pacifica*. *Candelaria concolor*, common in e and sw N Am, is much less frequent than *C. pacifica* in our area.

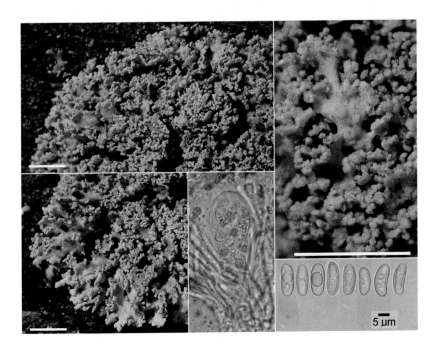

Cavernularia—*see* Hypogymnia

Cetraria

Thallus foliose or fruticose, erect, suberect, or prostrate, but always ascending from the substrate; lobes narrow or medium broad, often channeled, often with protruding marginal pycnidia that often resemble isidia; upper surface yellow, yellowish green, pale yellow, green, reddish brown, brown, black, or gray; lower surface variously colored but generally similar in color to the upper surface; rhizines sparse or lacking; pseudocyphellae common, often marginal or on the lower surface of foliose species; soredia or isidia present or not; lichen substances diverse as indicated by color, the bright yellow species having cortex and medulla K-, C-, KC-, P- and containing pulvinic acid derivatives, including pinastric and vulpinic acids; the pale yellowish green species with usnic acid, KC+Y; brown species usually with spot tests negative, with no lichen substances or only fatty acids, but some containing fumarprotocetraric acid (medulla P+R) and two with alectoronic acid (medulla KC+R, UV+ white); apothecia common on some species, lecanorine, the disk brown in most species, less often blackish; many habitats and substrates.

Sources include: Divakar et al. (2017), Esslinger (1973), Glew (1994), Goward (1985), Kärnefelt (1979, 1986), Mattson (1993), Nelsen et al. (2011), Printzen et al. (2013), Saag et al. (2014), Thell (1995), Thell et al. (1995, 2002, 2018), Thell and Goward (1996).

ID tips: Although species concepts within *Cetraria* are relatively well settled, generic placement is not; this will generate some confusion when you consult other sources. We use a broad concept of *Cetraria* because the numerous segregate genera (e.g., *Ahtiana, Allocetraria, Cetrariella, Coelocaulon, Flavocetraria, Kaernefeltia, Tuckermanella, Tuckermannopsis,* and *Vulpicida*) are not well supported by phylogenetic analysis of molecular and morphological data, as demonstrated by Thell et al. (2002), Nelsen et al. (2011), and Divakar et al. (2017). The only proposed classification that actually fits the available data is from Divakar et al. (2017), which places most cetrarioid species in either *Cetraria* or *Nephromopsis*. This classification has, however, met with resistance from other N Am lichenologists. The only other proposed solution that is consistent with the data is to use *Cetraria* in the broad sense, including both *Nephromopsis* and *Cetraria*. We adopt this classification here. We do, however, recognize *Esslingeriana*, because it clearly separates from *Cetraria* in phylogenetic trees.

1a Thallus on soil, rock, or alpine sod; thallus erect, some shade of brown, yellow, or greenish yellow
  2a Thallus yellow, pale greenish, or pale greenish yellow
    3a Thallus deep golden yellow. On soil or tundra sod; arctic s through RM to NM, w to OP...*C. juniperina*
    3b Thallus pale yellowish to greenish yellow or pale greenish (when shaded)
      4a Lobes roundish to irregular in cross section. Thallus fruticose, occasionally whitish pruinose toward the branch tips; podetia isotomically branched, with few short lateral branches, pycnidia rare; medulla P-, C-, K-, KC- (usnic acid and fatty acids). Arctic s in RM to NM, rare.. *C. madreporiformis*
      (See photo with *Dactylina*, p. 111.)
      4b Lobes flat to strongly channeled and inrolled
        5a Thallus margins inrolled; lobes channeled; surface smooth to weakly pitted and rugose. Arctic-alpine s to Wash, NM .... *C. cucullata*
        5b Thallus margins not or weakly inrolled; lobes ± flat to weakly channeled; surface rugose and pitted. Arctic-alpine, rarely subalpine, s to Wash, NM ....*C. nivalis*
  2b Thallus pale to dark brown, reddish brown, or olive
    6a Thallus of branches that are ± round or angular in cross section
      7a Pseudocyphellae generally < 0.3 mm long, round or oval, often sparse; branches about 0.5 mm wide, ± terete, ± smooth; thallus generally 1-3 cm tall. Alas to Cal and Colo .... *C. muricata*
      7b Pseudocyphellae generally 0.3-1 mm long, round or elongate, often abundant; branches about 1 mm wide, terete or angular, often longitudinally furrowed; thallus to 4(10) cm tall, erect.........
....***C. aculeata***
    6b Thallus of erect, ± flattened lobes (though sometimes channeled or tubular); mainly at mid to high elevations
      8a On rock; thallus suberect or more often prostrate, < 1 cm tall
        9a Lower surface pale to brown; conidia cylindrical, the ends not swollen; medulla K-, P-, KC+R, UV+. Arctic-alpine s to Wash and Colo; rare, increasingly common n
....*C. commixta*
        9b Lower surface dark; conidia with the center narrow, the ends swollen; medulla K+Y to O, P+O, KC-, UV-
.... *Melanelia hepatizon*
      8b On alpine sod, moss mats, or on tree or shrub bases; thallus suberect to erect, mostly 2-10 cm tall
        10a Medulla P+R (containing fumarprotocetraric acid)
          11a Pseudocyphellae laminal and marginal. Widespread boreal and subalpine to arctic-alpine; occasional in Ore and Wash
....*C. islandica*
          11b Pseudocyphellae primarily marginal in continuous lines.

Widespread and frequent in N Am arctic to boreal, s to BC
....[*C. laevigata*]
10b Medulla P-, K- or spot test not performed
  12a Pseudocyphellae both marginal and laminal, conspicuous;
     lobes almost flat to weakly channeled *C. islandica* (Lead 11a)
  12b Pseudocyphellae mostly marginal or pseudocyphellae
     indistinct; lobes almost flat to strongly channeled
    13a Lobes channeled, reticulately ridged and wrinkled
       .... ***C. ericetorum***
    13b Lobes flat to weakly channeled........................ ***C. subalpina***
1b On bark or wood; thallus suberect or prostrate, some shade of brown,
  yellow, gray, green, or black
  14a Upper surface some shade of yellow or gray
    15a Upper surface whitish to greenish gray above; lower surface dark
      brown to black.. ***Esslingeriana idahoensis***
    15b Upper surface pale to deep yellow, yellowish green, or pale green
      above; lower surface yellowish, pale tan, or whitish
      16a Thallus with marginal soredia; medulla yellow.............***C. pinastri***
      16b Thallus lacking soredia; medulla yellow or white
        17a Upper surface bright to deep yellow; medulla yellow
          .... ***C. canadensis***
        17b Upper surface very pale yellow to pale greenish; medulla white
          18a Thallus suberect, pale greenish; pycnidia mainly marginal
           and submarginal, black; at low elevations
            ....***C. pallidula***
          18b Thallus closely appressed, pale greenish in shade to nearly
           black where exposed; pycnidia laminal, black; mainly
           subalpine..***C. sphaerosporella***
  14b Upper surface olive green to brown or black
    19a Soredia or isidia present
      20a Thallus with whitish to gray or brownish marginal soredia; lobes
      narrow, mostly < 3 mm wide; apothecia very rare
        ....***C. chlorophylla***
      20b Thallus isidiate or warty, occasionally becoming lobulate (but
      often isidia lacking); lobes broader, mostly > 3 mm wide;
      apothecia common.. ***C. platyphylla***
    19b Soredia and isidia lacking (although prominent pycnidial warts are
      commonly present)
      21a Lobes tightly appressed to bark; upper surface yellowish green,
      greenish, or blackening; habitat usually subalpine
        .... ***C. sphaerosporella***
      21b Lobes suberect to erect; upper surface various in color; habitat
      various
        22a Thallus medium sized (to 10 cm broad), dark to pale brown to
        pale olive brown

23a Lobes elongate and dichotomously branched and
    rebranched; on shrub and tree bases, mainly subalpine
    .... ***C. subalpina***
23b Lobes short, not distinctly dichotomously branched and
    rebranched; habitat various
    24a Upper surface wrinkled, warty, occasionally isidiate or
        lobulate; lower surface brown to dark brown, sparsely
        rhizinate; pycnidia mainly laminal .... ***C. platyphylla***
    24b Upper surface smooth to warty or slightly wrinkled,
        without isidia or lobules; lower surface white to brown,
        moderately rhizinate; pycnidia prominent, mainly
        marginal, glossy, dark, shaped like small papillae or
        barrels
        25a Medulla UV-, KC-; common in PNW .... ***C. orbata***
        25b Medulla UV+ white, KC+R; boreal and e N Am
        .... [*C. halei*]
        (See notes under *C. orbata*)
22b Thallus small (generally < 2 cm broad), dark greenish black,
    brown black, or brown
26a Thallus dark olive green to green black or black
    27a Lobes ± appressed; c RM and e .... [*C. fendleri*]
    27b Lobes subfruticose or fruticose, finely dissected; w RM
        28a On the immediate coast; lobes (branches) roundish in
            cross section ....***C. californica***
        28b Not on the coast; lobes flat or channeled .....***C. merrillii***
26b Thallus light to dark brown or pale olive
    29a Medulla UV+ white, KC+R (alectoronic acid). Apothecia
        common; thallus either compact like the following species
        or more expanded like *C. orbata*; boreal forests of e N
        Am, occasionally w to Alta and e BC...................[*C. halei*]
    29b Medulla UV-, KC-. Apothecia generally numerous,
        dominating the thallus; lobes 2-3 mm wide, the margins
        undulating; generally on shrub twigs, especially in bogs;
        Alas s to n Cal and nw Mont........................... *C. sepincola*
        (Small individuals of *C. orbata* will key here, but that
        species is much more common, not restricted to bogs,
        has prominent marginal pycnidia, and is less abundantly
        apotheciate.)

Cetraria aculeata

Description: Thallus fruticose, erect, moderately to highly branched, reddish brown, shiny, brittle, mostly 2-4(10) cm tall; pseudocyphellae generally 0.3-1 mm long, round or elongate, often abundant; branches about 1 mm wide, terete or angular, often longitudinally furrowed, with numerous short, sharp projections (spinules); apothecia occasional; spot tests negative.

Range: Alas s to Colo and Ore, coastal mountains inland to the RM.

Substrate: Soil, humus over rock, and alpine sod.

Habitat: Mostly in open subalpine to alpine habitats, but occasionally at low elevations (talus or outcrop areas).

Notes: Cetraria aculeata and *C. muricata* are distinctive among our *Cetraria* species in having branches that are nearly round in cross section, rather than flat with a differentiated upper and lower surface. *C. californica* is also truly fruticose, but has a pale olive brown to blackish thallus, while *C. aculeata* and *C. muricata* are reddish brown. *C. muricata* is similar to *C. aculeata* but is more compact and has smaller and less abundant pseudocyphellae. Some of our material is intermediate between these two spp, but most is more similar to *C. muricata* than *C. aculeata*. Lower-elevation collections are often *C. aculeata* while high-elevation collections are more likely to be *C. muricata*. Recent work on the population genetics of these species revealed unsolved problems in w N Am.

Cetraria californica

Description: Thallus tufted fruticose, some shade of pale olive brown, pale grayish olive, to olive black, small (generally < 2(3) cm diam); branches roundish to irregular in cross section, mostly < 1(2) mm diam; soredia and isidia lacking but with short pointed branches; apothecia fairly common; spot tests negative.

Range: On the coast, se Alas and BC to Cal.

Substrate: Usually on conifer twigs, less often on shrubs.

Habitat: Most common on open edges of *Pinus contorta* or *Picea sitchensis* forests within sight or sound of the ocean, quickly diminishing inland, not occurring in the Willamette-Puget trough, Cas, or Sierra Nevada.

Notes: Frequently confused with *Cetraria merrillii* and *Nodobryoria abbreviata* but easily distinguished from these in both habitat (neither of these occurs in oceanside forests) and form. *C. merrillii* usually has flatter, darker (greenish black) lobes. *N. abbreviata* is always reddish brown, never with olive tints. Despite these differences, *C. californica* has frequently been misreported from inland areas, and those records persist in herbaria, so that herbarium-based distribution maps are wildly incorrect, instead showing the combined distribution of these three species.

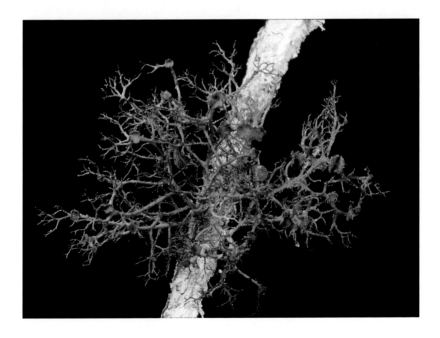

Cetraria canadensis

Description: Thallus foliose, with ascending lobes, to 4(6) cm diam; both upper and lower surfaces bright to deep yellow; lobes to 1 cm broad; medulla yellow; soredia and isidia lacking; apothecia usually present, the disk brown; spot tests negative or still yellow (pinastric, vulpinic, and usnic acids).

Range: BC to Cal, inland to w Mont, mainly e Cas.

Substrate: Bark and wood, especially frequent on *Pinus* and *Larix*, but also on shrubs and other conifers.

Habitat: Common, especially on twigs in conifer forests near water (floodplains, lakeshores, valley bottoms, and riparian habitats), but occasionally in upland habitats.

Notes: Cetraria canadensis is a distinctive species unlikely to be confused with any other. *Cetraria pinastri* is also bright yellow, but is sorediate. *Cetraria pallidula* is similar in form, but is a pale yellowish green or pale greenish.

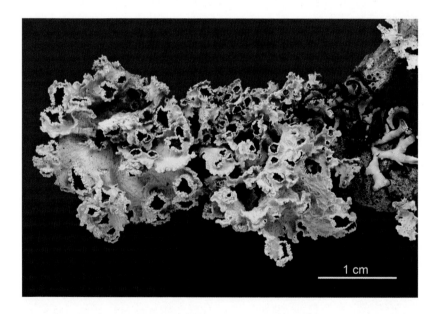

1 cm

Cetraria chlorophylla

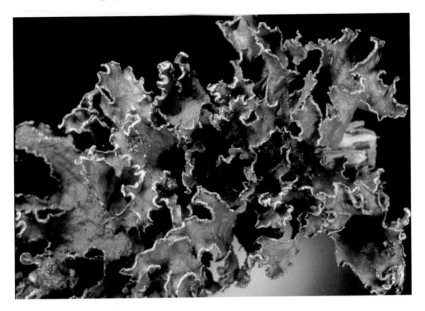

Description: Thallus foliose, small to medium sized (generally < 5(8) cm diam); lobes suberect, generally narrow (1-2 mm), ± channeled, with whitish, gray, or brown marginal soredia (sometimes poorly developed), sometimes granular or nearly isidiate; upper and lower surfaces light to dark brown, occasionally pale greenish brown in shade, usually paler below; apothecia very rare; all spot tests negative (fatty acids).

Range: Circumpolar, subarctic s through Cas and RM to Cal and Colo; common throughout the PNW.

Substrate: Bark and wood, fallen logs, fences, roofs, rarely on rock.

Habitat: Within the PNW, this is one of the most versatile foliose lichens, frequenting a range of habitats matched by few plants or lichens. Its habitats range from extreme oceanic sites to dry interior forests in continental climates. It tolerates both full sun and deep shade. One wonders if this versatility in habitat is matched by high genetic variability; its asexuality would suggest otherwise.

Notes: Despite the great variation in color and size of the thallus, and degree of development of soredia, the species can be consistently recognized by looking for the narrow, slightly channeled lobes with marginal soredia. In shaded sites the color is often more greenish than brown, while in sunny sites it is a rich coffee brown. Use a hand lens to inspect carefully for the relatively subtle soredia on specimens from shaded sites.

Cetraria ericetorum

Description: Thallus of upright to decumbent, dichotomously branched lobes, generally 2-3(5) cm high; lobes channeled, reticulately ridged and wrinkled, 2-4 mm broad; pseudocyphellae submarginal, whitish, visible with naked eye or hand lens as a delicate border on the underside of the lobes; pycnidia usually present on isidia-like projections from the margins; soredia and isidia lacking; apothecia brown, infrequent; spot tests negative.

Range: Circumpolar, s in the mountains to Ore and Colo.

Substrate: On soil and moss, often where these form a thin layer over rock; frequently a component of moss and *Selaginella* mats.

Habitat: Montane talus slopes and outcrop areas, more frequent in subalpine to alpine sods in exposed to sheltered microsites.

Notes: This is the most common brown terricolous *Cetraria* in our region. The next most common species in this group, *C. islandica*, is P+R and has frequent laminal pseudocyphellae on the backs of the lobes. Do not confuse the erect projections on the lobe margins with isidia. In contrast to isidia, these projections bear pycnidia at their tips.

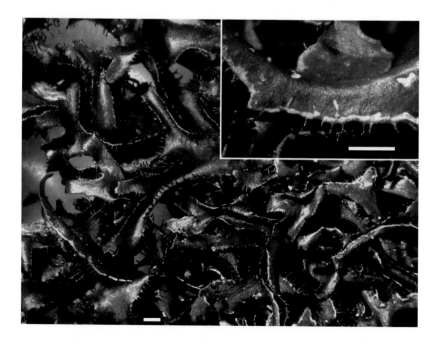

Cetraria merrillii

Description: Thallus small (generally < 2 cm diam), dark greenish black, dark olive green or black; lobes subfruticose, generally finely dissected to < 1 mm broad but often some lobes > 1 mm broad are present; soredia and isidia lacking; apothecia common, similar in color to the thallus; spot tests negative.

Range: Yukon and se Alas, s to Cal, inland to w Mont; most common between the Cas crest and Cont Divide, but also common in the Siskiyou Mts of s Ore and n Cal; not present on the immediate coast.

Substrate: Bark and wood, mainly on twigs of conifers (especially *Pinus albicaulis, P. contorta, P. ponderosa,* and *Larix occidentalis*); also on shrubs, especially *Arctostaphylos* (manzanita).

Habitat: Well-lit branches in forests and on more open, exposed trees, such as *Pinus albicaulis* on rocky ridges; also in treetops of moist low-elevation forests; present at all forested elevations.

Notes: Most easily confused with *Cetraria californica,* a species that is restricted to the immediate coast; *C. merrillii* is not found on the coast. *Cetraria fendleri* and *C. coralligera* are two species found in the central RM that can have a similar olive-black color. *Cetraria merrillii* is a conspicuous component of heavily lichenized shrubs on serpentine barrens and sparse *Pinus jeffreyi* woodlands in n Cal and s Ore. The species is also noticeable on fallen pine branches e Cas, where it forms dense colorful communities with *C. canadensis, C. platyphylla, Letharia vulpina,* and *Nodobryoria abbreviata.*

Cetraria orbata

Description: Thallus foliose, medium sized, to 5(10) cm diam, light to dark brown, olive brown to greenish in shade; lobes 1-5 mm broad, with few or no marginal cilia; upper surface smooth to slightly wrinkled; lower surface whitish, pale brown, or mid brown, moderately rhizinate; pycnidia mainly marginal, glossy, dark, papilla-like or barrel shaped; soredia and isidia lacking; spot tests negative (fatty acids).

Range: BC to Cal, inland to w Mont; rare e Cont Div, common in areas of strongest Pacific coastal influence.

Substrate: On bark and wood, mainly conifers but also on hardwoods.

Habitat: Mainly in low-elevation moist forests, fairly shade tolerant but also in exposed microsites.

Notes: In exposed sites the darker forms of *C. orbata* sometimes resemble *C. platyphylla*. That species, however, will have a wrinkled and warty upper surface, a dark brown lower surface, and almost no rhizines. See that species for additional comparisons. *Cetraria halei* has more marginal cilia and contains alectoronic acid (UV+ bright white, KC+R). Broadly distributed across boreal N Am, *C. halei* is not yet known from Ore and Wash. The prominent marginal pycnidia of *C. orbata* are commonly mistaken for isidia.

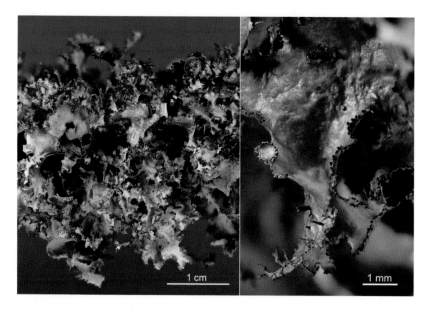

Cetraria pallidula

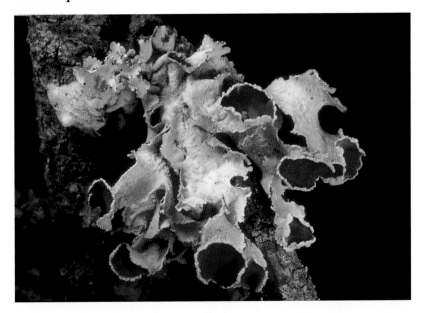

Description: Thallus foliose, small, to 3 cm diam, the lobes short and broad, with few lobes per thallus; upper surface very pale yellow or pale greenish with a faint yellowish tinge; lower surface similarly colored or somewhat paler; medulla white; soredia and isidia lacking; pycnidia marginal to submarginal, dark, often ± stalked; apothecia common; spot tests negative or cortex KC+Y, UV+ dark orange-red (usnic, caperatic, and unknown fatty acids).

Range: Yukon to Cal, e to Mont; generally uncommon but perhaps most common on the e slope Cas, sparse w Cas; in RM restricted to areas of strongest coastal influence.

Substrate: On bark and wood of conifers, especially on *Larix* and *Pinus contorta*.

Habitat: On well-lit branches beneath an open canopy or on branches in the upper crown; mesic forests at low to mid elevations.

Notes: This species has the appearance of a very pale version of the more common *C. canadensis*. Usually only a few thalli are found at a site. *Cetraria sphaerosporella* can be similar in color but is closely appressed and has black laminal pycnidia.

Cetraria pinastri

Description: Thallus foliose, mostly < 3 cm diam, with ascending lobes; both upper and lower surfaces yellow to greenish yellow; lobes to 1 cm broad but mostly narrower; medulla yellow; soredia present, marginal, yellow; apothecia rare; spot tests negative or still yellow (pinastric, vulpinic, and usnic acids).

Range: Alas s to Ore, e to Mont and s to NM, most frequent in continental mountain climates; common but seldom abundant.

Substrate: Tree bases, shrubs, and dead wood.

Habitat: Mountain forests, generally at medium to high elevations, often in cold sites and more tolerant of snow burial than other epiphytic *Cetraria*-like species.

Notes: This is the only bright yellow sorediate *Cetraria* in the PNW. Like the other bright yellow *Cetraria* species, it contains the poisonous yellow pigment, vulpinic acid.

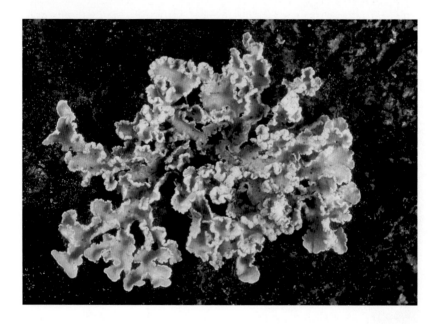

Cetraria platyphylla

Description: Thallus foliose, medium sized, to 5(10) cm diam, dark to pale brown to pale olive brown; lobes 3-12 mm broad; upper surface wrinkled, warty, these projections occasionally isidiate or lobulate; lower surface pale to dark brown, generally sparsely rhizinate; pycnidia not strongly marginal; spot tests negative or with spots of medulla that are K+Y and KC+Y or O.

Range: BC to Cal, inland to Alta and Mont; common between Cas and RM crests, occasional w Cas.

Substrate: Bark and wood, mainly conifers.

Habitat: In the crowns of trees in low-elevation moist forests, drier inland forests, and open-grown trees to timberline.

Notes: This is the most common dark brown epiphytic *Cetraria* in the PNW. The combination of the color, the strongly wrinkled thallus, and the relatively broad lobes (for a *Cetraria*) are distinctive. Some difficulties will, however, be encountered separating shade forms of *C. platyphylla* from *C. orbata*. That species has a smoother upper surface, primarily marginal pycnidia, and a pale (often white) lower surface with rhizines. *Cetraria orbata* is also more common w Cas and more shade tolerant than *C. platyphylla*.

1 cm

Cetraria sphaerosporella

Description: Thallus foliose, medium sized, yellow green, often blackening in exposed habitats, frequently tightly wrapped around small twigs or tightly appressed to larger stems; lower surface pale, rhizinate; soredia and isidia lacking; pycnidia black, laminal, immersed or slightly protruding; apothecia generally present, the disk pale brownish or greenish or darkening; spot tests negative except cortex KC+Y and UV+ dark orange-red.

Range: NWT s to Cal, inland to Mont and Alta.

Substrate: Bark and wood of conifers.

Habitat: Subalpine forests to isolated trees at timberline, occasionally in lower-elevation montane forests.

Notes: *Cetraria pallidula* can be similar in color, but is suberect, has mainly marginal and submarginal pycnidia, and is not known to occur in the subalpine. *Cetraria sphaerosporella* is unique among our *Cetraria* species in having lobes that tightly wrap around branches. Because of this, it was originally considered a *Parmelia* rather than a *Cetraria*. Thell et al. (2009) clearly demonstrated that the species does not belong with other species previously assigned to *Ahtiana* (*A. pallidula* in our biota), nor did they find that *Ahtiana* had merit as a monotypic genus.

Cetraria subalpina

Description: Thallus of upright to decumbent, dichotomously branched lobes, generally 2-3(5) cm high; lobes flat to weakly channeled, ± smooth, 2-4 mm broad; pseudocyphellae submarginal, whitish, visible with naked eye or hand lens as a delicate border on the underside of the lobes; pycnidia usually present on slender isidia-like projections from the margins; apothecia brown, occasional; spot tests negative.

Range: Coastal Alas s to Ore, mainly w Cas, rarely inland to n Id and nw Mont.

Substrate: Most common on bases of ericaceous shrubs and other woody plants; occasionally on bases of conifers or spreading onto the ground.

Habitat: Typically in semi-open to open subalpine forests, rarely to middle elevations in the mountains, mostly not in true alpine areas.

Notes: The distinctive substrate and habitat are good clues for naming this species; it can be verified by the nearly flat, smooth lobes, as opposed to the more channeled, wrinkled lobes of *C. ericetorum*. *Cetraria islandica* has prominent white to gray laminal pseudocyphellae on the lower surface, while in *C. subalpina* they are restricted to near the margins.

Cetrelia

Cetrelia cetrarioides

Description: Thallus foliose, medium sized to large (generally 5-20 cm diam); lobes broad (7-15 mm), rounded; upper surface whitish gray to greenish gray; lower surface black with brown margin; rhizines black, simple, sparse; pseudocyphellae present as small white spots on both upper and lower surfaces; soredia present, in elongate marginal soralia; apothecia not seen in the PNW; cortex K+Y (atranorin); medulla K-, P-, C-, KC- or + pinkish, UV+ blue-white (perlatolic acid).

Range: Coastal Alas to w Ore, mainly in the Coast Ranges, not known e Cas in w US but rarely so in BC. Sporadic throughout its range in w N Am.

Substrate: Bark, mainly on *Alnus rubra* and other hardwoods, rarely on mossy rock.

Habitat: Moist riparian and valley bottom forests, especially older *Alnus rubra* stands over seepy or swampy ground.

Notes: With the appearance of a *Parmotrema* at first glance, *C. cetrarioides* can be distinguished by its pseudocyphellae and elongate marginal soralia, which often follow the edge of a lobe for 5-10 mm or more. *Cetrelia cetrarioides* is broader lobed than *Parmelia* species, and is often distinguishable at a distance by its very broad thalli on trunks.

Sources include: Culberson and Culberson (1968).

Cladina—*see* Cladonia

Species in *Cladonia* subgenus *Cladina*, also known as reindeer lichens, were formerly placed in the familiar genus *Cladina*. Molecular evidence shows reindeer lichens to be part of *Cladonia* in its current sense (Stenroos et al. 2002). Reindeer lichens are richly branched, lack a persistent primary thallus, and have a cobwebby surface texture because they lack a cortex. See Group 1 in the *Cladonia* key.

Cladonia

Thallus of two parts, but sometimes only one of the parts: primary squamules and secondary erect fruticose podetia; podetia simple to richly branched, sometimes the branching in whorls, corticate or not, often with soredia or squamules, terminating in blunt to sharp points, cups, or apothecia; cups often proliferating from their margins, the proliferations often supporting apothecia or additional tiers of proliferations, sometimes proliferating centrally; apothecia at the tips of the podetia or sessile on the primary squamules, pale to dark brown or red, lecideine; pycnidia often at the apices of the podetia, black, brown, or red; spot tests various but often P+O or R (fumarprotocetraric acid); on a wide range of substrates and occurring in most habitats.

Sources include: Ahti (1961, 2007), Athukorala et al. (2016), Brodo and Ahti (1996), Fontaine et al. (2010), Hammer (1993a, 1993b, 1995), Pino-Bodas et al. (2011, 2012, 2013, 2015), Smith et al. (2012), Spribille et al. (2010), Stenroos (1989a, 1989b, 1990), Stenroos et al. (2002, 2019), Thomson (1967b), Timdal and Tønsberg (2012), Tønsberg and Ahti (1980).

ID tips: Cladonia is a big, beautiful genus, abundant in the PNW, and enticing to beginners. While some species are utterly distinctive, many have perplexing variability among sites or even within a colony. Your struggles will be lessened if you make good use of K and P spot tests, as well as UV fluorescence. Take a little extra time to make these tests to greatly reduce the possibilities for difficult specimens, ultimately saving time and increasing accuracy. If spot tests are new to you, be sure to read details of the methods in the Collecting and Identifying Lichens chapter (pp. 20–22).

Introductory Key

1a Stalks richly branched (generally branched more than 5 times over the length of a mature individual), the branch tips pointed
.... **Group 1, reindeer lichens** (Lead 7)
1b Stalks simple to few-branched, pointed, blunt or tipped with cups
2a Apothecia red ....**Group 2** (Lead 15)
2b Apothecia brown or lacking

3a Podetia lacking, only primary squamules present (spp keyed here
   often or generally lack podetia, however all spp have the potential
   to lack podetia)..**Group 3** (Lead 32, p. 119)
3b Podetia present
   4a Podetia with definite cups
      5a Podetia sorediate....................................**Group 4** (Lead 43, p. 120)
      5b Podetia not sorediate ....**Group 5** (Lead 61, p. 122)
   4b Podetia lacking definite cups
      6a Podetia sorediate....................................**Group 6** (Lead 74, p. 124)
      6b Podetia not sorediate ....**Group 7** (Lead 86, p. 125)

Group 1

Richly branched (reindeer lichens)

7a Surface smooth, corticate at least in part; primary squamules often
   persistent...(*Cladonia* **Group 7**)
7b Surface dull and fibrous, like matted cobwebs (use lens), cortex lacking;
   primary squamules absent (*Cladonia* subgenus *Cladina*, the reindeer
   lichens)
   8a Thallus mats often subdivided into discrete rounded heads; thallus P-,
      K-, UV+ whitish, usually KC+Y
      9a Branching in 4s and 5s; habitat boreal forest, rarely southward
         ....*C. stellaris*
      9b Branching in 2s and 3s; habitat old coastal dunes and roadcuts
         through dunes, rarely farther inland .... **C. portentosa**
   8b Thallus forming cushions but not subdivided into discrete heads;
      thallus spot tests various but always UV-
      10a Podetia grayish, whitish, or brownish gray, lacking usnic acid (KC-
         or not detectable because of K+Y reaction), K- or K+Y, always P+R
         11a Base of podetia blackened or with pale
            patches on a blackish background
            (melanized) and the inner wall blackish.
            Thallus K+Y; Alas to Wash inland to Alta
            ....*C. stygia*
         11b Base of podetia grayish or brownish or with
            only scattered blackish patches and the
            inner wall pale or brown
            12a Thallus K+Y (atranorin present); branch
               clusters mostly in 3s and 4s
               ....**C. rangiferina**
            12b Thallus K- (lacking atranorin); branch
               clusters mostly in 2s and 3s]
               ....**C. ciliata** var. **ciliata**

C. stygia

      10b Podetia pale yellowish, containing usnic acid
         (KC+Y), K-, P- or P+R
         13a Podetia P- ....**C. mitis**
         13b Podetia P+R (fumarprotocetraric acid)

14a Branching in 3s and 4s; pycnidial jelly clear. On soil or soil over rock; circumpolar s to Wash, Mont, and Colo; uncommon, mostly in cool rocky sites at low elevations............. *C. arbuscula*

14b Branching in 2s and 3s (several successive pairs on older axes are diagnostic); pycnidial jelly red (usually difficult to find). On soil or soil over rock; se Alas to near Olympia, Wash) .... ***C. ciliata* var. *tenuis***

Table 4. Spot tests and lichen substances for reindeer lichens in the PNW

species	atranorin (K+Y)	fumar-protocetraric acid (P+R, K+brownish)	perlatolic acid (UV+)	usnic acid (KC+Y)
arbuscula	-	+	-	+
ciliata var. *ciliata*	-	+	-	–
ciliata var. *tenuis*	-	+	-	+
mitis	-	-	-	+
portentosa	-	-	+	+/-
rangiferina	+	+	-	-
stellaris	-	-	+	+
stygia	+	+	-	-

Group 2

Apothecia red

15a Podetia lacking soredia (do not confuse a warty cortex or small squamules with soredia)

16a Podetia with abundant small squamules or squamules plus warts or granules; medulla UV- or UV+ strong bluish white

17a Medulla UV-, K+Y, P+O ....thamnolic acid chemotype of ***C. bellidiflora***

17b Medulla UV+ strong blue-white (squamatic acid), K-, P-

18a Podetia with small, dense squamules and roundish corticate granules, the granules mainly on the upper half of the podetia; primary squamules small, to 4 mm long; on soil or moss over rock. Uncommon from Alas s to Ore and Sask .... *C. straminea*

18b Podetia heavily squamulose, cup forming but cups often irregular or absent; on mossy rock, bark, and rotten wood .... ***C. bellidiflora***

16b Podetia generally platy-areolate to warty; medulla UV- or UV+ weak

19a Lower surface of squamules distinctly yellowish, the squamules large. Squamules recurved when dry, exposing the lower surface; often sterile or intimately associated with podetia of *C. borealis*, *C. coccifera*, or *C. straminea*; rarely making its own podetia which are cupless or nearly so, ecorticate, and with a thin, yellow, cottony surface layer, similar to the lower surface of the squamules

.... *C. luteoalba* (Lead 33a)
(Keyed here according to the podetia of its common and intimate associate, *C. borealis*)

19b Lower surface of squamules white or discolored but not a clear pale lemon yellow, the squamules small to medium. Cortex of podetia of mostly flat, scaly plates to warty; squamules or granules occasionally present in cups or on outside of podetia; primary squamules small or up to 10 × 5 mm; on soil, soil over rock, less often on rotten wood; widespread, common at all elevations in Ore and Wash ....*C. borealis*
(Most specimens formerly named *C. coccifera* from w N Am are actually *C. borealis*; *C. coccifera* is rare in PNW and contains zeorin instead of barbatic acid.)

15b Podetia sorediate
  20a Medulla UV+, P-
    21a Podetia thick (generally 1-3 mm diam), yellowish (usnic acid present), the sides often fissured and split, completely covered with yellowish soredia ....**C. sulphurina**
    21b Podetia slender (generally 0.5-1.5 mm diam), not yellowish (usnic acid lacking), the sides not fissured and split, not yellowish sorediate
      22a Podetia generally tapering to narrow points; medulla weakly UV+ or UV- (barbatic acid). On rotten logs, humus, and tree bases; widespread, uncommon in PNW .... *C. bacillaris*
(Morphologically similar to *C. macilenta*, *C. bacillaris* is considered by some authors to be the barbatic acid chemotype of *C. macilenta*.)
      22b Podetia generally branched, pointed or blunt; medulla UV+ intense bluish white (squamatic acid) .... **C. umbricola**
  20b Medulla UV-, P- or P+
    23a Podetia with definite cups
      24a Thallus K-, P-
        25a Podetia tall (2-8 cm) and slender, gradually to abruptly flaring to cups, densely covered with powdery yellowish soredia. On rotten logs, bark, and humus-rich soil, occasionally on humus over rock; widespread, uncommon, mainly e Cas .... *C. deformis*

C. pleurota

        25b Podetia shorter (generally < 2(4) cm tall), flaring more abruptly from the base, with granular (rarely powdery) soredia (though soredia often scarce when fertile). On soil and soil over rock; widespread, uncommon in PNW..*C. pleurota*
      24b Thallus K+Y, P+ deep Y to O (thamnolic acid)
        26a Primary squamules large (5-15 mm broad). Podetia 15-40 mm tall, corticate at the base, the upper part with fine whitish to

yellowish or greenish soredia; on humus-rich soil, rotten logs, and tree bases; widespread but uncommon in N Am, rare in PNW ....*C. digitata*
26b Primary squamules small to medium (usually 1-4 mm broad) .... **C. transcendens**
23b Podetia lacking cups or with irregular, partial, or tiny cups
27a Podetia yellowish tinged; containing usnic acid
28a Thallus K-, P-...*C. deformis* (Lead 25a)
28b Thallus K+ deep Y or orangish, P+ deep Y or O (thamnolic acid) .... **C. transcendens**
27b Podetia pale greenish, greenish gray, or whitish; lacking usnic acid
29a Medulla K-; podetia unbranched, cupless ....*C. bacillaris* (Lead 22a)
29b Medulla K+ deep Y (thamnolic acid); podetia various
30a Primary squamules large (5-15 mm broad) .... *C. digitata* (Lead 26a)
30b Primary squamules small to medium sized (< 5 mm broad)
31a Podetia pointed or blunt, never with cups, unbranched. On soil, tree trunks, and rotten wood; widespread, uncommon, coastal Alas to Cal...................... *C. macilenta*
31b Podetia typically cup bearing, but cups often poorly developed or lacking, podetia often sparingly branched .... **C. transcendens**

Group 3

Podetia lacking (all spp may lack podetia; only those chronically so are keyed here). TLC is often needed for identification, so chemical characters are included.

32a Squamules yellowish tinged (usnic acid present)
33a Primary squamules with a yellowish lower surface, large. Squamules strongly curled back when dry; widespread, occasional in many different open habitats, from alpine to low elevations; often intimately associated with red-fruited spp ....*C. luteoalba*
33b Primary squamules with whitish lower surface, small to large
34a Containing barbatic acid; common.................*C. borealis* (Lead 19b)
34b Containing zeorin; very rare in PNW (TLC required) .... [*C. coccifera* (see Lead 19b)]
32b Squamules not yellowish tinged (usnic acid lacking)
35a Medulla P- or P+ strong Y; cortex K+Y (atranorin)
36a Medulla P+ strong Y (psoromic acid); squamules medium to large, often > 10 mm long....................................... *C. dahliana* (Lead 92b)
36b Medulla P-; squamules small to medium, rarely to 7 mm long ....*C. cariosa*
35b Medulla P+O or P+R (fumarprotocetraric or norstictic acid, with or without atranorin), K reaction various
37a Thallus K+Y to R (norstictic acid and atranorin)

38a Squamules convex or flat, not much of lower surface exposed; steppe to moist forests, central Wash to n Id .... *C. andereggii*

38b Squamules variably reflexed, often exposing much of the lower surface when dry; northern and boreal .... *C. symphycarpa*

37b Thallus K- or K+ brownish (fumarprotocetraric acid, with or without atranorin)

39a Squamules small (2-6 × 2-4 mm), generally thick and rounded

40a Squamules closely contiguous, occasionally forming a rosette .... *C. pocillum* (Lead 51a)

40b Squamules ± dispersed, not rosette forming .... ***C. pyxidata***

39b Squamules larger (mostly 2-25 × 1-8 mm), generally elongate, thick or thin

41a Squamules very large, to 50 mm long and 5 mm broad; lower surface of squamules partially corticate; podetia when present not proliferating from the cup centers. On soil or rock; coastal Alas and BC .... [*C. schofieldii*]

41b Squamules large, to 25 mm long but usually much shorter; lower surface of squamules ecorticate; podetia when present proliferating from the cup centers

42a Squamules usually without whitish cortical thickenings; lower surface K+ brownish (atranorin lacking) .... ***C. verticillata***

42b Squamules with whitish cortical thickenings forming a distinct texture on the upper surface; lower surface K+Y or brownish Y (atranorin) .... ***C. macrophyllodes***

Group 4

Cup-forming; sorediate

43a Medulla strongly UV+

44a Podetia short, mostly < 2 cm tall, cups goblet shaped ..... UV+ members of *C. chlorophaea* group (***C. albonigra***, *C. grayi, C. imbricarica, C. merochlorophaea, C. novochlorophaea*)

44b Podetia medium to tall, mostly > 2 cm; cups not goblet shaped

45a Podetia yellowish (usnic acid); primary squamules large (> 3 mm), persistent; cups seldom proliferating from the margins .... ***C. sulphurina***

45b Podetia greenish to gray green or whitish green, lacking usnic acid; primary squamules small (1-3 × 1 mm) or disappearing; cups usually proliferating from the margins .... ***C. cenotea***

43b Medulla UV- or weakly UV+

46a Thallus distinctly yellowish, generally containing usnic acid

47a Thallus K+Y, P+ deep Y to O or R .... ***C. transcendens*** (Lead 31b)

47b Thallus K-, P-

48a Apothecia yellowish to pale brown (K-) but often lacking; cups often strongly dentate .... ***C. carneola***

48b Apothecia and pycnidia red (K+ purple), generally found with careful searching; cups generally not markedly dentate

49a Podetia tall (2-8 cm) and slender, gradually to abruptly flaring to cups, densely covered with powdery yellowish soredia ....*C. deformis* (Lead 25a)

49b Podetia short (generally < 2 cm tall), flaring more abruptly from the base, with granular soredia externally and on the inside of the cups ....*C. pleurota* (Lead 25b)

46b Thallus not yellowish, lacking usnic acid

  50a Interior of the cups with small peltate squamules and round corticate granules

    51a Primary squamules closely packed or subcontinuous, rosette-forming when growing freely; in drier habitats than the next sp. Squamules often brownish, thick; podetia, chemistry, and substrates similar to the next sp .... *C. pocillum*

    51b Primary squamules separate, not closely packed or forming rosettes; in more mesic sites than the preceding sp....**C. pyxidata**

  50b Interior of the cups with granular to powdery soredia

    52a Thallus K+Y, P+ deep Y .... **C. transcendens** (Lead 31b)

    52b Thallus K- or brownish, P- or P+R

      53a Thallus P- .... *C. carneola*

      (Specimens of *C. carneola* with low amounts of usnic acid will key here.)

      53b Thallus P+R (fumarprotocetraric acid)

        54a Cups generally slender, shallow to moderately deep

          55a Soredia coarse, granular, generally grading into squamules; podetia proliferating from margins and centers of cups...**C. verruculosa**

          55b Soredia fine, powdery, not grading into squamules (but squamules sometimes present on podetia); podetia proliferating from margins of cups

            56a Podetia corticate toward base .... **C. coniocraea**

            56b Podetia lacking a cortex toward base

              57a Cups narrow, with elongate (> 2 mm) marginal subulate proliferations; cup interiors occasionally perforate....................................*C. subulata* (Lead 85a)

              57b Cups of various widths, often with short marginal proliferations (generally < 2 mm) or giving rise to further cups; cup interiors closed...........**C. fimbriata**

        54b Cups generally deep goblet shaped

          58a Podetia granular or powdery sorediate on outside; medulla UV-

            59a Cups containing granular soredia, often broadly flaring or deep goblet shaped.............................**C. chlorophaea**

            (See text for related species.)

            59b Cups containing powdery soredia, mostly narrow ....**C. fimbriata**

          58b Podetia warty or with soredia compounding into numerous tiny projections; medulla UV- or UV+

60a Lower half of podetia not blackened between the granules or squamules, similar in color to the upper half; medulla UV- or weakly UV+, with fatty acids. On rotten wood or soil; most common w Cas but inland to w Mont and ne Ore ....*C. asahinae*
(TLC is required to distinguish *C. asahinae* from *C. chlorophaea* with confidence.)

60b Lower half (or more) of podetia blackened between the granules or squamules; medulla UV+........***C. albonigra***

Group 5

Cup forming; soredia lacking

61a Medulla UV+ blue-white, P-

  62a Cups imperforate; podetia usually < 2 cm tall; cortex areolate to verrucose

    63a Podetia olive, gray-green, or brownish, rarely yellowish green (usnic acid lacking); medulla UV+ weak blue white (sekikaic and homosekikaic acids). On soil or moss over rock; widespread but rare in PNW ....*C. novochlorophaea*

    63b Podetia yellowish (usnic acid present); medulla UV+ strong blue-white (squamatic acid) .... *C. straminea* (Lead 18a)

  62b Cups perforate if present; podetia medium sized to tall (1-11 cm); cortex continuous or nearly so, smooth or with slightly developed areoles. Podetia greenish to brownish (usnic acid lacking), cupless or with perforate cups, the margins of the cups becoming repeatedly proliferate; branch axils often perforate; containing squamatic acid; on soil, soil over rock, and rotten wood; widespread in northern N Am, rare in southern PNW (two locations in Mont) .... *C. crispata*

61b Medulla UV-, P+R

  64a Podetia coarsely granular to squamulose, becoming decorticate with age; proliferations arising from the margins of the cups and from the centers. Podetia 3-10 cm tall, often tipped with cups, the cups shallow, often irregular or lacking .... ***C. verruculosa***

  64b Podetia smooth or squamulose, mostly retaining the cortex with age; proliferations lacking or arising from the margins or centers

    65a Podetia proliferating centrally from the cups

      66a Thallus K- or K+ brownish; podetia generally well developed, often to 5 cm tall; squamules generally without a mosaic of whitish cortical thickenings

        67a Climate oceanic; podetia mostly < 1 mm diam; cups mostly < 4 mm diam..***C. concinna***

        67b Climate suboceanic to continental; podetia mostly > 1 mm diam; cups often > 4 mm diam; distribution mainly e Cas .... ***C. verticillata***

66b Thallus K+Y (atranorin); podetia generally poorly developed, mostly < 2 cm tall; squamules textured with a mosaic of pale cortical thickenings.......................................*C. macrophyllodes*

65b Podetia unbranched or proliferating marginally (rarely centrally) from the cups

68a Cups with a single large, often gaping hole; thallus K+ deep Y, P+ deep Y or O (thamnolic acid); coastal

69a Cups regular to oblique; lateral perforations and splits in podetia uncommon; podetial cortex smooth, continuous to areolate. Uncommon on soil and moss on the immediate coast, Alas to Cal......*C. japonica* (*C. japonica* is considered by some authors to be the thamnolic acid chemotype of *C. crispata*. Our use of the name *C. japonica* is a convenience for tracking chemotypes in this group.)

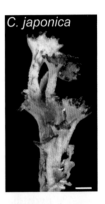

C. japonica

69b Cups irregular, formed by dilated axils; lateral perforations in podetia common ....*C. artuata*

68b Cups imperforate or with multiple small perforations; thallus K- or brownish or only cortex K+Y

C. artuata

70a Podetia short, < 3(5) cm; cups often goblet shaped; interior of cups usually with sparse to abundant peltate squamules

71a Cups with elongate, flattened, fissured, fertile proliferations; podetia < 5 cm, squamulose. Rare w Cas, very rare inland to n Id...................................... *C. dimorpha* (*Cladonia prolifica* will also key here. Unlike *C. dimorpha* and *C. pyxidata*, *C. prolifica* is a rare coastal species that lacks squamules inside the cups; it has oblique cups with marginal proliferations and squamules.)

71b Cups lacking elongate marginal proliferations; podetia short, generally < 3 cm..*C. pyxidata*

70b Podetia taller, to 10 cm; cups generally small, occasionally expanded, sometimes perforate; interior of cups lacking squamules

72a Cortex K+Y (atranorin), smooth, occasionally squamulose; podetia 2-10 cm tall, cupless or with narrow cups, with marginal (occasionally central) proliferations from the cups ....*C. ecmocyna*

90a Apothecia pale brown, generally present, capitate and terminal; medulla weakly UV+ (barbatic acid); podetia generally < 1 cm tall. On rotten wood, occasionally on humus-rich soil; widespread, northern, s to s BC and nw Mont, disjuncts to Colo .... [*C. botrytes*]

90b Apothecia dark brown, generally lacking, not capitate; medulla strongly UV+ blue-white (squamatic acid); podetia usually > 20 mm tall; many substrates; mainly boreal .... *C. crispata*

89b Medulla UV-, P- or P+

91a Podetia with terminal capitate apothecia

92a Thallus K+Y (atranorin), P- or P+ pale Y, rarely P+R; primary squamules small to medium, to 7 × 2 mm but generally much smaller ....**C. cariosa**

92b Thallus K+Y, P+ deep Y (atranorin + psoromic acid); primary squamules medium to large, often > 10 mm long. On soil or humus; occasionally in moist valley forests to subalpine forests; widespread but scattered in boreal and montane areas, s to S Dak, Mont, and Ore..*C. dahliana* (Psoromic-containing specimens in this group are polyphyletic in the PNW, in part the psoromic acid chemotype of *C. symphycarpa*, and in part at least two species outside of *C. symphycarpa*.)

91b Podetia lacking terminal capitate apothecia

93a Cortex or medulla K+Y, smooth or sparsely squamulose

94a Cortex K+Y, medulla K-

95a Basal parts of podetia melanotic, with pale patches against a blackish background; Arctic s to Wash and RM .... *C. stricta*

95b Basal parts discoloring to yellowish or brown, not melanotic; common e Cas .... **C. ecmocyna**

94b Cortex K-, medulla K+ strong Y; coastal

96a Cups absent; axils narrowly and irregularly open to closed; podetia branching 1-3 times. Podetia to 45 mm tall and 2 mm diam, uniform and slender, grayish, browned only at the tips, lateral perforations absent. On soil or soil over rock, near the coast and in the Coast Range, Cal to BC; rare.....................................*C. poroscypha*

96b Cups present, though often irregular or oblique; axils wide open; podetia unbranched except from the cup margins ....**C. artuata** (Lead 69b) and *C. japonica* (Lead 69a)

93b Cortex K-, smooth, warty, or squamulose

97a Branch axils and cups (if present) perforate; podetia sparingly to richly branched

98a Fertile branches often slightly flattened and grooved; never forming cups but branch axils gaping; w Cas .... *C. furcata*

98b Fertile branches terete; usually forming distinct or irregular, multiperforate cups; e Cas. On soil or soil over rock ....*C. multiformis*

97b Branch axils and cups (if present) closed; podetia simple to moderately branched

99a Base of podetia whitish spotted on melanized background (whitish patches of cortex on blackening base of podetia)

C. multiformis

100a Podetia and cups rather broad, podetia (1)1.5-2.5 mm diam below cups; tomentum on podetia very fine, thin, in the upper half; upper parts of podetia not melanotic ... **C. phyllophora**

100b Podetia and cups rather narrow, 0.5-1.5 mm diam below cups; tomentum on podetia restricted to the tips; upper parts of podetia often melanotic. Arctic s to Wash and RM .... *C. stricta*

99b Base of podetia not distinctly whitish spotted

101a Podetia becoming decorticate with age, generally minutely warty, squamulose, or even coarsely sorediate, sometimes moderately squamulose ....**C. verruculosa**

101b Podetia persistently corticate, smooth with continuous to slightly dispersed areoles, sometimes squamulose but never granular sorediate; cups present or not; branches generally proliferating from the margins of cups; on soil, humus, soil over rock, or rotten logs; widespread in N Am, occasional in PNW ....*C. gracilis*

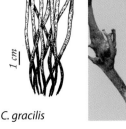

Cladonia gracilis subsp. *turbinata*

(Highly variable in form, *C. gracilis* subsp. *turbinata* (left) is predominantly cup forming, subsp. *vulnerata* is tall and slender with few cups, and subsp. *elongata* is similar to subsp. *vulnerata* but blackened at the base, as illustrated at far left.)

C. gracilis

1 cm

Cladonia albonigra

Description: Podetia mostly 1-2(4) cm tall, pale greenish gray, forming distinct cups, proliferating from the margins or centers of the cups, sometimes into more than one tier of cups; similar in form to *C. chlorophaea*, but the podetia are finely warty or isidioid-squamulose to sorediate, these structures compounding into numerous tiny projections; base of podetia decorticate and blackening, contrasting with the pale granules, squamules, or areoles; apothecia brown, occasional; medulla K- or K+ dingy Y, P+R, sw UV+ (often weak; fumarprotocetraric and 4-*O*-methylcryptochlorophaeic acid; other chemotypes farther north).

Range: Alas to n Cal, inland to n Id and w Mont.

Substrate: Humus over rock or soil, occasionally on bark.

Habitat: Mainly in areas of moist conifer forests, often on mossy tree trunks in old forests, humus over rock, and roadcuts, mainly at low to mid elevations.

Notes: This member of the *C. chlorophaea* group is distinguished by chemistry and morphology. The blackening base and isidia-like projections on the podetia of *C. albonigra* distinguish it from most other members of the group. *Cladonia asahinae* has similar projections but differs in chemistry (see **Table 5**, p. 135) and lacks the blackened base.

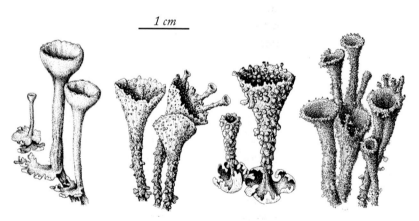

1 cm

Podetia of, left to right: *C. Fimbriata, C. chlorophaea, C. pyxidata, C. albonigra*

Cladonia artuata

Description: Podetia pale gray-green (to brownish in exposed sites), mostly 1.5-7 cm tall and 2-6 mm diam, proliferating from the margins of irregular, often asymmetric cups; podetial tips and cups often opening to the interior; podetial sides with short to elongate lacerations; podetial surface smooth, the cortex continuous or becoming areolate at the base, sometimes sparsely squamulose; primary squamules to 11 × 1-3 mm wide; soredia and isidia lacking; apothecia light to dark brown, sessile or on short stipes; thallus P+ deep Y, persistent, K+ deep Y to O, UV- (thamnolic acid with accessory barbatic acid).

Range: Narrowly distributed along the Pacific Coast, BC to central Cal.

Substrate: Soil, less often on rotten wood.

Habitat: Full sun to partial shade, most often in semi-open vegetation on stabilized sandy roadcuts, dunes, and deflation plains, rarely slightly inland at low elevations in the Coast Range.

Notes: This species is occasional along the northern California and Oregon coasts, usually occurring in *Cladonia*-rich mats. It can be recognized by its lacerate podetia, irregular cups, and the strong K and P reactions (above) indicating presence of thamnolic acid. Some authors consider this a synonym of the thamnolic chemotype of the widespread northern species, *C. crispata*, but preliminary DNA sequencing suggests that *C. artuata* is distinct at least from northern European *C. crispata*. Sequences are not, however, currently available for related Japanese and North American material. The very similar *C. japonica*, also considered a synonym of *C. crispata* by some authors, differs from *C. artuata* in having more regular cups with few lateral perforations in the podetia.

Cladonia bellidiflora

Description: Podetia heavily squamulose, mostly 1-3 cm tall, cup forming but cups often irregular or absent; yellowish tinged from usnic acid; cortex thick but becoming disrupted, leaving plates and squamules on a decorticate surface; primary squamules to 8(12) mm long and 4(7) mm broad; apothecia red (or discoloring to blackish red), common; medulla UV+ strong bluish white, P-, K- (squamatic acid chemotype) or UV-, P+O, K+ strong Y (thamnolic acid chemotype).

Range: In and w Cas, rare inland in BC and Wash.

Substrate: On mossy rocks, less often on bark, wood, and soil.

Habitat: Most frequent on cool, moist talus slopes, outcrops, and old lava flows; occasional in forests and sand dunes.

Notes: Cladonia bellidiflora is distinguished by the combination of red fruits, a yellowish tinge, especially at the base of the podetia, and squamulose podetia. *Cladonia straminea* is a more northern species that is similar but the squamules grade into granular soredia. Sterile material is distinguished by the yellowish color from the more grayish, brown-fruited, heavily squamulose

species, *C. squamosa.* The squamatic acid chemotype is most often found on moss over rock, especially old lava flows. The thamnolic acid chemotype is most common on shaded tree trunks. Another species concept holds the latter as an extremely squamulose and esorediate form of *C. transcendens.*

Cladonia cariosa

Description: Podetia pale greenish gray to whitish, to 2.5 cm tall, with distinct longitudinal fissures, ± branched towards the top, often tipped with brown apothecia; cortex continuous at first, soon warty and fissured; primary squamules small to medium, to 7 × 2 mm but generally much smaller; thallus K+Y (atranorin), P- or P+ pale Y, rarely P+R.

Range: Widespread, common throughout the PNW.

Substrate: On soil, rarely on rotten wood.

Habitat: Roadcuts, stream banks, and other disturbed sites, exposed to somewhat sheltered; valleys to subalpine, occasionally alpine.

Notes: The lacerate podetia tipped with capitate brown apothecia are characteristic. *Cladonia dahliana* can form similar podetia, but that species is P+ deep Y (psoromic acid) and has larger squamules. Another related species, *C. symphycarpa*, is K+Y to R (norstictic acid) and also has larger squamules. It is widespread but is not yet known from Ore and Wash. Another norstictic-containing species, *Cladonia andereggii*, is typically sterile, with squamules only. The squamules are similar to *C. cariosa* but usually larger; it is also distinguished from *C. cariosa* by K+R and P+O reactions from norstictic acid.

Cladonia carneola

Description: Podetia short (generally < 2 cm, up to 3 cm), forming distinct cups, covered with fine yellowish green soredia; cups often strongly dentate; primary squamules middle sized to large (to 13 × 10 mm but generally much smaller); apothecia pale brown or yellowish but usually lacking; medulla K-, P-, UV- (or UV+ pale whitish); containing usnic acid and zeorin, ± barbatic acid.

Range: Widespread in N Am, fairly common throughout the PNW.

Substrate: Humus-rich soil, stumps, and rotten wood, occasionally on tree trunks.

Habitat: Mainly in low- to mid-elevation forests or previously forested areas.

Notes: No other small consistently cup-forming *Cladonia* in the PNW is both P- and K- and yellowish tinged from usnic acid. The yellow tinge is stronger in more exposed habitats. *C. carneola* is similar in form to the *C. chlorophaea* group. With practice, *C. carneola* can usually be distinguished in the field by the short, regular, often dentate cups, pale brown fruits, and the yellowish tinge.

Cladonia cenotea

Description: Podetia tall, to 2-4(10) cm, greenish to gray green or whitish green (lacking usnic acid), fissured or not, completely sorediate or not, sparingly branched; not clearly cup forming, but irregular cups can be formed at branch points; branch axils perforate to gaping; soredia powdery or grading into small squamules; primary squamules persistent or disappearing, small (1-3 × 1 mm); apothecia uncommon, pale brown to brown; medulla K-, P-, UV+ strong blue-white (squamatic acid).

Range: Widespread, common e Cas, rare w Cas.

Substrate: Rotten logs, tree bases, and humus-rich soil.

Habitat: Cool, moist forests from valleys to subalpine, in both shaded and semi-open habitats.

Notes: This highly variable species, common in cool continental forests, is distinguished by its tall, sparingly branched podetia that are sorediate and UV+ strong blue-white. *C. squamosa* is similarly UV+ but is squamulose, not sorediate.

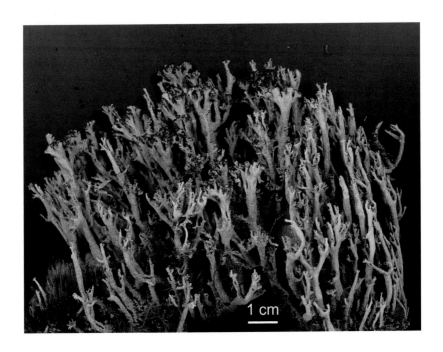

1 cm

Cladonia chlorophaea

Description: Podetia mostly 1-3(6) cm tall, gray-green, forming goblet-shaped cups, often broadly flaring, the transition gradual or abrupt; cups often proliferating from the margins; interior of the cups containing granular soredia, with granular or powdery soredia on outside of the cups and podetia; apothecia brown, fairly common; thallus P+R (fumarprotocetraric acid), K-, UV-.

Range: Cosmopolitan, common in the PNW.

Substrate: On a wide variety of substrates, including soil, humus, bark, rotten wood, and mosses or detritus over rock.

Habitat: Common in many habitats, most often where partly shaded or exposed.

Notes: C. chlorophaea is distinguished by its granular-sorediate cups. *C. fimbriata* has powdery sorediate cups, both inside and out, and usually a narrower profile, the cups seldom flaring broadly. *C. pocillum* and *C. pyxidata* occupy the other end of the size-of-propagule spectrum, having small squamules inside the cups (use hand lens). Adding to the difficulty of this group, a number of closely related and largely chemically defined species can be distinguished, but these usually require the use of TLC. These are described in **Table 5** (p. 135).

Table 5. The *Cladonia chlorophaea* **group in the Pacific Northwest**
(Most are P+R and contain granular soredia in the cups.)

Species	Positive spot tests[1]	Chemical content	Notes
albonigra	UV+, K- or K+ dingy Y	4-*O*-methylcryptochlorophaeic acid (other chemotypes farther north)	base blackening
asahinae	none	a. rangiformic or norrangiformic acid or both b. lichesterinic or protolichester- inic acid or both	erect pointy gran- ules on podetia; podetia often grayish
chlorophaea	none	none	the commonest member of this group in PNW
conista	none	bourgeanic acid	rare in PNW; podetia corticate at base, powdery sorediate above
[*crypto- chloro- phaea*]	UV+, K+ wine R, KC+ purplish R	cryptochlorophaeic acid	e Cont Div
[*grayi*]	UV+	grayanic acid	e Cont Div
humilis	K+Y	atranorin	rare, w Cas; podetia as in C. conista
imbricarica	P-, UV+	sphaerophorin; fumarprotoce- traric acid lacking	short podetia; dry sites e Cas
merochloro- pahea	UV+	merochlorophaeic, 4-*O*-methylcrypto-chlorophaeic, cryptochlorophaeic	rare, e Cas
novochloro- phaea	P-, UV+ weak, rarely P+R	sekikaic and homosekikaic acids; fumarprotocetraric acid lacking, protocetraric sometimes present	erect pointy or rounded corti- cate granules on podetia; melanotic; broad cups; rare
pulvinella	P+R, K+Y	atranorin, bourgeanic, and fumarprotocetraric acid	short cups; infre- quent, e Cas in Wash and Ore to coastal Ore, s to s Cal

[1] In addition to P+O or R from fumarprotocetraric acid, unless otherwise noted

Cladonia ciliata

Description: Thallus fruticose, richly branched, terminal branches mainly as pairs, erect or drooping, grayish brown to brownish (var. *ciliata*) or pale yellowish green (var. *tenuis*), to 15 cm or more tall, main branches mostly 0.8-1.3 mm diam (dry); apothecia usually lacking, terminal, brown; pycnidial jelly red (hard to find); thallus K-, P+R (fumarprotocetraric acid), with or without usnic acid, UV-.

Range: Coastal Alas s to w Wash, where it is rare; infrequent northward. Although several older herbarium records exist for coastal Ore, these have not been confirmed genetically.

Substrate: On soil, humus, or those substrates over rock.

Habitat: Mima mounds in Thurston Co., Wash; farther north primarily on thin soil over rock outcrops and talus.

Notes: Identification of this species can be challenging, so that the reported range in N Am is overly broad; many herbarium database records are likely incorrect. Although readily distinguished from other richly branched *Cladonia* species by genetic markers, a combination of TLC with morphology can be very useful in identification. The variety *tenuis* is similar in color and chemically identical to the much more common *C. arbuscula*, but that species has terminal branches mainly in 3s and 4s, while in *C. ciliata* terminal branches are mainly in 2s. The var. *ciliata* is similar in appearance to *C. rangiferina*, but that species has atranorin, giving a K+Y reaction, in addition to the P+O or R reaction of fumarprotocetraric acid. In contrast, while also P+R, *C. ciliata* lacks atranorin and is thus K- or brownish. Separation of the two varieties of C. *ciliata* is often easy by color (yellowish tint of usnic acid in var. *tenuis*), or by spot test (KC+Y in var. *tenuis*, KC- in var. *ciliata*).

C. ciliata var. tenuis   C. ciliata var. ciliata

Cladonia concinna

Description: Podetia gray-green (to brownish in exposed sites), often to 5 cm tall, mostly < 1 mm diam; cups mostly < 4 mm diam, proliferating from the centers of the cups; squamules to 8 mm long by 4 mm wide, the podetial surface usually smooth; podetia sometimes sparsely squamulose; soredia and isidia lacking; apothecia brown; thallus P+R, K- or K+ brownish, UV-.

Range: Narrowly distributed along the Pacific Coast, Alas to s Cal, where it appears to replace the similar *C. cervicornis.*

Substrate: Soil or humus over rock.

Habitat: Full sun to partial shade at all elevations, most common on sand and outcrops very near the coast.

Notes: The multiple tiers of stalks coming from the centers of the cups are distinctive, but very similar to *C. verticillata, C. cervicornis,* and *C. macrophyllodes. Cladonia concinna* is distinguished from those by its very slender podetia and cups and strictly coastal distribution, while *C. verticillata* and *C. macrophyllodes* are more often found in boreal and montane areas, respectively. In general structure and secondary chemistry it is similar to *C. verticillata,* a relatively coarse lichen compared to the delicate *C. concinna.* This species was only recently described; older specimens in herbaria are commonly named *C. verticillata* or *C. cervicornis.*

Cladonia coniocraea

Description: Podetia elongated, mostly 1-3 cm tall, greenish or gray green, variable in shape (slenderly pointed, apiculate, club shaped, or nearly cylindrical), often forming narrow cups that are rarely broader than the podetia; podetia corticate toward the base, otherwise greenish sorediate, a few squamules occasionally present on the base; primary squamules often large (to about 3-8 mm diam), entire to divided, sorediate on the margins; apothecia brown, uncommon; thallus K-, P+R (fumarprotocetraric acid), UV-.

Range: Widespread, very common throughout PNW forests.

Substrate: Rotten wood, tree bases, humus, and occasionally soil.

Habitat: Relatively indiscriminate, occurring in a wide array of forests, in both shaded and exposed habitats.

Notes: The distinction between this species and *C. ochrochlora* was troublesome for a long time. Molecular studies have shown, however, that the struggle is over: we can consider them one polymorphic species, now going under the older name, *C. coniocraea.* One form has large, almost undivided squamules and podetia that often end in narrow, poorly formed cups. Another form has finely dissected squamules and only pointed podetia. And you don't need to look very long before you find strange forms with branched podetia, narrow cups, and largely corticate bases. All three forms appear to smoothly intergrade. Individuals with squamules on the podetia and small cups can be confused with *C verruculosa*, a species most frequent on soil or soil or moss over rock. *Cladonia coniocraea* has extensive powdery soredia while *C. verruculosa* has primarily granular soredia.

Cladonia cornuta

Description: Podetia 2-16 cm tall, 1-4 mm diam, simple or occasionally sparingly branched; pale greenish gray, olive green, or brownish; generally tapering to points, cupless, occasionally with narrow cups; soredia in rounded to irregular patches on the upper part of the podetia but otherwise podetia mostly corticate; primary squamules small, to 3 mm, roundish; apothecia rare, brown; cortex K-, medulla K-, P+R (fumarprotocetraric acid), UV-.

Range: Widespread, occasional throughout the PNW.

Substrate: Peaty soil, mosses, occasionally on rotten wood, most frequent on moss mats over rock.

Habitat: Cool, moist, montane habitats, in both oceanic and continental areas but more frequent in the latter, especially frequent on cool, moist talus slopes and outcrop areas.

Notes: This species, when well developed, is easily recognized by tall, simple podetia with roundish patches of soredia on the upper parts. Shorter specimens may be confused with *C. coniocraea*, which is more heavily and continuously sorediate and often has larger, divided squamules. Some forms of *C. coniocraea* on rotten wood w Cas have soredia in patches and lots of cortex on the lower half, but these are more prone to form cups, branches, and apothecia.

Cladonia ecmocyna

Description: Podetia 2-10 cm tall, simple to sparingly branched, not lacerate or perforate, whitish to olive green or brownish, often orange brown or yellowish brown at the base, cupless and pointed or with narrow cups, with marginal (occasionally central) proliferations from the cups; cortex smooth or with sparse squamules; primary squamules small, to 3 mm; apothecia common, brown; cortex K+Y (atranorin), medulla K- or brownish, P+R (fumarprotocetraric acid), UV-.

Range: Widespread, throughout the PNW, common e Cas.

Substrate: Soil, humus, or moss.

Habitat: Abundant in many exposed to lightly shaded habitats, especially talus slopes and outcrop areas, also forming extensive mats in some open subalpine forests.

Notes: The podetia are highly variable in outward form, varying from simple and pointed to tall and cup forming, sometimes with multiple tiers of branching. This same range of variation is found in the very similar species, *C. gracilis*, which can be differentiated by its K-cortex. In contrast, *C. ecmocyna* gives an instant K+ clear yellow reaction (test on the upper third of a podetium). Furthermore, *C. ecmocyna* is separated by a yellowish or orangish discolored base and pruinose young parts of the podetia.

Cladonia fimbriata

Description: Podetia to 2.5(4) cm tall, completely covered with powdery sore-dia (unless soredia have eroded off, leaving whitish decorticate areas); cups of various widths, but characteristically narrow, often with short marginal proliferations (generally < 2 mm) or giving rise to further cups; cup interiors closed; primary squamules 2-10 mm long; apothecia occasional, brown; me-dulla K-, P+R (fumarprotocetraric acid), UV-.

Range: Widespread, very common throughout the PNW.

Substrate: On soil, rotten wood, and bark; common on stumps, fenceposts, roadcuts, and tree bases; w Cas on virtually every available substrate includ-ing roofing, fiberglass, brick, etc.

Habitat: In a very broad array of habitats, exposed to shaded, mainly at low to mid elevations.

Notes: Frequently confused with *C. chlorophaea* and related species, *C. fimbri-ata* is distinguished by its powdery soredia, both inside and outside the cups. Also, the cups are typically narrower than in *C. chlorophaea*. *C. subulata* is similar to *C. fimbriata*, but has deeper cups and forms long (1 cm or more) pointed marginal proliferations from the cups.

Cladonia furcata

Description: Podetia moderately to richly branched, tall (to 10 cm), gray green, brownish, or pale greenish, sometimes blackening at the base similar to *C. phyllophora*; axils open, the edges inrolled, and often with a longitudinal inrolled fissure extending down the podetium from the axil; fertile branches often ± flattened and fissured; squamules usually scattered on the podetia; primary squamules usually disappearing soon; soredia and isidia lacking; apothecia brown, small, terminal, common; cortex K-, medulla K-, P+R (fumarprotocetraric acid), UV-.

Range: Alas to Cal, very common w Cas.

Substrate: On soil, moss, and humus, occasionally on rotten wood or tree bases.

Habitat: Moist coastal forests at low to mid elevations, frequent on partly shaded roadcuts.

Notes: Although perforated cups found in most forms of *C. multiformis* readily

distinguish it from *C. furcata*, the occasional cupless forms of *C. multiformis* are more challenging. In these cases one must rely on finding the somewhat flattened fertile branches of *C. furcata*, in contrast to the round fertile branches of *C. multiformis*. They may also be separated by range, *C. multiformis* being common e Cas and *C. furcata* largely restricted to w Cas. The taxonomy of this group is unsettled, *C. furcata* in the PNW bearing little resemblance to *C. furcata* in e N Am and Europe. The western form has been called *C. herrei* by some authors.

Cladonia macrophyllodes

Description: Podetia proliferating centrally from the cups (but often the po-detia are poorly formed or lacking), sometimes producing several tiers of cups but usually < 2 cm tall; squamules large, 5-20 mm long by 1-8 mm wide, tex-tured with a mosaic of pale cortical thickenings (insert photo); lower surface of squamules pure white or blackening at the base; cortex K+Y (atranorin); medulla K-, P+R (fumarprotocetraric acid), UV-.

Range: In and e Cas, Alas s to Cal, inland throughout the RM.

Substrate: On soil and soil over rock.

Habitat: Common at middle to high elevations, occasionally at low elevations, open forests to exposed grasslands and tundra.

Notes: Cladonia macrophyllodes is similar to *C. verticillata,* but that species is K- or K+ brownish and only rarely shows a well-developed mosaic of cortical thickenings on the squamules. *Cladonia macrophyllodes* is most frequently found as just squamules or with a few poorly developed podetia, while *C. ver-ticillata* usually has well-developed podetia. *Cladonia macrophyllode*s is most commonly seen in cold or dry habitats e Cas.

Cladonia mitis

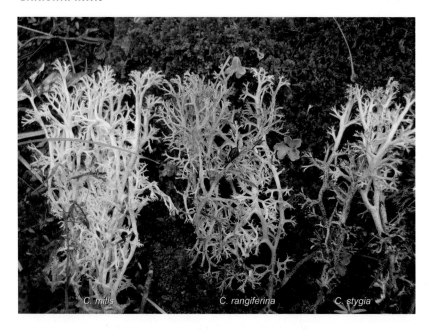

Description: Thallus fruticose, richly branched in 3s and 4s, erect, pale yellowish or pale yellowish green or yellowish white, to 8(10) cm tall, thicker branches 1-1.5(4) mm diam; apothecia lacking; spot tests negative except cortex KC+Y (usnic and rangiformic acids); thallus UV-.

Range: Circumpolar s to Ore, Id, Wyo, and e US; uncommon in Ore and Wash.

Substrate: On soil, humus, or soil over rock.

Habitat: Outcrops and talus slopes, mainly cool, moist slopes at low elevations.

Notes: Very similar to *C. arbuscula*, from which it must be distinguished by a P test or, with some practice, a taste test will work. *Cladonia mitis* has a mild flavor ("*mitis*" means "mild"), while *C. arbuscula* and *C. rangiferina* have a somewhat bitter taste from fumarprotocetraric acid, the P+R constituent. These three species of reindeer lichen become rare toward the southern part of the PNW. Photos and indoor lighting often do not show the yellowish tint given by usnic acid. Presence of usnic acid distinguishes *C. arbuscula* and *C. mitis* from *C. rangiferina*. When identifying these indoors, check the KC reaction, which should be yellow to strongly yellow if usnic acid is present, negative or just greenish if usnic acid is lacking.

Cladonia norvegica

Description: Podetia slender, mostly unbranched, to 1.5(2) cm tall and about 1 mm diam, pale greenish, gray green, or whitish, slenderly pointed or blunt, cupless or with tiny shallow cups; podetia partly corticate at the base but becoming ecorticate and covered with soredia, a few squamules occasionally present near the base; primary squamules small (mostly < 2 mm diam), finely incised or narrowly lobed; apothecia subspherical or turban-like, pale brown, pinkish tan, or pale tan; thallus K-, P-, medulla UV- or UV+ weak (barbatic acid).

Range: Alas to Ore, mostly w Cas in oceanic forests, but also inland to s BC.

Substrate: Rotten wood, tree bases, and tree trunks.

Habitat: Most frequent in mature to old conifer forests at low to middle elevations in the mountains.

Notes: Most populations of *C. norvegica* are sparse and mixed with other species, most commonly *C. coniocraea* and *C. transcendens*. If apothecia are lacking, *C. norvegica* is difficult to distinguish from associated species without chemical tests. Furthermore, *C. bacillaris* is morphologically similar to *C. norvegica* and shares the same chemistry (barbatic acid) but *C. bacillaris* is red fruited, while *C. norvegica* has pale brown or pinkish tan apothecia. Much more common than either of those two is *C. coniocraea*, which usually has larger squamules and is P+R.

Cladonia phyllophora

Description: Podetia simple to moderately branched, to 8 cm tall and 4 mm diam, cupless or with irregular cups, usually squamulose; upper part of podetia with a very fine, thin tomentum; base of podetia melanotic, i.e., whitish spotted on a brown to black melanized background (inset photo); cup margins proliferating; cortex continuous to areolate, the part between the areoles decorticate; apothecia brown, common, small to medium sized, on the tips of proliferations from the cups; thallus K-, P+R (fumarprotocetraric acid), UV-.

Range: Northern circumpolar s through RM, only sporadically w Cas.

Substrate: On soil, or soil over rock.

Habitat: Low-elevation talus and outcrops, usually fairly open sites on cool, moist slopes.

Notes: This species is variable in form (thickness of the podetia, degree of branching, and presence of cups), but is easy to recognize on close inspection by the black-spotted bases of the podetia. Overlooking this character, one is likely to mistake it as an atypical form of *C. ecmocyna*, *C. gracilis*, or *C. multiformis*. A closely related species to *C. phyllophora* is *C. stricta*. This primarily arctic species has been reported s to Wash and through the RM. *Cladonia stricta* sometimes contains atranorin, while *C. phyllophora* does not, but they also differ morphologically. *C. phyllophora* has broader cups and podetia (1-2.5 mm just below the cups) vs. narrower cups and podetia (0.5-1.5 mm) in *C. stricta*.

Cladonia portentosa

Description: Thallus fruticose, yellowish to pale gray green, to 7(10) cm high, richly branched, mainly with 3 (occasionally 2 or 4) branches arising from a point; mats often subdivided into rounded heads; squamules lacking; apothecia brown, not seen; thallus P-, K-, UV+ whitish (perlatolic acid, with or without usnic acid).

Range: Within several km of the coast from se Alas to Cal, rarely inland to Cas, Siskiyou lowland, and s Willamette Valley.

Substrate: Sandy soil and soil, humus, or moss over rock outcrops.

Habitat: Old sand dunes, dry areas in deflation plains, stabilized roadcuts through dunes, and rock outcrop areas, usually in exposed to partly shaded sites; rarely on hummocks in grassy wetlands.

Notes: Ours is the subspecies *pacifica*. Although it is usually separable from the other reindeer lichen species by habitat, the identification can be confirmed by checking for branches typically arising in threes. Some forms contain usnic acid and are yellowish tinted; others lack usnic acid and are grayish, similar to *C. rangiferina*. Apart from the branching pattern, *C. mitis* is similar in morphology and spot tests, but is UV-. *Cladonia portentosa* is more translucent than other reindeer lichen species in this area. Another species with the tendency to subdivide into rounded heads like cotton puffs, but more consistently so than *C. portentosa*, is *C. stellaris*, which occurs northward and more inland. It also contains perlatolic acid, so is UV+.

Cladonia pyxidata

Description: Podetia to 2(4) cm tall, gray-green to olive green or brownish; cups deep goblet shaped, the interior of the cups with small roundish squamules and round corticate granules; outside of podetia coarsely areolate and granular to minutely squamulose; primary squamules small to medium sized (2-7 × 4 mm), appressed or ascending, thick, olive green to brownish, separate, not closely packed or forming rosettes; apothecia brown, fairly common; thallus K- or brownish, P+R (fumarprotocetraric acid), UV-.

Range: Widespread.

Substrate: Mineral soil or soil or moss over rock, occasional on humus.

Habitat: In a wide array of open or semi-open habitats, in some cases where recently disturbed, including roadcuts and clearcuts, but also on undisturbed, open, rocky sites; at all elevations.

Notes: Both *C. pyxidata* and *C. pocillum* have tiny squamules inside the cups. *Cladonia pyxidata* has separate primary squamules, unlike the densely packed, contiguous or shingled primary squamules of *C. pocillum*. *Cladonia pyxidata* occupies more mesic sites than *C. pocillum*, while *C. pocillum* is best developed on dry or calcareous sites. Extreme forms are easy to separate, but some intermediates will be found.

Cladonia rangiferina

Description: Thallus fruticose, richly branched (mainly in 3s and 4s, occasionally in 2s), erect, grayish, whitish, or brownish gray, to 10(15) cm tall, thicker branches 1-1.5(3) mm diam; apothecia usually lacking, terminal, brown; thallus K+Y (atranorin), P+R (fumarprotocetraric acid), UV-.

Range: Circumpolar s to Ore and Mont; infrequent in the southern part of its range; abundant northward.

Substrate: On humus, or soil over rock.

Habitat: Talus slopes and outcrop areas, often on cool, moist slopes and in narrow valleys and canyons.

Notes: Although rather easily distinguished from other reindeer lichens when growing side by side in the field, the distinctive grayish color of *C. rangiferina* is subtle enough to make field identification difficult when it is growing alone, particularly when wet. *C. mitis* and *C. arbuscula* have a yellowish cast. *Cladonia portentosa* is primarily coastal and forms aggregations of rounded heads. *Cladonia furcata* in the PNW splits down from the axils, is corticate, and is never as richly branched as *C. rangiferina*. *Cladonia stygia* is similar in appearance and chemistry to *C. rangiferina* but blackens at the base (see photo in key, p. 116).

Cladonia squamosa

Description: Podetia simple to few-branched, mostly 1-5 cm tall, densely squamulose, pale gray green (brownish in exposed sites); branch axils commonly opening into the interior of the podetia; cortex often disintegrating, exposing the white (or darkening) ± cobwebby medulla; soredia absent; primary squamules similar to the lower podetial squamules; apothecia dark brown, uncommon; medulla K-, P-, UV+ blue-white (var. *squamosa*, with squamatic acid) or K+ deep Y, P+O, UV- (var. *subsquamosa*, with thamnolic acid).

Range: Widespread in N Am, occasional throughout much of the PNW.

Substrate: Soil, soil over rock, rotten logs, and tree trunks.

Habitat: Moist forests at low to mid elevations.

Notes: Although highly variable in overall size and density and size of podetial squamules, *C. squamosa* is easily recognized by the combination of the UV+ medulla and squamulose, gray-green podetia. *Cladonia squamosa* var. *subsquamosa* is morphologically similar but UV-, K+ deep Y, P+Y to O, is common in and w Cas, but is not yet known e Cas. *Cladonia singularis*, also w Cas, is similar to *C. squamosa*, but has pointed podetia and lateral branches. The axils and apices are closed or with only a small perforation, as opposed to the open, often gaping axils and apices of *C. squamosa*. Although described from w Cas in Ore and Wash, *C. singularis* has surprisingly been reported as common in heaths, talus, krummholz, and alpine in some areas of se Alas.

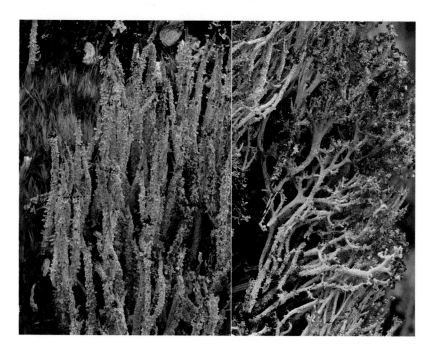

Cladonia sulphurina

Description: Podetia 2-8 cm tall, greenish with a yellow tinge from usnic acid, completely sorediate, with irregular, narrow cups; primary squamules large, 5-10 × 2-5 mm; apothecia red, fairly common (if absent, inspect carefully for red apothecial initials); medulla K-, P-, UV+ blue-white (squamatic acid).

Range: Widespread, most common in cool continental forests, sparse w Cas, frequent in n RM.

Substrate: On rotten wood or bark or humus-rich soil.

Habitat: Lowland to subalpine forests and outcrop or talus areas.

Notes: The fairly large, lacerate and irregular yellowish podetia are characteristic. The UV+ medulla distinguishes it from the similar but somewhat smaller *C. deformis*. With practice, the large, sorediate, UV+ squamules make even poorly developed material identifiable. While the medulla is UV+ bright blue-white, the usnic acid in the cortex and soredia absorbs UV light. This absorption makes the surface appear dark, so one must expose the medulla to see the UV+ reaction.

Cladonia transcendens

Description: Podetia to 4 cm tall and 2.5 mm diam, usually arising near the centers of the squamules, yellowish green to gray green, unbranched or sparingly branched, cupless or just as often forming irregular, poorly developed cups; podetia partially corticate or becoming completely covered with powdery to granular soredia, often squamulose; primary squamules 2-7 mm long and 1-3(5) mm broad, sorediate; apothecia red, common; medulla K+Y, KC- or KC+Y, P+O (thamnolic acid), UV-.

Range: Alas s to Cal, Id, and Mont; in the PNW mainly w Cas with disjuncts to Id and Mont.

Substrate: On bark and wood; commonly on fenceposts, stumps, tree bases, and fallen logs.

Habitat: In both shaded and open sites, including both clearcuts and old-growth forests, at low to mid elevations; e Cas rare and largely restricted to moist, partly shaded locations, often near streams or in cool canyons.

Notes: Easily the most common red-fruited *Cladonia* w Cas. *Cladonia belli-diflora* has fine warts or squamules on the podetia rather than soredia and is usually UV+. *Cladonia bacillaris* is relatively rare, unbranched, cupless, and P- and K-. *Cladonia macilenta* lacks usnic acid (and is therefore never yellowish tinged), is unbranched, and cupless. In contrast *C. transcendens* has a distinct yellowish cast when growing in exposed sites and often forms irregular cups. When sterile, *C. transcendens* can be confused with *C. coniocraea*, but that is K- or K+ brownish (from fumarprotocetraric acid), has larger primary squamules, and lacks the yellowish tinge sometimes seen in *C. transcendens*.

Cladonia umbricola

Description: Podetia usually greenish gray, sometimes yellowish green, to 3 cm high, generally 0.5-1.5(4) mm diam, simple or sparingly branched, occasionally ± cup-forming, lacking a cortex, lacking squamules or occasionally with scattered squamules; podetia partly to wholly sorediate, the soredia usually powdery, occasionally granulose; basal squamules 3-4(6) mm long and 3(4) mm wide, sorediate or not; apothecia red, common; medulla UV+ blue-white (squamatic acid), K-, P-, KC-, rarely with usnic acid and then KC+Y.

Range: Alas to n Cal, immediate coast inland to w Mont.

Substrate: Bark or decaying wood.

Habitat: Uncommon in low- to mid-elevation moist forests in canyons, stream bottoms, and valleys.

Notes: Cladonia transcendens can be similar, but is usually yellowish tinged from usnic acid, is UV-, and is more prone to make cups and granular soredia than *C. umbricola. C. macilenta* is morphologically similar to *C. umbricola,* but is cupless, unbranched, and UV-, K+Y, P+O. *Cladonia bacillaris* is similar to *C. umbricola* in its negative spot tests and red fruits, but is only weakly UV+ and is much rarer than *C. umbricola.*

Cladonia verruculosa

Description: Podetia simple to moderately branched, highly variable in form, pointed or tipped with cups, 2-12 cm tall, gray green, or with whitish or brownish parts; branch axils and cups closed; cups shallow, often irregular; branches proliferating from the margins and the interiors of the cups; podetia becoming decorticate with age, the decorticate parts sometimes blackening; podetia generally minutely warty, squamulose, coarsely sorediate, or with soredia partly fine and partly coarse, sometimes richly squamulose; apothecia frequent, dark to pale brown; thallus K-, P+R (fumarprotocetraric acid), UV-.

Range: BC to Cal, very common w Cas, fairly common inland to w Mont.

Substrate: Usually on soil or mossy rock, less often on rotten wood.

Habitat: A wide variety of open to partly open disturbed sites, including roadcuts, clearcuts, and stabilized dunes, in and around areas of low- to mid-elevation moist forests, occasional in the steppe fringes of forested ecosystems.

Notes: Although variable in development of cups and podetial squamules, this species consistently produces a gradation of squamules to small granules on the podetia. *C. coniocraea* can be similar in appearance but has finer soredia and usually occurs on bark and wood.

Cladonia verticillata

Description: Podetia gray-green (to brownish in exposed sites), often to 5 cm tall, mostly > 1 mm diam, proliferating from the centers of the cups; cups often > 4 mm diam; squamules to 8 mm long by 4 mm wide, the upper surface usually without a mosaic of whitish cortical thickenings (use hand lens); surface of podetia smooth or sparsely squamulose; soredia and isidia lacking; apothecia brown; thallus P+R, K- or K+ brownish, UV-.

Range: Cosmopolitan (but highly variable in form in different parts of the world), occasional throughout the PNW.

Substrate: Soil or soil over rock, rarely on rotten wood.

Habitat: Exposed habitats at all elevations, except near the outer coast.

Notes: The multiple tiers of stalks coming from the centers of the cups are distinctive, but also found in *C. macrophyllodes* and *C. concinna*. Partially developed forms of *C. verticillata* are common and must be carefully distinguished from *C. macrophyllodes*. That species is K+Y (from atranorin), has poorly developed podetia that are coarser and less proliferous, and has a mosaic of cortical thickenings or bumps on the squamules. In contrast, *C. verticillata* is K- or K+ brownish, has poorly to well-developed podetia, and usually lacks the cortical thickenings on the squamules. *C. concinna*, a more slender species with narrow cups, replaces *C. verticillata* along the coast from BC to Cal, and is fairly common in dune ecosystems in Ore and n Cal. *Cladonia verticillata* is a polyphyletic taxon and because no sequences are available for material from the PNW the taxonomy is especially uncertain. We use it here in the broad sense.

saxicolous *"Collema,"* Alas s to Utah and Colo
....*Lathagrium fuscovirens*
15b On noncalcareous rock where periodically inundated by
streams, lakes, or seepage. Isidia globose to cylindrical,
coarse (generally > 0.1 mm broad), sometimes forming
a layer that obscures the upper surface; arctic-alpine,
uncommon, reported from Colo to S Dak, w to Wash and
n ....*C. glebulentum*
9b Apothecia present; isidia and lobules present or not
16a Spores 2-celled. Thallus similar in appearance to *Enchylium tenax*;
on soil; e Cas ....*Enchylium coccophorum*
16b Spores > 2-celled
17a Spores becoming submuriform or muriform
18a Thallus crustose, in small areolate patches; lobes < 0.3 mm
wide or indistinct; upper surface occasionally granular; on
calcareous rock and mortar; w Cas ....*Scytinium callopismum*
18b Thallus foliose; lobes to 3 mm wide, elongate, ± adnate to
substrate; on calcareous rock or soil or moss over rock
.... *Lathagrium cristatum* (Lead 14a)
17b Spores 4-celled (occasionally more)
19a Lobes tips not or only slightly swollen
20a On noncalcareous seacoast rock; spores 8-13 μm broad.
Spores 3(5)-septate, rarely with a longitudinal septum,
(18)20-28 μm long; se Alas to Wash; uncommon
.... *Blennothallia fecunda*
20b On calcareous rock not on the immediate coast; spores 13-15
μm broad ....*Blennothallia crispa* (Lead 11b)
19b Lobes tips ± swollen (thallus of *tenax* type)
21a On rock; apothecia often elevated on the ends of short erect
branches, the thalline margin almost disappearing, rolling
under the apothecia. On calcareous rock, rarely on siliceous
rock or moss over rock, generally in sunny, periodically wet
habitats (e.g., seepage and lakeshore rocks); widespread,
montane to alpine.................................*Enchylium polycarpon*
21b On soil; apothecia otherwise
22a Apothecial margin crenulate to lobulate; spores pale
yellowish brown. Thallus sometime lobulate; apothecia
generally numerous, sessile with a constricted base; on
calcareous soil, low elevations to alpine; Colo n to Arctic
....*Enchylium bachmanianum*
22b Apothecial margin ± even; spores hyaline. Widespread
and common, especially on calcareous soil, especially in
steppe and tundra, low elevations to alpine
.... *Enchylium tenax*

Collema curtisporum

Description: Thallus foliose, to 2(4) cm broad, black, gelatinous when wet, with broadly rounded lobes; isidia and soredia lacking; spores 4-septate, hyaline, 20-40 × 3-4.5 μm, with blunt ends, straight or curved; apothecia generally present; spot tests negative.

Range: E Wash, n Id, w Mont, and ne Ore, n to Alas.

Substrate: Almost always on *Populus trichocarpa*.

Habitat: Moist riparian forests, often in narrow sheltered valleys, commonly associated with *Collema furfuraceum, Leptogium saturninum, Physconia* spp, *Lobaria pulmonaria*, and *L. hallii*.

Notes: The thallus is very similar to *C. nigrescens*, from which it is reliably distinguished only by examination of the spores. Fortunately, apothecia are almost always present. The two species also differ ecologically and geographically, *C. nigrescens* being found on a variety of hardwoods w Cas, while *C. curtisporum* occurs mainly on *Populus* between the Cascade and Rocky Mountain crests.

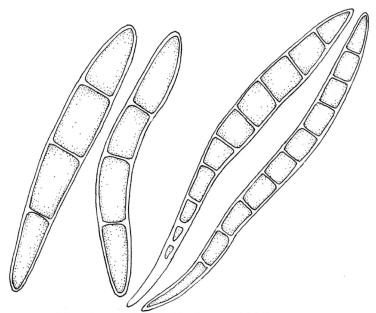

Spores of *C. curtisporum* (left) and *C. nigrescens* (right)

Collema furfuraceum

Description: Thallus foliose, to 5 cm broad, black, gelatinous when wet, with broadly rounded lobes, distinctly ridged, of the *nigrescens* type; isidia mainly on ridges, 0.05-0.15 mm diam, globular or slender, simple or branched; apothecia occasionally found; spot tests negative.

Range: Widespread in N Am, found throughout the forested part of the PNW, in both oceanic and continental climates.

Substrate: Bark (especially broad-leaved trees), occasionally on rock.

Habitat: In Ore and Wash most common in moist low-elevation riparian forests and in oak savannas.

Notes: Distinguished by the distinctive ridged and blistered appearance of the thallus along with abundant isidia, usually without apothecia. Variety *luzonense* is usually fertile with pruinose apothecia and occurs on hardwoods and shrubs on the immediate coast from BC to Cal.

Collema nigrescens

Description: Thallus foliose, to 2(5) cm broad, olive black, brownish, or black, gelatinous when wet, small to medium sized but with broadly rounded lobes; isidia present or not, soredia lacking; spores 6-13-celled, hyaline, 50-90 × 3-5 μm, with elongate pointed ends, straight or curved; apothecia generally present and numerous, to 1 mm diam; spot tests negative.

Range: Alas to Cal, w Cas, fairly common in sw Ore and n Cal.

Substrate: Bark of broad-leaved trees and shrubs; on many species but most frequent on oak (*Quercus*).

Habitat: Low-elevation hardwood forests and savanna, mainly in the valleys and the foothills.

Notes: Very similar in appearance to *C. curtisporum*, from which it is distinguished by spore characters visible only with a compound microscope. See comparison of spores under *C. curtisporum. Collema nigrescens* also differs in sometimes having isidia.

Cornicularia

Cornicularia normoerica

Description: Thallus fruticose, erect to suberect in dense tufts, brownish black (paler below), stiff, shiny, (0.5)1-2(2.5) cm tall, sparingly branched; branches about 0.3-1(3) mm diam, knotty, strap-like or wiry; isidia and soredia lacking; pseudocyphellae lacking; apothecia common, blackish, terminal, the margin dentate or fibrillate; spot tests negative.

Range: Circumpolar arctic-alpine to subalpine or upper montane with coastal affinities; s Alas to n Cal, e to nw Mont, rare e Cas, occasional on w slope Cas and OP.

Substrate: Noncalcareous rocks.

Habitat: Exposed rocky summits and ridges; one of the few erect-fruticose lichens that grows on exposed edges, corners, and faces of rocks, as opposed to occupying more sheltered faces and crevices.

Notes: A distinctive lichen, both in form and habitat. All other species formerly placed in *Cornicularia* are now referred to *Bryocaulon* and *Cetraria*. The stubble-like growths on exposed rock seem well suited for trapping and col-

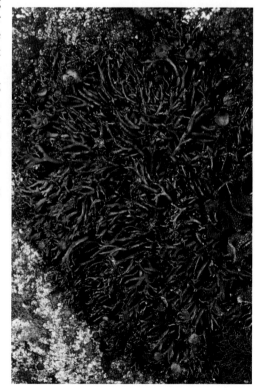

lecting water from the fog that frequently sweeps over ridges. These microsites are also the first to be swept bare of snow, making photosynthesis possible during much of the wet season. The photo shows a prostrate growth form but the species is just as often suberect to erect.

Sources include: Kärnefelt (1986).

Dactylina

Thallus fruticose, yellowish, pale greenish, brownish yellow, or brown, of erect stalks, simple to moderately branched, corticate, lacking rhizines; stalks hollow or with a loose central medulla; apothecia rare; thallus P+ or P-, UV+ or UV-, C+ or C-; on arctic-alpine sod.

Sources include: Kärnefelt and Thell (1996), Thomson and Bird (1978).

ID tips: The species below are readily distinguished morphologically, but you might find spot tests helpful in ambiguous cases. All of these species are restricted to alpine habitats.

1a Thallus generally > 2 cm tall, unbranched or nearly so, appearing inflated, hollow, yellow to brownish yellow, C+R, KC+R (usnic and gyrophoric acids). Arctic to s BC, s Alta, and Wash Cas .... *D. arctica*
   (*D. arctica* subsp. *beringica* contains physodic and physodalic acids in addition to usnic and gyrophoric acids and occurs throughout most of the range of *D. arctica* in N Am.)
1b Thallus generally < 2 cm tall, ± branched, C-, KC+Y or R
   2a Stalks with few short lateral branches, yellow or yellowish green, occasionally whitish pruinose toward the tips; pycnidia rare; medulla P-, C-, K-, KC- (usnic acid and fatty acids). Arctic s in RM to NM
   .... *Cetraria madreporiformis*
   2b Stalks knobby with short side branches, yellowish or brown, generally with a light violet pruina toward the tips; pycnidia common; medulla P+R (P- chemotype farther n), K-, KC+O or R (usnic, physodic, physodalic, protocetraric, and virensic acids); arctic s in RM to Wyo
   .... *D. ramulosa*

Dendriscocaulon—*see* Lobaria

Dermatocarpon

Thallus squamulose or foliose, to 5 cm or more, appressed to ascending; often umbilicate; upper surface gray, brown, or dark brown, often whitish or grayish pruinose; lower surface pale tan, orange brown, brown, or black, smooth, papillose, veined, or irregularly rugose; rhizines absent except in a few spp; photobiont green; soredia and isidia lacking; perithecia common, immersed, showing from the upper surface as dark spots and from the lower surface as roundish bulges; spores nonseptate or occasionally 1-septate, rarely 2-4-septate; thallus K-, C-, KC-, P-; usually on rock, occasional on soil or gravel.

Sources include: Breuss (2003), Glavich and Geiser (2004), Heidmarsson (1998, 2001, 2003), Heidmarsson and Breuss (2004), McCune and Rosentreter (2007), McCune et al. (2007), Rosentreter and McCune (1992).

ID tips: *Dermatocarpon* species can be challenging to identify, but ascospores can be very useful for this, if you have access to a compound microscope with a micrometer. Be sure to measure fully developed spores and ignore the smaller immature spores that are usually present.

Melzer's iodine reactions of the medulla are sometimes used to separate species, but the usual I (IKI) reagent will often give different results. Melzer's reagent (Common 1991) is as follows: 1 g elemental iodine; 3 g potassium iodide, 40 g chloral hydrate, and 40 ml water. Unfortunately, Melzer's reagent is difficult to get, because one ingredient, chloral hydrate, is a controlled substance in the United States. To use Melzer's, transfer a thallus section to a thin smear of the reagent on a microscope slide or spot plate and examine it under the dissecting microscope. The positive reaction is a deep red brown, while a negative reaction is pale yellowish.

Introductory Key

1a Thallus almost crustose, of small, closely packed squamules, the
    squamules 1-3 mm broad ....**Group 1** (Lead 3)
1b Thallus squamulose or foliose; if squamulose then the squamules mostly
    > 3 mm broad
    2a Lower surface rhizinate ....**Group 2** (Lead 6)
    2b Lower surface smooth, wrinkled, or papillose but not rhizinate,
        sometimes with scattered peg-like holdfasts ....**Group 3** (Lead 8)

Group 1

Almost crustose to squamulose

3a Perithecia mostly external; spores subspherical; thallus gray; central Ore,
  Colo ....*Verrucaria sphaerospora*
3b Perithecia mostly internal, externally visible only as a dark spot; spores
  broadly ellipsoidal to elongate-ellipsoidal
  4a Spores broadly ellipsoidal (length to width usually about 2:1 or
    broader), 12-17 × 6-9 µm; thallus dark brown. On dry calcareous rock
    in exposed sites; widespread and common in w N Am
    ....*Heteroplacidium compactum*
  4b Spores elongate-ellipsoid (length to width usually > 2:1), 13-22 × 5-7
    µm; thallus color various
    5a Medulla with elongate cells (LM); lower cortex with gelatinized
      hyphae arranged perpendicular to the surface. Thallus dark brown
      to grayish brown, often grayish pruinose; squamulose-colonial, the
      squamules to several mm diam; lower surface blackish; infrequent;
      nw Wash to Wyo s to Ariz and Cal; on outcrops and stones (esp.
      basalt), often in seepage tracks in otherwise dry habitats
      ....*D. leptophyllodes*
    5b Medulla of roundish cells; lower cortex also of roundish cells. Baja
      Cal and Ariz n to e Ore ....*Placopyrenium conforme*

Group 2

Foliose; rhizinate

6a Lower surface granular to warty. Rhizines sparse to dense, short and
  stubby, blunt, sometimes flattened laterally; spores 13-17 × 6.5-8 µm; on
  calcareous rock; e Cas, BC and Mont ....*D. atrogranulosum*
6b Lower surface smooth, apart from the rhizines
  7a Rhizines coralloid-branched, often elongate; lower surface black. Poorly
    known, apparently rare...*D. vellereum*
  7b Rhizines simple to sparingly branched, short; lower surface orangish
    brown to dark brown. Spores 10-13 × 5-8 µm; on dry often calcareous
    rock; widespread in w N Am but uncommon....................*D. moulinsii*
    (Nash et al. [2002] included *D. schaechtelinii*, separating it from
    D. moulinsii by its thicker rhizines that contain a medulla, a
    thicker thallus, and slightly longer spores (10-15 × 4-7 µm in *D.
    schaechtelinii*). See also *Umbilicaria vellea* and *U. americana*, with
    rhizines black from granular-sooty thalloconidia.)

Group 3

Foliose; not rhizinate

8a Thallus growing loose on soil or gravel (occasionally found attached to rock at the same sites where it is vagrant), light to dark brown or grayish brown; margins tightly downrolled

9a Lower surface finely papillose or granular.........................**D. reticulatum**

9b Lower surface smooth to rugose, not papillose

10a Lower surface dark brown to black, except where stained by soil; upper surface brown to gray brown or gray; pycnidia and/or perithecia usually present and easy to find (look for blackish ostioles); lower cortex and medulla only occasionally infused with oil drops.. **D. deminuens** group

10b Lower surface brown, except often blackish brown near the lobe tips; upper surface dark brownish gray; pycnidia and perithecia absent (mature perithecia and/or pycnidia not yet seen); lower cortex and medulla almost always densely infused with oil drops....
....*D. "oleosum"*

(This unpublished species is so far known from central and eastern Oregon. Vagrant populations of the *deminuens* group seems more widespread (n Cal, s Ore) than "*oleosum*." Both occur on basaltic rocks and soil, often associated with *Artemisia rigida*. *Dermatocarpon oleosum* has not yet been found with either perithecia or mature pycnidia. We hypothesize that the energy-rich oil drops in the medulla and lower cortex might result from diverting energetic allocation from sexual reproduction to survival of the individual.)

8b Thallus firmly attached to rock or rock crevices; thallus form various

11a Lower surface finely papillose; upper surface gray

12a Lower surface distinctly papillose, the texture easily seen with a hand lens, brown to black; thallus commonly > 2 cm broad; spores 12-17 × 5-9 μm; on various rocks; habitat streamside to strictly terrestrial very common... **D. reticulatum**

12b Lower smooth to very finely papillose (DM), the texture not easily seen with a hand lens, black; thallus usually < 2 cm broad; spores 14–26(28) × 6–9(11) μm; on calcareous rock; habitat streamside rock in flood zone; apparently rare, e Wash and w Mont
.... *D. "micropapillatum"*

11b Lower surface not finely papillose, variously colored, smooth, warty, veined, or wrinkled, rarely granular; upper surface gray or brown; if lower surface finely granular then upper surface brown

13a Upper surface light to dark brown, seldom pruinose but sometimes with a grayish necrotic layer; lower surface brown to black; on rocks in or near perennial streams or cold lakes; spores long (> 15 μm)

14a Lower surface often slightly veined, rarely distinctly veined, brown to dark brown, matte to subglossy; thallus often colonial, with many small, crowded individuals....................**D. deminuens**

14b Lower surface distinctly veined, ± glossy; thallus generally ±
   single-lobed, but individuals often massed together. Thallus
   0.1-0.3 mm thick; spores 14-22 × 6.5-8.5 μm; on siliceous rocks;
   Alas s to Ore and Colo; occasional; alpine, subalpine, and in cold
   canyon-bottom streams... *D. rivulorum*
13b Upper surface brownish gray to gray when dry, slightly to strongly
   pruinose; lower surface various; habitat various; spores various
   15a Substrate calcareous rock; medulla Melzer's I+ reddish brown;
      habitat montane to alpine. Thallus to 3 cm diam; upper surface
      light gray to dark gray; lower surface beige, tan, pinkish tan, or
      orangish tan; spores 10-15 × 6-8 μm; uncommon; e Cas
      .... ***D. polyphyllizum***
   15b Substrate various; medulla Melzer's I- or yellowish; habitat alpine
      to lowlands
      16a Lower surface light to dark brown; upper surface light to dark
         brown or grayish, epruinose but sometimes with a thin, hazy
         epinecral layer. Lower surface smooth to rugose, hardly veined;
         spores 13-16 x 6-8 μm; on various rocks, usually calcareous
         .... *D. miniatum* (broad sense)
         (*D. miniatum* is a widely misapplied name; it is infrequent in
         the PNW.)
      16b Lower surface with an orangish or reddish tinge; upper surface
         light gray to gray, becoming pruinose. Lower surface smooth to
         rugose or slightly veined; spores 11-17(18) x 5.5-8.5 μm long;
         Thallus to 5 cm diam; on both calcareous and noncalcareous
         rock; low elevations to montane, sw N Am n at least to se Ore
         .... *D. taminium*

Dermatocarpon deminuens

Description: Thallus foliose to 3D foliose when vagrant, individuals mostly < 3.5 cm broad, often closely packed or shingled; upper surface light gray, gray brown, dark brown, or blackish, often with a hazy grayish epinecral layer; lower surface brown to black, smooth to rugose, matte to subglossy; rhizines lacking; perithecia common, nearly invisible or conspicuous; spores 14-21(25) × 6-9 µm, occasionally 1-septate; pycnidia common.

Range: Commonly collected from NWT s through Cas and RM.

Substrate: Rocks, often basalt, andesite, and serpentine.

Habitat: Seasonal seepage over rock and streamside, lakeside, or seashore rocks where frequently wetted, low to middle elevations; also on rock and vagrant in poorly drained basalt scablands.

Notes: Dermatocarpon meiophyllizum is a similar semi-aquatic species from Europe that does not occur here. Most specimens from the PNW that were previously assigned to that name belong to the *D. deminuens* complex, which also includes *D. bachmannii*, a name applied to a variant with slightly larger thalli. *D. rivulorum* is closely related and also frequents semi-aquatic habitats, but it has a veined and slightly shiny lower surface and is more common at higher elevations. The aquatic *D. luridum*, a common species in e N Am, bright green when wet, does not occur in the PNW.

Dermatocarpon polyphyllizum

Description: Thallus foliose, to 17–26 mm, monophyllous to polyphyllous, with 1-several holdfasts; upper surface light to dark gray, smooth or becoming scabrid-pruinose with an epinecral layer; lower surface beige, tan, orangish brown, or pinkish tan, matte, not papillose, smooth to rugose; veins sometimes present; rhizines lacking; medulla Melzer's I+ red-brown.

Range: Widespread e Cas, but distribution and abundance uncertain.

Substrate: Calcareous rock.

Habitat: On generally dry rock or with occasional seepage.

Notes: Confident identification requires Melzer's reagent or DNA sequencing; in lieu of that we suggest tentatively assigning material to *D. polyphyllizum* if it keys to couplet 15, was growing on calcareous rock, has a gray upper surface, a tan to orangish or pinkish tan lower surface, and small spores. *Dermatocarpon taminium* can be similar in appearance to *D. polyphyllizum* but is not a strict calciphile and is more associated with warm to hot climates, rather than the cold, dry habitats of *D. polyphyllizum*.

Dermatocarpon reticulatum

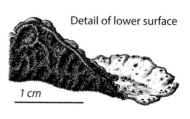

Detail of lower surface

Description: Thallus foliose, umbilicate (rarely unattached), 2-6(8) cm broad; upper surface some shade of gray, sometimes brownish gray, smooth to pruinose; lower surface finely papillose (use lens), orangish brown to dark brown or black, coarsely wrinkled or not, occasionally veined or coarsely warted or rugose, rarely with a few stubby rhizines; perithecia common; spores 12-17 × 5-9 μm; medulla Melzer's I-.

Range: Widespread in w N Am, common in the PNW.

Substrate: On noncalcareous or somewhat calcareous rock.

Habitat: On rock faces and in crevices in rather dry habitats and rocky streamside areas where occasionally flooded.

Notes: Like most *Dermatocarpon* species, this species is highly variable in outward form, and reliable identification is based on the lower surface. Although frequently misidentified, this species is easily distinguished by the continuously papillose texture of the lower surface. These papillae are clearly visible with a good hand lens but are mostly invisible to the naked eye. Occasionally unattached individuals are found in poorly drained steppe with thin, rocky soils. These "vagrant" individuals have tightly rolled narrow lobes in a contorted, strongly three-dimensional form.

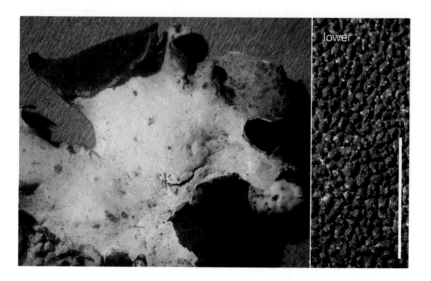

lower

Erioderma

Erioderma sorediatum

Description: Thallus foliose, small, mostly 0.5-3 cm diam; upper surface gray, with short, erect, stiff hairs or tomentum on the lobe tips; lower surface whitish, veinless, tomentose, sparsely rhizinate; lobes 2-4 mm wide; photobiont blue green; soredia grayish; soralia elongate, marginal; apothecia not seen; medulla P+O (eriodermin).

Range: Coastal se Alas to s Ore, always within a few km of the ocean, except farther inland on the OP; rare; also in s hemisphere.

Substrate: Most frequent on a thin layer of bryophytes over stems of ericaceous shrubs (*Vaccinium, Rhododendron*); also on conifer branches.

Habitat: Semi-open habitats on the coast, most often in dune woodlands and deflation plains, also in coastal *Picea* forests.

Notes: See comparison under *Leioderma sorediatum*, a species of similar color, size, and form, with which *Erioderma* occurs. *Erioderma pedicellatum* (see photo in key to genera, p. 56) is a very rare species, known in the PNW only from Alas, more common in the N Atlantic oceanic region. It is easily distinguished from *E. sorediatum* by the absence of soredia and presence of apothecia.

Sources include: Galloway and Jørgensen (1975, 1987), Glavich et al. (2005a), McCune et al. (1997), Nelson et al. (2009).

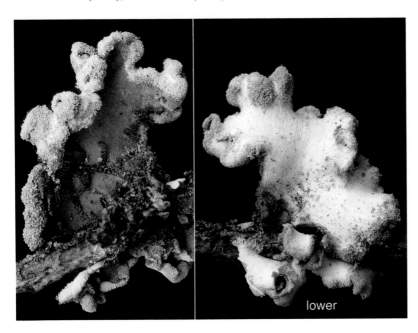

Esslingeriana

Esslingeriana idahoensis

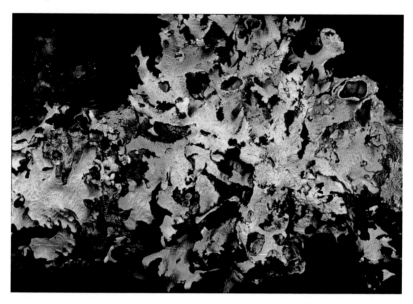

Description: Thallus foliose, suberect or appressed with upturned edges; upper surface whitish to pale greenish gray; lower surface dark brown to jet black; soredia and isidia lacking; pycnidia black; apothecia usually present; cortex K+Y (atranorin), medulla white to yellowish, lw UV+Y or O (anthraquinones).

Range: BC to Cal e to w Mont.

Substrate: Bark and wood, mainly conifers but also on shrubs.

Habitat: Occasional in low- to mid-elevation forests w Cas; more common on e slope Cas and in moist forests on the w slope RM.

Notes: This species is frequently overlooked because of its superficial resemblance to a smallish specimen of *Platismatia glauca*. That species, however, has soredia or isidia. The lower surface of *P. glauca* is often mottled with white or brown, while the lower surface of *E. idahoensis* is wholly black except for a narrow brown marginal zone. An odd variant with a lower surface white to mottled with brown and black, similar to *Platismatia*, is known from one site in the Columbia Gorge (Skamania Co., Wash, 200 m). But like *E. idahoensis* the medulla is yellowish in some areas and lw UV+O.

Sources include: Esslinger (1971).

Evernia

Thallus truly fruticose or fruticose-like but upper and lower surfaces differentiated, erect or pendent, greenish to yellowish, ± soft, attached by a single holdfast or draped over the substrate in one sp; soredia or isidia present in most spp; cortex KC+Y (usnic acid), medulla sw UV+ blue white (divaricatic or evernic acid); apothecia rare, lecanorine, the disk red brown; epiphytic, but one species on alpine sod.

Sources include: Bird (1974).

ID tips: An abundant species in the PNW, *E. prunastri* is commonly confused by beginners with *Ramalina*. See notes under *E. prunastri* below. Other species of *Evernia* are extremely rare in the PNW.

1a Branches flattened; soredia present (though often sparse) .....*E. prunastri*
1b Branches angular, not flattened; soredia present or not
  2a Soredia or isidia present, produced in cracks along ridges. Thallus to 10 cm long, suberect; branches to 1 mm thick; containing usnic and divaricatic acids; apothecia rare; generally on conifers; boreal; very rare w Cont Div, but more frequent n into Can.........[*E. mesomorpha*]
  2b Soredia and isidia lacking
    3a Thallus pendulous, to 30 cm long; on trees in conifer forests. Thallus cream yellow when fresh, turning reddish brown in the herbarium; cortex frequently broken, exposing the medulla; medulla loose; thallus containing usnic and divaricatic acids; apothecia rare; on conifers in mesic forests; RM from BC and Alta to NM and Ariz; in Ore and Wash it is known only from Wallowa Mts.......*E. divaricata*
    3b Thallus prostrate, to 5 cm long; on tundra. Thallus cream yellow; cortex often broken; medulla fairly dense; apothecia lacking; on rocky soil; arctic-alpine s through RM to Ariz and NM; rare ....
    ....[*E. divaricata* (alpine variant)]

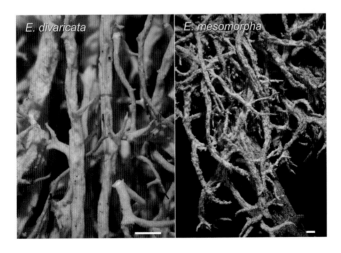

E. divaricata   E. mesomorpha

Evernia prunastri

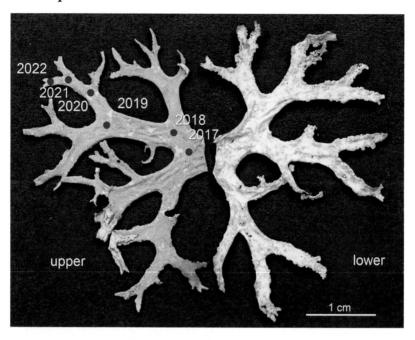

Description: Thallus appearing fruticose, but foliose on close inspection (by virtue of having differentiated upper and lower surfaces), erect or drooping, to 7(13) cm long, dichotomously branched; branches flattened, 1-5(10) mm wide, ± soft; upper surface greenish or yellowish green; lower surface whitish; soredia present (though soredia occasionally very sparse); apothecia not seen; cortex KC+Y (usnic acid), K+Y (atranorin); medulla sw UV+ white (evernic acid).

Range: Common w Cas, sporadic e Cas to w Mont.

Substrate: Wood or bark, especially hardwood trees and shrubs, occasionally on conifers.

Habitat: Ubiquitous in most habitats at low elevations w Cas, especially hardwood forests, savannas, and in urban and agricultural areas; less abundant in lower mountain forests, where it is best developed in pockets of hardwoods; e Cas almost entirely restricted to riparian forests and shrubby areas.

Notes: Shade forms have narrow lobes and will be confused with *Ramalina* species. Look carefully at these for the subtle contrast between the greenish upper surface and slightly paler lower surface. With practice you will also recognize the difference between the stiffer, cartilaginous texture of *Ramalina* when dry and the softer feel of *Evernia*. Because *Evernia prunastri* in the PNW rebranches once per year, the age of an individual can be estimated by counting a series of branch points, as illustrated here.

Flavoparmelia

Flavoparmelia caperata

Description: Thallus foliose, ± adnate, to 15(20) cm broad; lobes medium to broad, 3-8 mm wide; upper surface pale yellowish green when dry, often finely wrinkled; pseudocyphellae lacking; lower surface black except brown near the margin; soredia present, primarily laminal, erupting from the upper cortex in discrete soralia, initially small, becoming large and confluent; rhizines black, simple; apothecia rare; cortex K±Y, KC+Y (usnic acid, atranorin); medulla C-, KC+ pink, P+O or R (protocetraric and caperatic acids).

Range: One of the most common macrolichens in e N Am, occasional in Cal and sw US, but rare in the PNW, where it is known from only a few locations in coastal (Waldport, Ore), San Juan Islands, and urban (Seattle) areas, possibly as an import on nursery plants. The species appears to be rapidly expanding its distribution in w Wash.

Substrate: Bark and wood.

Habitat: In the PNW in low-elevation urban and suburban areas.

Notes: This distinctive species stands out in the field as one of our few largish pale yellowish green foliose species. Most likely to be confused with *Flavopunctelia*, that genus has tiny white pseudocyphellae, and a P-, C+R medulla. *Flavopunctelia soredica* can have only weakly developed pseudocyphellae, but it has primarily marginal rather than laminal soredia, so it is easy to distinguish from *Flavoparmelia caperata*, even without a spot test. In the US, you can almost always find *F. caperata*, but in the PNW it is still rare.

Flavopunctelia

Thallus foliose, ± adnate, to 10 cm broad; lobes medium to broad; upper surface yellowish green; pseudocyphellae often present; lower surface black or dark brown, brown near the margin; rhizines simple; soredia present; apothecia rare; cortex K+Y (atranorin), KC+Y (usnic acid); medulla C+R, KC+R (lecanoric acid); on bark and wood.

Sources include: Hale (1984).

ID tips: Although these two *Flavopunctelia* species are rare in the PNW, they can be quite common and conspicuous elsewhere (e N Am, Cal). Typical specimens are readily identified to species, but occasional difficult specimens may at first seem intermediate in morphology.

1a Soredia laminal and marginal; upper cortex with distinct but small white marks (pseudocyphellae), especially near the lobe tips, the pseudocyphellae developing into laminal soredia. Boreal forests to Alta, s to Cal and Mex, very rare n into BC; rare in PNW; mainly in low-elevation riparian forests .... *F. flaventior*
1b Soredia primarily marginal, in linear soralia on suberect crescent-shaped lobes; upper cortex often with faint white angular markings (not visible with naked eye) but generally lacking prominent pores. Cal and NM, n to Wash and Mont, mainly e of RM, also in e N Am .... *F. soredica*

Flavopunctelia flaventior

Flavopunctelia soredica

Fuscopannaria, with Steineropsis and Vahliella

Thallus squamulose to lobate and placodioid, occasionally crustose, small, gray, brown, blue gray, or less often blackish, with lobe margins often whitish felty with a frosted appearance; attached to substrate by a well-developed blue to black hypothallus or by a tomentum of dark rhizohyphae; lobes generally < 2(3) mm wide; lower surface lacking a cortex; isidia or soredia present or not; apothecia at first immersed, eventually sessile, with thalloid margin or not; disk reddish brown, orange-brown, dark brown, or black; hymenium I+ blue green turning red brown; hypothecium pale or hyaline; spores simple, hyaline, smooth or uneven, the outer spore wall (epispore) sometimes thickened, warty, or pointed; thallus P-; very fine superficial terpenoid crystals often apparent after storage; photobiont blue green (*Nostoc* or *Scytonema*); on rocks and trees, often overgrowing bryophytes, mostly in sheltered humid habitats.

Sources include: Jørgensen (1994b, 2000a, 2000b, 2008), Spribille et al. (2020), Carlsen et al. (2012).

ID tips: Spot tests of are of almost no use in separating *Fuscopannaria* species; however, the presence or absence of terpenoid crystals can be diagnostic. In many species terpenoids develop into very fine external crystalline needles after storage in the herbarium (use 40× magnification with DM). The key includes all genera formerly in *Pannaria* in the broad sense, except for *Pannaria* itself.

Introductory Key

1a Thallus with distinct marginal lobes; cortex and/or medulla P+O; apothecia with thalline margin.. see ***Pannaria***

1b Thallus with well developed marginal lobes or squamulose or crustose, lobate or not; cortex and medulla P-; apothecia with or without a thalline margin

2a Apothecia nearly always present; spores warty, large (25-30 × 9-12 µm); apothecia with beaded thalline margin

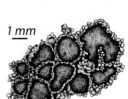

3a Photobiont blue green. Thallus pale bluish gray to dark brown, fresh material dark bluish when wet; on shady soil, logs, tree bases, and mossy rocks; widespread, common e Cas.....................*Protopannaria pezizoides*

Protopannaria pezizoides

3b Photobiont green (except for cephalodia) .... *Psoroma hypnorum*

2b Apothecia present or not; spores uneven to smooth, mostly < 25 µm long; apothecia various

4a Photobiont green; so far known only from n Alas.......[*F. viridescens* and *F. globigera*]

4b Photobiont blue green; widely distributed

Psoroma hypnorum

5a Soredia, isidia, or lobules present
  6a Thallus completely granular, lacking squamulose, areolate, and lobate parts .... **Group 1** (Lead 9)
  6b Thallus squamulose, areolate, or placodioid-lobate, though often the thallus partially dissolved into granules
    7a Thallus becoming densely isidiate and lobulate, so as to obscure the lobes; hypothallus black, usually prominent; thallus dark gray or brownish gray; thallus lacking terpenoids (not developing super-fine needle-like external crystals in storage). Spores 10-17 x 5-8 μm; on bark, wood, and rock; widespread in temperate n hemisphere, but rare in Ore, Id, and Mont, more frequent n to BC and se Alas ....*Parmeliella triptophylla*
    7b Thallus isidiate or lobulate or neither of those; hypothallus conspicuous or not; thallus color various; thallus with or without terpenoids
      8a Thallus with flattened lobules .... **Group 2** (Lead 11)
      8b Thallus isidiate or sorediate .... **Group 3** (Lead 14)
  5b Soredia, isidia, and lobules absent (though with narrowly incised lobes, knobby margins, or isidioid warts) .... **Group 4** (Lead 22)

Group 1

Completely granular

9a Thallus P+O; photobiont green. On rock and soil; Sierra Nevada and Coast Ranges of Cal n to s Ore
  .... *Leprocaulon adhaerens* and *L. santamonicae* (not treated here)
9b Thallus P-; photobiont blue green
  10a Granules corticate, pale bluish gray; growing on soil or on moss over soil or rock; terpenoids lacking; apothecia rare. Spores 10-15 × 5-8 μm. ....*Moelleropsis nebulosa*
  (*Fuscopannaria cyanolepra* can appear similar but has brownish squamulose parts with bluish gray marginal true soralia.)
  10b Granules lacking a cortex, totally dissolved into soredia, pale lead gray to bluish gray; growing on trees or mossy rock; terpenoids present; apothecia unknown. Wash and Ore, w Cas............................*F. leprosa*

(The general habit approaches *F. mediterranea* and *F. cyanolepra*, but those species have corticate parts; check carefully with dissecting scope or hand lens.)

Group 2

Flattened lobulate

11a Lobules and isidia laminal, often thickly developed and obscuring the thallus; marginal lobes not much enlarged
....*Parmeliella triptophylla* (Lead 7a)
11b Lobules arising from the edges and lower surface of the thallus
  12a Lobules flattened but roundish or ear-shaped in outline; thallus containing terpenoids (look for very fine needle-like crystals on herbarium specimens; key fresh specimens both ways)
    13a Habitat rocks very near the coast, including upper rocky intertidal, rarely farther inland; lobules not produced consistently
....*F. maritima* (Lead 26a)
    13b Habitat never coastal rocks, lobules produced consistently..**F. aurita**
  12b Lobules strap shaped, elongate and flattened, sometimes becoming coralloid; thallus lacking terpenoids, containing atranorin. Thallus usually brownish but sometimes gray, incised-squamulose, often with distinctly elongate marginal lobes to 1(2) mm broad, forming extensive colonies with a distinct hypothallus; lobes often shingled; photobiont *Scytonema*-like, with large (to 10 μm) cells that are emerald green and in short chains; apothecia occasional, flat, to 1.5 mm diam, usually brownish, usually with a thalline margin, the margin becoming lobulate, or thalline margin lacking; spores (12)15-20(23) × 7-9 μm; usually on mossy conifer branches and tree bases or wood, occasionally on hardwoods (*Alnus, Salix*), or bryophytes on the ground or rock; n Cal to se Alas where it is fairly common, usually near the coast...*Steineropsis laceratula*

Group 3

Isidiate or sorediate

14a Growing on soil, moss over soil, rock, or moss over rock
  15a Thallus nearly completely sorediate, composed of brown corticate granules or small squamules that soon burst into bluish gray soredia, usually coalescing into a sorediate crust. Apothecia rare, with a sorediate margin that turns under and brown disk; spores 12-14(17.5) × 7-8 μm; fairly common but usually overlooked, most common on soil and moss over rock on steep roadcuts and cliffs, often overgrowing adjacent substrates, including bare rock and roots; low to mid elevation, Wash to Cal, inland to n Id ....*F. cyanolepra*

F. cyanolepra

(This epithet has been applied to several different taxa having largely or completely granular thalli, including *F. leprosa* and *Moelleropsis nebulosa*—see Group 1.)

15b Thallus with persistent, extensive corticate areas

16a Isidia merely warty; terpenoid crystals developing on herbarium specimens. On mosses over rock and rock .... *F. praetermissa* and *F. thiersii* (See Group 4.)

16b Isidia well developed, cylindrical to coralloid or somewhat flattened; terpenoid crystals lacking

17a Thallus more gray than brown; hypothallus conspicuous .... *Parmeliella triptophylla* (Lead 7a)

17b Thallus brownish; hypothallus inconspicuous or not apparent. Thallus of deeply divided squamules with elongate coralloid lobes, often with swollen apices; apothecia rare, with poorly developed thalline margin or thalline margin completely lacking; disk brown, convex; spores (16)19-23(27) × 8-10 μm; coastal Cal to Ore and OP; on mossy soil, soil over rock, and rock; possibly a short-lived pioneer .... *F. coralloidea*

14b Growing on bark or wood (or moss over those substrates), rarely on rock

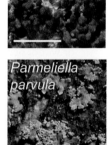

F. coralloidea

Parmeliella parvula

18a Thallus lacking terpenoids, not developing tiny needle-like crystals of terpenoids in the herbarium. Thallus with elongated, incised lobes; isidia mostly globular or compound globular, though some cylindrical isidia can be found; apothecia rare, small, brown, the proper margin blackening, lacking a thalline exciple; in the PNW most frequent in old-growth forests; usually on conifer twigs or *Rhododendron*; distribution poorly known because of its inconspicuous appearance, but probably widespread in oceanic areas of coastal states and provinces .... *Parmeliella parvula*

18b Thallus containing terpenoids, usually developing tiny needle-like crystals after several years in the herbarium

19a Upper surface pruinose or scabrid; thallus with elongate lobes that branch more than once, often with expanded lobe tips

20a Isidia or soredia tending to aggregate in marginal or terminal hemispherical or lip-shaped groups; atranorin lacking. Thallus with elongate lobes, to 2 cm diam; lobes 0.5-1.5 mm wide, weakly to distinctly scabrid; apothecia rare in N Am; BC and Alas; on bark .... [*F. ahlneri*]

(*Fuscopannaria alaskana* appears like a robust specimen of the closely related *F. ahlneri*. Spores of *F. ahlneri* are 20-25 × 10-11 μm, while those of *F. alaskana* are reportedly 9-13 × 7-8 μm. However, spore characters are practically useless in these primarily sterile spp.)

20b Soredia marginal, not aggregated into hemispherical soralia; atranorin present. Thallus distinctly lobate, to 2 cm diam; lobes scabrid or pruinose and sometimes yellowish gray to bluish gray pruinose or frosted near the margins; apothecia uncommon, to 1 mm diam, with a sorediate bluish thalline margin; spores about 15 × 8.5 μm; w Ore to s Cal; few collections, but fairly frequent on *Quercus*; w Cas, valleys and foothills ......*F. pulveracea*

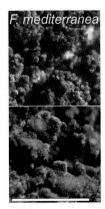

F. pulveracea

19b Upper surface not pruinose; thallus squamulose to placodioid-lobate

21a Thallus of convex squamules, olive brown with olive brown or contrasting lead-gray felty granular isidia or soredia. Squamules to 1-3 mm wide, often irregularly rounded and incised, often obscured by propagules, white felted to tomentose marginally; granules olive brown to lead gray or bluish gray, developing by repeated budding (like yeast), initially on the upturned margins; apothecia unknown in N Am; on bark or wood, rarely on rock; se Alas to Cal, with rare disjuncts inland to Ariz and Colo......... *F. mediterranea* (*F. cyanolepra* is very similar but differs in substrate and color. Its soredia have a bluish tinge while those of *F. mediterranea* are often grayer.)

F. mediterranea

21b Thallus not appearing swollen, gray or brownish, without paler, felty soralia. Thallus to 1 cm diam; lobes to 3 mm long, rebranching, incised; upper surface uneven, matt but glabrous, mostly blue-gray with marginal granular soralia; propagules mainly appearing dull and ecorticate (sorediate) but developing from corticate granules; hypothallus distinct; apothecia rare, to 1 cm diam, the disk concave to flat, without a thalline margin, but sometimes granules or soredia clinging to the proper margin; spores 15-20 × 10-12 μm; so far known from *Picea* and *Salix* in moist habitats; se Alas to Wash ....*F. ramulina*

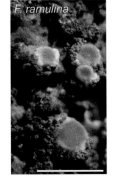

F. ramulina

(Appears like a sorediate *F. pacifica* and easily confused with large specimens of *Parmeliella parvula* (Lead 18a), which is similar but lacks terpenoid crystals.)

Group 4

No isidia or soredia, but often with knobby margins

22a On bark or wood

   23a Thallus bluish gray to light brownish gray, whitish scabrid or pruinose (use lens); lobes 0.5-2 mm wide; apothecia with thalline margin .... *F. leucostictoides*

   23b Thallus gray to brown, never whitish scabrid or pruinose; lobes mostly < 0.5 mm wide; apothecia lacking thalline margin or thin and disappearing... *F. pacifica*

22b On rock, moss over rock, soil over rock, or alpine sod

   24a Lobes with whitish margins that are often knobby; terpenoid crystals present after storage; apothecia dark brown to black

      25a On coastal rocks, rarely in Coast Range or Cascades. Thallus

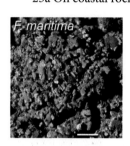

squamulose to short-lobed placodioid, coalescing to form congested, often extensive mats; lobes light to dark brown with whitish or blue-gray pruinose knobby margins, often convex, with lobe tips expanding to small fans mostly < 0.5(1) mm wide; upper surface smooth to slightly and minutely rugose; thallus containing atranorin and terpenoids; apothecia brown, to 1 mm diam, common, convex at maturity, the thalline margin variably developed; spores 13-16(20) × 7-10 μm.................... *F. maritima*

      25b On rock or mossy rock in other habitats

         26a Occurring directly on rock, usually ultramafic (serpentine or peridotite). Thallus to 6 cm diam or more, forming rosettes of imbricate lobes with pale knobs on the margins, centrally with an areolate to squamulose appearance; apothecia common, brown to dark brown, to 2 mm diam, becoming convex, often with a whitish-beaded, discontinuous thalline margin; spores 17-22 × 7.5-9.5 μm; w Cas, Cal to w Ore.. *F. thiersii*

         26b Occurring on organic matter over rock or soil, calcareous or not, rarely on partially rotted or charred stumps or logs. Thallus to 3 cm diam or more, lobes becoming densely imbricate, sometimes becoming complete covered with knobby outgrowths that become whitish; apothecia often numerous when other propagules are absent, to 1.5 mm diam, convex, with or without a thalline margin; spores (15)18-22 × 8-11 μm, often with one large oil drop; widespread in northern regions; arctic-alpine, subalpine, and montane .... *F. praetermissa* (Previously applied broadly to many species.)

   24b Lobes uniform in color or faintly whitish margined; terpenoid crystals lacking; apothecia various

      27a Habitat arctic-alpine; spores small, 10-12 × 6-7 μm. Thallus forming small greenish brown cushions to 2 cm broad; apothecial disk

blackish with a strong thalline margin, beaded but not surrounded
by lobules; on rock; Colo and Arctic .... [*F. hookerioides*]
27b Habitat lowland to montane or subalpine; spores large (> 15 μm long)
   28a Thallus forming a thin monolayer (or nearly so) over the
      hypothallus, lobate-squamulose to placodioid, often forming
      small rosettes; lobes < 0.5 mm broad; apothecia common, disk
      pale tan, orangish, or brown, thalline margin lacking or thin and
      disappearing; on bark, wood, and rock ....***F. pacifica***
   28b Thallus imbricate, cushion forming; lobes often > 0.5 mm broad;
      apothecia various

      29a Lobe tips fan shaped, commonly 0.5-3
         mm broad. Thallus brownish to buff, dull
         but not pruinose or scabrid, central parts
         with congested smaller convex lobes that
         becoming knobby or granulose; apothecia
         brown to dark brown, convex at maturity
         and excluding the margin, but often with
         thick lobes surrounding the disk; spores
         15-20 × 8-10 μm without epispore, 18-23
         × 9-11 μm with epispore; epispore forming
         short knobs or tapered points on the spore
         ends; rare; Mont, Alta, BC, Ore, and Wash
         ....*F. cheiroloba*
         (This species can be mistaken for *Vahliella californica, F.
         praetermissa,* or even *Massalongia carnosa. F. praetermissa*
         has bluish gray margins contrasting with the brown lobes,
         larger spores, and terpenoid crystals. *Vahliella californica*
         forms a cushion of similar-sized lobes that are not particularly
         fan tipped, while *F. cheiroloba* has tightly packed lobes in the
         centers, but with expanded peripheral lobes. *Massalongia
         carnosa* has more marginal granules, the peripheral lobes are
         dark brown, and the spores are septate. Also, *M. carnosa* has a
         tan lower surface with sparse rhizines in contrast with the felty
         blue-black hypothallus of *Fuscopannaria*.)
      29b Lobe tips only slightly expanded, usually < 1 mm broad. Lobes
         convex, thick (to 0.5 mm), congested. Thallus brownish,

uniform in color to faintly
whitish margined, becoming
cushion-like; apothecia
brown to black, to 1.5 mm
diam, becoming compound
and convex, with a variable
thalline margin, sometimes
with flat lobules; spores
17-25 × 7-10 μm; on dry
noncalcareous rock; BC to s
Cal, inland to RM
................*Vahliella californica*

Fuscopannaria aurita

Description: Thallus lobate or placodioid, medium to dark brown or bluish gray from "frosted" lobules, to 4 cm diam or coalescing over larger areas, forming rosettes or congested mats; lobes thick, to 300 μm; margins ascending and covered with blue-gray frosted ear-shaped lobules; lobe tips expanding paw-like, plane to slightly concave fans mostly 1-3 mm broad; upper surface minutely but distinctly rugose; lower cortex absent, the medulla gradually merging into the blackish rhizohyphae; apothecia rare, to 1.5 mm diam; disk brown, finally convex, with a lobulate thalline margin; spores 14-15 × 7-8 μm; spot tests negative; minute needle-like terpenoid crystals developing in storage.

Range: Cal to BC inland to w Mont, endemic to the PNW.

Substrate: Usually on mossy soil or mossy rock, rarely on mossy wood.

Habitat: Moist low-elevation valleys and mountains.

Notes: This lobulate species is one of the most common *Fuscopannaria* species w Cas. The somewhat flattened, frosted marginal lobules differ from the ascending knobs sometimes found in *F. praetermissa* and *F. thiersii.* In some habitats the frosted lobules dominate, giving a sorediate impression, and suggesting *F. cyanolepra,* but *F. aurita* never forms marginal soredia. It also resembles *Massalongia,* but that genus lacks terpenoid crystals; has narrower, septate spores; and has a more uniformly brown thallus.

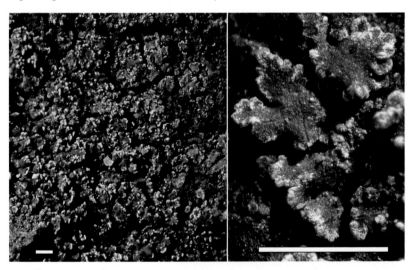

Fuscopannaria leucostictoides

Description: Thallus squamulose to placodioid, mostly < 2 cm diam; lobes generally < 1(2) mm broad and usually < 3 mm long, margins smooth or knobby; upper surface bluish gray to brown, white-scabrid to pruinose or thinly felted tomentose toward the edges, giving the thallus a bluish gray color; hypothallus usually well developed, blackish; apothecia common, the disk brown, to 1.5 mm diam, usually with a smooth to knobby persistent thalline margin; spores 14-18(20) × 7-10 μm, including the epispore; epispore sometimes thickened at the ends and thin at the sides, or evenly thickened, or sometimes lacking; spot tests negative; atranorin and terpenoids present.

Range: BC to Cal, w Cas but with rare disjuncts in Id.

Substrate: Bark and wood of hardwoods and conifers, rarely on rock.

Habitat: Edges of coastal forests, oak savannas, valley and foothill woodlands, rarely inland to n Id in suboceanic climates.

Notes: The bluish gray tinge, lack of isidia and soredia, and the epiphytic habitat are characteristic of this species. Well-developed vigorous individuals may approach *Pannaria oregonensis* in appearance, but that species can be separated by its P+O reactions. If the lobes are finely dissected or lobulate, see *Steineropsis laceratula*; if ± sorediate, see *F. pulveracea*. *Fuscopannaria convexa* appears like a small version of *F. leucostictoides*, but without distinctly elongate lobes. Both *F. convexa* and *F. leucostictoides* contain atranorin. *Fuscopannaria convexa* is known only from coastal se Alas. It has a brown to black disk with a thalline margin, a prominent black hypothallus, spores 13-16(20) × 7-10 μm, and favors *Populus* bark.

Heterodermia sitchensis

Description: Thallus foliose, tiny (< 2 cm diam), whitish gray to pale greenish gray, sometimes discoloring to blue black; lobes suberect, mostly 1-2 mm wide, with marginal cilia; soredia present in urn-shaped lobe tips; lower surface white, partly or wholly lacking a cortex; apothecia occasional, becoming sorediate on the margins; cortex K+Y; medulla K+Y, P+Y (atranorin and zeorin).

Range: se Alas to Ore, on the immediate coast.

Substrate: On bark of *Picea sitchensis*, probably on other conifers as well, especially where nutrient enriched.

Habitat: Cool, moist oceanic forests in exposed positions on the coast.

Notes: This very rare species, endemic to the PNW, is so far known from only a handful of sites in BC and se Alas and one site in Ore (Cape Lookout). This species is so small and inconspicuous that it is likely to be overlooked as an odd little *Physcia tenella*. Close inspection of *H. sitchensis* reveals, however, the distinctive urn-shaped soralia.

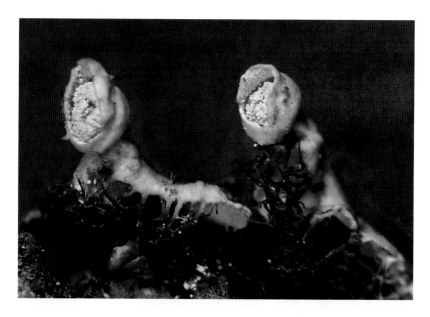

Hypogymnia

Thallus medium sized to small, foliose, erect, appressed, or pendulous; lobes hollow (rarely filled in a few spp), mostly 1-6 mm wide, whitish gray to pale greenish gray or brown-gray above with black pycnidia; lower surface black and typically strongly wrinkled, often with a hole at the lobe tip or axil; rhizines lacking; soredia or lobules sometimes present; apothecia brown, with thalline margin, often short-stalked; spot tests diverse, but all have a K+Y cortex (atranorin), most have a KC+ pink, orange, or red medulla (physodic acid), and can contain often one or more of the following: physodalic, protocetraric, diffractaic, hypoprotocetraric, 2'-O-methylphysodic, and 3-hydroxyphysodic (= conphysodic) acids; abundant on bark and wood, occasionally rock, low elevations to alpine.

Sources include: Ahti and Henssen (1965), Goward (1988), Goward and McCune (1993, 2007), Goward et al. (2010, 2012), Lumbsch et al. (2011), McCune et al. (1997), McCune and Rosentreter (1997), McCune et al. (2011), McCune and Conway (2022), Miadlikowska et al. (2011), Ohlsson (1973a), Pike and Hale (1982).

ID tips: The most common cause of misidentification of *Hypogymnia* is not taking the time to verify a morphological identification by spot tests, in particular the P test. The results of this test are usually clear. By studying the morphology and checking the spot tests you will increasingly be able to accurately identify to species on the basis of morphology and to predict spot test results. But the morphology is variable enough within species that even the experts routinely apply spot tests to aid identification. Note that while spot tests here often refer to the medulla, because the upper cortex is transparent when wet, the color reaction of the thallus is similar to the color reaction of the medulla. So when making spot tests you can simply break off about 1 cm of a lobe tip and test that.

Introductory Key

1a Lobes < 1 mm wide; lower surface densely pock-marked with tiny pits
(use lens); mainly w Cas ....**Group 1** (Lead 3)
1b Lobes usually > 1 mm wide; lower surface not densely pitted
  2a Soredia present...**Group 2** (Lead 4)
  2b Soredia lacking...**Group 3** (Lead 12)

Group 1

Narrow lobed; lower surface densely pitted

3a Soredia present ....*H. hultenii*
3b Soredia absent..*H. lophyrea*

Group 2

Sorediate

4a Soredia on the inside of the lobe tips, which burst open, exposing often
lip-shaped soralia
  5a Lobes lacking narrow, short adventitious lobes; lobe cavities white
above, dark below; medulla P+R; lower surface lacking holes except
where soralia are forming; common and widespread.......**H. physodes**
  5b Lobes with narrow, short adventitious lobes; lobe cavities dark above
and below; medulla P-; lower surface with large holes near the lobe
tips. Epiphytic or on mossy rock or tundra sod; arctic, boreal, and
coastal Alas to BC, not yet known from Ore and Wash ..... [*H. vittata*]
4b Soredia laminal or terminal but not forming lip-shaped soralia
  6a Lobes mostly not hollow..**H. pulverata**
  6b Lobes hollow
    7a Medulla P+R; soredia laminal to terminal on upper surface
.... *H. oceanica*
    7b Medulla P-; soredia various
      8a Soredia generally laminal
        9a Lobes separate. Common in the s hemisphere, but known in N
Am from a single site in the old dunes near Clear Lake, Lane
Co., Ore..*H. subphysodes*
        9b Lobes contiguous to subcontiguous, forming an appressed
rosette
          10a Upper surface brownish or greenish brown in partly to
exposed environments; soredia coarsely granular, eroding to
powdery, often becoming confluent over much of the upper
surface.. **H. austerodes**
          10b Upper surface whitish gray in partly to exposed
environments; soredia powdery in discrete roundish soralia.
Epiphytic, uncommon in montane conifer forests
.... *H. tuckerae*
      8b Soredia primarily terminal or subterminal, but occasionally
laminal on older lobes
        11a Lobes appressed, the branching pattern various; soralia small
(< 1 mm) on ends of narrow, lateral, often upturned lobes
and forming larger soralia on the blunt ends of main lobes,
becoming more laminal on old thalli. On bark, wood, or rock;
arctic and boreal, rarely s in RM to Ariz.................... [*H. bitteri*]
        11b Lobes erect or suberect, dichotomously branched; soralia
capitate, surrounding the lobe tips...........................**H. tubulosa**

Group 3

Soredia lacking

12a Interior of lobes snow white ....**H. imshaugii**
12b Interior of lobes grayish to dark brown, at least in part (ceiling of cavity
may be white and floor of cavity dark)

13a Medulla P+R throughout
  14a Lobes distinctly trailing and pendulous from the substrate
    15a Lobes slender (about 1 mm), cascading in arcs from the substrate, nodulose; ceilings of lobe interiors usually white .... *H. duplicata*
    15b Lobes broad (2-5 mm), pendulous but rarely in a series of arcs, nodulose or not; ceilings of lobe interiors dark ....*H. enteromorpha*
  14b Lobes erect or appressed to the substrate
    16a Lobes mostly 2-5 mm wide.................................*H. enteromorpha*
    16b Lobes mostly < 2 mm wide
      17a Thallus usually erect to suberect; lobes often spreading from each other, bearing narrow perpendicular side lobes; growing very near the coast ....*H. heterophylla*
      17b Thallus appressed to the substrate; lobes contiguous, lacking narrow perpendicular side lobes; widespread .... *H. physodes* (Juveniles with very sparse or no soredia will key here.)
13b Medulla largely P- (or pale Y)
  18a Medullary cavity ceiling white to graying above, the floor of the cavity dark; lobes short and narrow
    19a On soil or mossy rocks, alpine; upper surface light to dark brown; isidia or lobules present on the edges and upper surface; common in Arctic, rare in RM s to Colo ....[*H. subobscura*]
    19b On bark or wood; upper surface whitish to pale greenish gray but sometimes brown at the tips; isidia and lobules lacking
      20a Lobes contiguous, closely appressed to the substrate; medulla K-; lobe tips at most slightly browned.............................*H. wilfiana*
      20b Lobes contiguous to imbricate, becoming loosely appressed to the substrate; medulla K+ reddish brown; lobe tips distinctly browning in exposed microsites...........................*H. canadensis*
  18b Medullary cavity ceiling and floor grayish to black; lobes various but not short and narrow
    21a Lobes broad, appressed, often neatly dichotomous and contiguous; lobe tips and axils without holes; thallus with a soft papery texture when dry; upper surface becoming strongly wrinkled; thallus often forming large (10-20 cm diam) rosettes............... *H. rugosa*
    21b Lobes various; thallus with a ± cartilaginous texture; lobe tips and axils with sparse to abundant holes (use hand lens); upper surface smooth or wrinkled; thallus rarely forming a rosette > 10 cm diam
      22a Branches erect or drooping, isotomic dichotomous ... *H. inactiva*
      22b Branches appressed to trailing, aniso- or isotomic dichotomous
        23a Medulla KC+O; lobes appressed to the substrate, rarely trailing, with short, swollen, bud-like side lobes .... *H. occidentalis*
        23b Medulla KC-; lobes trailing or appressed, often nodulose, usually lacking bud-like side lobes.....................*H. apinnata*

Table 6. Esorediate species of *Hypogymnia*. P+R reactions are deep yellow-orange to brick red (physodalic and protocetraric acids). KC+O reactions are pinkish red or orange (physodic acid). K+R reactions are slowly reddish brown (3-hydroxyphysodic acid).

Species	Spot tests (medulla)			Field characters
	P	KC	K	
apinnata	-	-	-	Appressed or trailing, puffy, often nodulose lobes
canadensis	-	O	R	Like *H. wilfiana* but lobes more imbricate, browner tips, K+
duplicata	R	-	-	Trailing, slender, arcuate, nodulose lobes; apothecia rare; w Cas only
enteromorpha	R	O	-	Like *H. apinnata* but with short, bud-like side lobes
heterophylla	R	O	-	Erect or semi-appressed, with narrow side lobes; immediate coast only
imshaugii	±R	±O	±R	White lobe interiors, erect or suberect lobes, similar in habit to *H. inactiva*, but sometimes appressed
inactiva	-	O	-	Like *H. imshaugii*, but with dark lobe interiors and perforations
occidentalis	-	O	-	Appressed, puffy lobes, with short bud-like side lobes
rugosa	- or O in lobe tips	O	-	Appressed, large, soft, and puffy, neatly dichotomously branched, mid to high elevations
wilfiana	-	O	-	Small, appressed, contiguous lobes

Hypogymnia apinnata

Description: Thallus foliose, medium sized (mostly 5-15 cm), whitish gray to pale greenish gray; lobes puffy, hollow, trailing or appressed, mostly 2-5 mm wide; often nodulose; lobe interiors dark; lobe tips swollen, often with a terminal hole; short perpendicular bud-like side lobes lacking or sparse; lower surface black; soredia and isidia lacking; apothecia common, brown; cortex K+Y; medulla K-, P- (or pale Y), KC- (atranorin and fatty acids).

Range: Alas to Cal, common w Cas, infrequent inland to n Id and nw Mont.

Substrate: Bark and wood, mainly on conifers.

Habitat: Moist conifer forests, low to mid elevations.

Notes: This species often grows intermixed with *H. enteromorpha*. In most cases they can be distinguished morphologically, *H. apinnata* having few or no short, swollen bud-like side lobes, while *H. enteromorpha* typically has sparse to abundant bud-like side lobes; ambiguous cases must be separated with a P test. *Hypogymnia rugosa*, another large, dichotomously branched species, always grows appressed to the substrate, has a papery texture and wrinkled upper surface, and grows at middle elevations in the mountains. *Hypogymnia apinnata* grows either appressed or pendulous, has a more cartilaginous texture, and grows mainly at lower elevations.

Hypogymnia austerodes

Description: Thallus foliose, small to medium sized (mostly 2-6(12) cm broad), gray or greenish gray, often brownish, occasionally strongly brown; lower surface black; lobes contiguous, appressed, hollow, < 2 mm wide, imperforate; lobe interiors with dark floors and often white ceilings; short bud-like side lobes often present; soredia present, arising by gradual disintegration of the cortex, either on laminal warts or a roughening upper surface, producing fine to coarsely granular soredia; apothecia occasional; cortex K+Y; medulla K-, P- (or pale Y), KC+O (physodic acid, usually with 3-hydroxyphysodic acid).

Range: Alas to Mex, e Cas.

Substrate: Most often on bark and wood, occasionally on rock.

Habitat: Most common in subalpine and boreal forests in continental climates but found from low-elevation cool riparian forests to alpine.

Notes: Hypogymnia austerodes is characterized by the small brownish tinged thallus with laminal soredia and contiguous appressed lobes. *Hypogymnia oceanica* is seldom brownish, has soredia more toward the lobe tips, and is P+R. *Hypogymnia austerodes* has been frequently confused with two much rarer sorediate species, *H. bitteri* and *H. farinacea*. The former has small soralia that are terminal on short side lobes and occasionally on the main lobes. *Hypogymnia farinacea*, like *H. austerodes*, has laminal soredia but in *H. farinacea* these arise from wrinkles that burst open into flakes and granules. In *H. austerodes* the soredia arise from a gradual breakdown of the cortex on a warty or rough surface, producing granular soredia. Although the form of the soralia in *H. austerodes* is quite variable, as illustrated by three forms recently described as species (*H. dichroma, H. salsa, H. verruculosa*), DNA sequences in *H. austerodes* show a high level of uniformity worldwide, regardless of the form of the soredia.

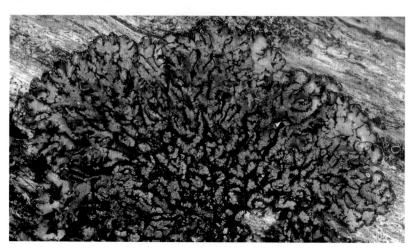

Hypogymnia canadensis

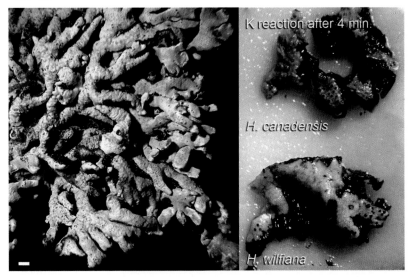

Description: Thallus foliose, small to medium sized, to 6(10) cm broad, pale gray or greenish gray but lobe tips brownish in exposed habitats; lower surface black; lobes contiguous to separate, appressed or shingled, hollow, 0.4-2.0 mm wide, imperforate; lobe interiors with dark floors and pale to dark brown ceilings; short bud-like side lobes occasional; soredia absent; apothecia common; cortex K+Y; medulla K+ slow reddish brown (3-hydroxyphysodic acid; see photo), P- (or pale Y), KC+O (physodic acid).

Range: Coastal se Alas to Ore Cas, mostly w Cas crest, inland to se BC.

Substrate: Bark and wood, mostly on conifers, rarely on rock.

Habitat: Cool, moist forests at low to mid elevations.

Notes: Small or poorly developed specimens of *H. canadensis* can easily be confused with *H. wilfiana* in the field. But well-developed specimens of *H. canadensis*, especially from the n part of its range, conspicuously differ from any other *Hypogymnia* species in w N Am. These specimens have lobes that are shingled in layers, rather than the contiguous, appressed lobes of *H. wilfiana*. In ambiguous cases, the presence of 3-hydroxyphysodic acid (K+ slow reddish brown) distinguishes *H. canadensis* from N Am specimens of *H. wilfiana*.

Hypogymnia duplicata

Description: Thallus foliose, medium sized to large (mostly 4-20(30) cm), whitish gray to pale greenish gray above; lobes nodulose, hollow, cascading in arcs, about 1 mm wide; lobe tips without holes but lower surface sometimes sparsely perforate; lobe interiors with a dark floor and often a white ceiling; lower surface black; soredia and isidia lacking; apothecia uncommon, brown; cortex K+Y; medulla K-, P+R, KC- (diffractaic, physodalic, and protocetraric acids).

Range: Alas to Ore, Coast Ranges and w slope Cas. At present not known s of Corvallis in the Coast Range nor s of Mt Hood in the Cas.

Substrate: Bark and wood of conifers.

Habitat: Cool moist coastal forests, low elevations to coastal ridges.

Notes: Readily recognized in the field by its distinctive cascading, arcuate lobes. Some drooping individuals of *H. inactiva* may be mistaken for *H. duplicata*, but *H. inactiva* is never P+R. The two other commonly pendulous species, *H. apinnata* and *H. enteromorpha*, both have broader lobes (about 2-5 mm) than *H. duplicata* (about 1 mm).

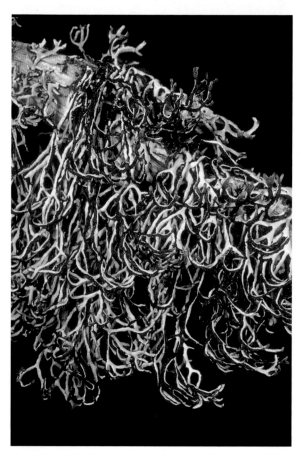

Hypogymnia enteromorpha

Description: Thallus foliose, medium sized (mostly 5-15 cm), whitish gray to pale greenish gray; lobes puffy, hollow, trailing or appressed, mostly 2-5 mm wide; often nodulose; lobe interiors dark; lobe tips swollen, often with a terminal hole; short perpendicular bud-like side lobes frequent or sparse; lower surface black; soredia and isidia lacking; apothecia common, brown; cortex K+Y; medulla K-, P+Y to R, KC+O (diffractaic, physodic, physodalic, and proto-cetraric acids).

Range: Alas to Cal, very common w Cas, uncommon to rare e to Cont Div.

Substrate: Wood and bark, mainly conifers.

Habitat: Moist low- to mid-elevation forests.

Notes: This species is very similar to *H. apinnata*, but *H. enteromorpha* usually has short, bud-like side lobes and is always P+, KC+. *Hypogymnia apinnata* usually lacks bud-like side lobes and is P- (or pale yellowish), KC-. *Hypogymnia enteromorpha* is the only member of this genus in the PNW that is both broad-lobed and P+R.

Hypogymnia heterophylla

Description: Thallus foliose, medium sized (mostly 3-10 cm), whitish gray to pale greenish gray; lobes hollow, erect (or somewhat appressed in the central part) to somewhat drooping, mostly 1-3 mm wide; lobe tips and axils often with holes; lobe interiors dark; narrow perpendicular side lobes usually frequent, occasionally sparse; lower surface black; soredia and isidia lacking; apothecia common; cortex K+Y; medulla K-, P+Y to R, KC+O (physodic, physodalic, and protocetraric acids).

Range: Coastal BC to Cal, seldom more than a few km inland. Abundant in Ore and Cal, rare in Wash and BC.

Substrate: Bark and wood, most often collected from *Pinus contorta* or *Picea sitchensis*, but occurring on hardwoods as well.

Habitat: Best developed in open conifer forests on tree and shrub branches, occupying both well-lit and shaded branches.

Notes: In its typical form, *H. heterophylla* is readily identified in the field once the slim perpendicular side lobes are recognized. Specimens with few such side lobes will require a P test to separate them from *H. inactiva* (which is P-). *H. imshaugii* is another common species with erect lobes, but it has white lobe interiors and is not usually found near the coast.

Hypogymnia hultenii

Description: Thallus foliose, small (generally < 2(5) cm diam); lobes < 1 mm wide, usually loosely appressed or extending off the substrate; upper surface pale gray to greenish gray; lower surface black or brown, thoroughly pock-marked with tiny pits (use lens); apothecia infrequent; soredia present; cortex K+Y, medulla K-, KC+R, P- (physodic acid).

Range: Alas to Cal, w Cas.

Substrate: Bark and wood, mainly conifers.

Habitat: Moist conifer forests at low to medium elevations.

Notes: This miniature *Hypogymnia* used to be placed in the genus *Cavernularia*. The minutely pitted lower surface is diagnostic for the species pair *H. hultenii* and *H. lophyrea*. *Hypogymnia lophyrea* is similar but lacks soredia and is more restricted to the immediate coast.

Hypogymnia imshaugii

Description: Thallus foliose, medium sized (mostly 2-8 cm), whitish gray to pale greenish gray, occasionally brownish or blackish in very exposed sites; lobes 1-2(3) mm wide, hollow (rarely solid), usually erect but appressed in harsh environments, imperforate; lobe interiors completely white (or the floor grayish where the black lower surface shows through the thin white medulla); perpendicular side lobes infrequent or lacking; lower surface black; soredia and isidia lacking; apothecia common; cortex K+Y; medulla K- or K+ brownish, rarely K+ slow reddish brown, P+Y to R or P-, KC+O (± physodic acid, diffractaic, physodalic, and protocetraric acids).

Range: BC to Mex, inland to Alta and Mont, common throughout the PNW except on the immediate coast, most common e Cas and w Cont Div.

Substrate: Bark and wood.

Habitat: Low-elevation to subalpine forests and savannas, occasionally in steppe (on fences and shrubs); in moist, dense forests, mostly restricted to the upper canopy.

Notes: Similar in general appearance to *H. imshaugii*, *H. inactiva* has dark lobe cavities and occasional holes in the lobe tips and axils. A wide variety of forms are presently included under *H. imshaugii*. Typical material has erect forking lobes and is P+R (physodalic and protocetraric acids). A common montane to subalpine form, particularly in exposed sites e Cas, has appressed lobes and is P- (physodic acid ± 3-hydroxyphysodic acid).

Hypogymnia inactiva

Description: Thallus foliose, medium sized (mostly 3-10(14) cm), whitish gray to pale greenish gray, occasionally brownish in very exposed sites; lobes mostly 1-3(4) mm wide, hollow, erect or often drooping in shaded sites; lobe tips and axils commonly with a hole; lobe interiors dusky or dark; perpendicular side lobes infrequent or lacking; lower surface black; soredia and isidia lacking; apothecia common; cortex K+Y; medulla K-, P-, KC+O (physodic acid).

Range: Alas to Cal, mainly w Cas, occasional inland to n Id, rare in w Mont.

Substrate: Bark and wood, on both conifers and hardwoods.

Habitat: Low- to mid-elevation moist conifer forests, more shade tolerant than most other *Hypogymnia* species.

Notes: Typical specimens with erect, forking lobes with dark interiors are easily recognized in the field. Drooping specimens are occasionally misidentified as *H. duplicata*, but that species is P+R. This close relative of *H. imshaugii* is one

of the most abundant *Hypogymnia* species w Cas, but it rapidly becomes hard to find on the e side. Students of *Hypogymnia* are frequently seen popping open lobes to check for the white interiors of *H. imshaugii* vs. *H. inactiva*. Another difference from *H. imshaugii* is that *H. inactiva* has sparse to abundant holes in the lobe tips and axils.

Hypogymnia lophyrea

Description: Thallus foliose, small (generally < 2(5) cm diam); lobes < 1 mm wide, usually loosely appressed or extending off the substrate; upper surface pale gray to greenish gray; lower surface black or brown, thoroughly pock-marked with tiny perforations (use lens); apothecia common; soredia and isidia lacking; cortex K+Y, medulla KC+ reddish.

Range: On the coast, Alas to Cal.

Substrate: Bark and wood, mainly conifers.

Habitat: Low- to mid-elevation conifer forests, fairly frequent on the immediate coast, becoming sporadic in the Coast Range. Most commonly found in semi-open coastal *Pinus contorta* and *Picea sitchensis* forests.

Notes: Similar to *H. hultenii* but lacking soredia and commonly fertile. These two species share a lower surface that is distinctly but finely pitted. They were formerly placed in the genus *Cavernularia*.

Hypogymnia occidentalis

Description: Thallus foliose, medium sized (mostly 2-8(15) cm), gray or greenish gray; lower surface black; lobes contiguous, appressed (rarely trailing off the substrate), hollow, < 3 mm wide, the tips puffy and often perforated with a hole; upper surface usually rugose in older parts of the thallus; lobe interiors with dark floors and ceilings; short bud-like side lobes present; soredia absent; apothecia common; cortex K+Y; medulla K-, P- (or pale Y), KC+O (physodic acid).

Range: Alas to Cal, inland to w Mont, rare e Cont Div.

Substrate: Bark and wood, usually on conifers, also on hardwoods.

Habitat: Moist low-elevation to subalpine forests; one of the most common *Hypogymnia* species in *Abies grandis*, moist *Pseudotsuga*, and *Thuja* forests e Cas and w RM; also common w Cas but less so than other *Hypogymnia* species (*H. apinnata, H. enteromorpha,* and *H. inactiva*).

Notes: Hypogymnia occidentalis is recognized in the field by its medium size, appressed, puffy lobes, and short bud-like side lobes. Its commonly rugose upper surface has resulted in frequent misidentifications as *H. rugosa*. That species, however, has regularly dichotomous branching, no bud-like side lobes, and a papery (rather than cartilaginous) texture. Small specimens of *H. enteromorpha* are very similar in form to *H. occidentalis*; these are separated by the P+ reaction of *H. enteromorpha*.

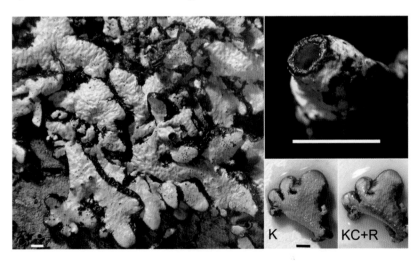

Hypogymnia oceanica

Description: Thallus foliose, small to medium sized (mostly 1-6 cm), gray or greenish gray; lower surface black; lobes not quite contiguous, appressed to slightly trailing, hollow, < 3 mm wide, puffy; lobe interiors with dark floors and ceilings; lobe tips often with a hole; short bud-like side lobes present; soredia present on the upper surfaces of the lobes, both on the tips of the lobes and sporadically back from the lobe tips; apothecia uncommon; cortex K+Y; medulla K-, P+R, KC+O (physodic, physodalic, and protocetraric acids).

Range: Common in coastal Alas, increasingly rare southward to s Ore, not known e Cas in Ore and Wash, rare in moist interior forests of BC.

Substrate: Bark and wood, mainly conifers.

Habitat: Moist coastal forests, including *Picea* and *Pinus contorta* forests on the immediate coast and *Pseudotsuga – Tsuga heterophylla* forests in the Coast Range and Cas; usually in small numbers.

Notes: This is the only local *Hypogymnia* that both is P+R and has soredia on the upper surfaces and tips of the lobes. *Hypogymnia tubulosa* has terminal soredia that are always restricted to the lobe tips and has erect forking branches, unlike the appressed lobes of *H. oceanica*. *Hypogymnia bitteri* and *H. austerodes* can have laminal soredia, but they live in more continental climates, are usually brownish, P-, and the lobes do not have the swollen, puffy appearance of *H. oceanica*. *Hypogymnia tuckerae* can be similar in appearance but is P- and imperforate or nearly so.

Hypogymnia physodes

Description: Thallus foliose, small to medium sized (mostly 2-6(10) cm), gray or greenish gray; lower surface black; lobes contiguous, appressed or with ascending tips, hollow, < 2(3) mm wide; lobe interiors with dark floors and usually white ceilings; short bud-like side lobes usually lacking; soredia on the inside of the lobe tips, which burst open into lip-shaped soralia; apothecia occasional; cortex K+Y; medulla K+ slow reddish brown, P+R, KC+O (physodic, 3-hydroxyphysodic, physodalic, and protocetraric acids).

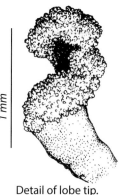

1 mm

Detail of lobe tip.

Range: Widespread, common throughout the PNW and n hemisphere.

Substrate: Bark and wood, occasionally on rock or mossy rock, rarely on tundra sod.

Habitat: Ubiquitous in forests at low to mid elevations, often in nonforested habitats (steppe, shrub thickets, farmlands, etc.), also common in urban and suburban areas.

Notes: The distinctive soredia lining the bursting lobe tips distinguish this species from all other local *Hypogymnia* species, except for *H. vittata*, a rare coastal species. However, *H. vittata* is P- and has characteristic slender adventitious side lobes that bear soredia.

Hypogymnia pulverata

Description: Thallus foliose, medium sized, to 10(15) cm diam, whitish to pale greenish gray; lower surface black; lobes contiguous or separate, usually appressed, less often suberect, mostly or completely solid, mostly < 3 mm wide; short bud-like side lobes sparse or lacking; soredia laminal, formed initially on edges of cracks or flakes of cortex; apothecia rare; cortex K+Y; medulla K+ slow reddish brown, P- or P+R, KC+O (physodic, 3-hydroxyphysodic, ± physodalic, and ± protocetraric acids)

Range: Alas to Ore on the immediate coast, very rare in the PNW, common in s hemisphere and e Asia.

Substrate: Bark and wood, especially conifers.

Habitat: In the PNW so far known only from *Picea sitchensis* forests on the immediate coast at low elevations.

Notes: This is the only *Hypogymnia* in the PNW with consistently solid lobes. *Hypogymnia imshaugii* is sometimes partly or wholly solid, but it lacks soredia. In the s hemisphere, where *H. pulverata* is much more common, both P+ and P- chemotypes are known. Very rare in N Am, *H. pulverata* is so far known from three areas, all coastal: Alas, Ore, and Quebec (Hudson Bay!). Another species, *H. subphysodes*, is also disjunct from the s hemisphere but very rare in the PNW. It is similar in overall appearance to *H. pulverata*, but *H. subphysodes* has hollow lobes with white interiors.

Hypogymnia rugosa

Description: Thallus foliose, medium sized to large (mostly 5-15(50) cm), whitish gray to pale greenish gray; lobes puffy, papery in texture when dry, hollow, appressed, mostly 2-6 mm wide, imperforate; upper surface wrinkled in the older parts; lobe interiors dark; short perpendicular bud-like side lobes lacking or sparse; lower surface black; soredia and isidia lacking; apothecia common; cortex K+Y; medulla K-, P- (sometimes P+O in cracks in lobe tips), KC+O (hypoprotocetraric acid).

Range: BC to Ore, inland to w Mont; uncommon except in *Abies* forests in the Cascades; rare e of 114°W longitude.

Substrate: Conifer bark.

Habitat: Moist mid-elevation to subalpine forests w of Cont Div; often in mixed *Abies* and *Tsuga mertensiana* forests, sporadically down into the *Pseudotsuga – Tsuga heterophylla* forests in cool sites.

Notes: With practice this large, puffy species is easily recognized in the field by its even dichotomous branching, large size, appressed habit, and papery texture. The species most often confused with *H. rugosa* is *H. occidentalis*, which is also usually appressed and has a rugose upper surface. That species, however, has short bud-like side lobes and lacks the even dichotomous branching of *H. rugosa*. Another large *Hypogymnia, H. apinnata*, occasionally has this regular dichotomous branching, but often shows some tendency to grow off of the substrate, has a more cartilaginous texture when dry, and usually does not have a wrinkled upper surface. Both *H. apinnata* and *H. occidentalis* have occasional to frequent perforations in the lobe tips and axils, while *H. rugosa* does not.

Hypogymnia tubulosa

Description: Thallus foliose, medium sized (mostly 1-6 cm), whitish gray to pale greenish gray, occasionally brownish near the lobe tips; lobes mostly 1-3 mm wide, hollow, erect, dichotomously branched, imperforate; lobe interiors white above and darkening below, or both ceilings and floors dark; perpendicular side lobes infrequent or lacking; lower surface black or brown; soredia present surrounding the swollen lobe tips; apothecia rare; cortex K+Y; medulla K+ slow reddish brown, P-, KC+O (physodic and 3-hydroxyphysodic acids).

Range: Widespread, common throughout the PNW.

Substrate: Bark and wood, rarely rock.

Habitat: Mostly in open or semi-open habitats at low to mid elevations, including riparian areas, farm trees, urban areas, savannas, and wooded wetlands; sporadic in closed forest.

Notes: The distinctive covering of soredia on the lobe tips is a characteristic and

reliable feature of *H. tubulosa.* Occasionally one will find an esorediate individual that might be confused with *H. inactiva* or *H. imshaugii.* Look for the incipient soredia on the lobe tips, which are at first slightly swollen, shinier and darker gray or browner than the rest of the thallus. Eventually these areas become dull, rough, and sorediate. *Hypogymnia oceanica* also has sorediate lobe tips but the soredia also extend back along the lobes. *Hypogymnia oceanica* is also differentiated by a more appressed habit and P+R medulla.

Hypogymnia wilfiana

Description: Thallus foliose, small to medium
sized (mostly 2-6(10)
cm), gray or greenish
gray; lower surface
black; lobes contigu-
ous, appressed, hollow,
< 2 mm wide, imperfo-
rate; lobe interiors with
dark floors and usually
white ceilings; short
bud-like side lobes
often present; sore-
dia absent; apothecia
common; cortex K+Y;
medulla K-, P- (or pale
Y), KC+O (physodic
acid).

Range: Alas to Cal,
inland to w Mont.

Substrate: Bark
and wood, most-
ly on conifers.

Habitat: Moist forests at low to mid elevations.

Notes: Distinguished by its small, appressed, contiguous lobes that lack soredia. *Hypogymnia canadensis* differs in its K+ slow reddish brown medulla and
imbricate lobes, rather than the K- medulla and contiguous lobes of *H. wilfiana*. Small individuals of *H. physodes* that have not yet developed soredia can
be confused with *H. wilfiana*, but *H. physodes* is P+R. Also check the underside of the lobe tips carefully for incipient soredia. Unusually small forms of
H. occidentalis can be similar to *H. wilfiana*; however, *H. occidentalis* has dark
ceilings of the lobe cavities and at least some perforate lobe tips.

Hypotrachyna

Thallus foliose, small to medium, mostly 1-4(10) cm wide, loosely appressed or with ascending lobe tips; lobes narrow, 0.5-2(4) mm wide; upper surface greenish, yellowish green, pale greenish gray, or grayish; lower surface black except brownish at the edges, densely rhizinate; rhizines dichotomously branched; soredia present (our species); apothecia rare; spot tests various, the thallus containing various substances including atranorin, usnic, gyrophoric, salacinic, and/or stictic acids.

Sources include: Hale (1975), Lendemer (2006a), Lendemer and Allen (2020), McCune (1998).

ID tips: Our three species are easily distinguished from each other, both morphologically and chemically. In very shaded environments, however, the characteristic yellowish usnic acid tinge of *H. sinuosa* may be very weak and cause confusion with the other two species.

1a Upper surface yellowish green or greenish; medulla K+Y to R, cortex KC+Y (usnic acid)..*H. sinuosa*
1b Upper surface gray, whitish gray, or bluish gray; medulla K- or K+Y to R, cortex KC-
  2a Rhizines dense, richly branched; soralia subterminal and terminal, powdery; medulla P+O or R, K+Y to R, C-, KC- .... *H. riparia*
  2b Rhizines sparse to moderately dense, sparsely branched; soralia laminal and marginal, becoming broad and diffuse, coarse; medulla P-, K-, C+R, KC+R..*H. afrorevoluta*

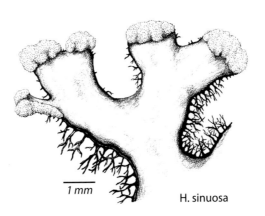

H. sinuosa

Hypotrachyna afrorevoluta

Description: Thallus foliose, medium sized (mostly < 8 cm diam), loosely appressed throughout; upper surface whitish gray to bluish gray; lobes to 4 mm broad, often downrolled at the edges, the tips broadly rounded; lobe axils often nearly circular; soralia laminal to subterminal, sometimes powdery but more often granular or pustular; rhizines simple to sparsely branched; apothecia and pycnidia not seen; cortex K+Y (atranorin), KC-; medulla K-, P-, C+R, KC+R (gyrophoric acid, 5-O-methylhiascic acid, and 4,5-di-O-methylhiascic acid), UV-.

Range: Rare on the w coast of N Am; coastal BC to Cal, rarely inland to w slope Cas.

Substrate: Trees (mainly) and rocks, on both conifers and hardwoods.

Habitat: Coastal and riparian forests.

Notes: Hypotrachyna revoluta is a closely related species that occurs along with *H. afrorevoluta* in e N Am, though the name *H. revoluta* was used more broadly in the past to include specimens from w N Am. *Hypotrachyna revoluta* produces soralia on the upper surface of the secondary lobes near the tips, while *H. afrorevoluta* produces laminal pustular soralia. Furthermore, in *H. revoluta* the secondary lobes are strongly ascending, giving the thallus a subfruticose appearance, while lobes of *H. afrorevoluta* are rather adnate throughout.

rhizines

Hypotrachyna riparia

Description: Thallus foliose, medium sized (mostly 3-8 cm), loosely appressed; upper surface whitish gray to mineral gray or bluish gray; lobes 1.5-5 mm broad, often downrolled just behind the lobe tips; lobe tips becoming hooded, often somewhat reflexed; soralia subterminal to capitate, the soredia powdery, on the upper surface of the lobes; rhizines richly dichotomously branched, to 0.5 mm long, often visible in top view on young, appressed lobes; apothecia and pycnidia not seen; cortex K+Y (atranorin), KC-; medulla K+Y to O, P+O (salacinic acid), C-, KC-, UV-.

Range: Foothills of Cas in w Ore; so far known from only a handful of sites in Lane Co. and Linn Co.; apparently a PNW endemic.

Substrate: Shrubs and hardwood trees

Habitat: Riparian forests and ash swamps with abundant cyanolichens. The known sites have wet to thinly flooded ground throughout the winter months.

Notes: Hypotrachyna riparia is distinguished from all of the other local gray parmelioid lichens by its dense, dichotomously branched rhizines. It is, however, easily overlooked because of its similarity in color and size to the ubiquitous *Parmelia sulcata. Hypotrachyna riparia* is similar in color to *H. afrorevoluta*, but that species has more sparsely branched rhizines, laminal to subterminal rather than terminal soralia, and has a medulla C+R, KC+R, K-.

Hypotrachyna sinuosa

Description: Thallus foliose, small (mostly 1-5 cm), loosely appressed or with lobe tips ascending; upper surface yellowish green or greenish; soralia roundish to somewhat elongate, on the lobe tips; rhizines dichotomously branched; apothecia rare; disk brown; cortex K-, KC+Y (usnic acid); medulla K+Y to R, P+O (salacinic ± norstictic and stictic acids), C-, KC-, UV-.

Range: Alas to Cal, frequent w Cas.

Substrate: Most frequent on *Alnus rubra*, but also on other hardwoods, occasional on conifers.

Habitat: Most common in moist riparian forests at low elevations, usually below 1000 m, from the coast to valleys and foothills in the Cas; n into BC and Alas; in more northern areas the species is increasingly restricted to near the coast.

Notes: Frequently overlooked by beginners, this species looks at a glance like a small, off-color *Hypogymnia* or *Parmelia*. However, the yellow-green tinge; flat, thin lobes tipped with soralia; and lobes thickly covered below with forked rhizines are diagnostic. Shade forms that are barely yellowish can be similar to *Parmelia*, but the upper surface lacks the whitish markings (pseudocyphellae) of *Parmelia*.

Imshaugia

Thallus foliose, whitish to gray or gray green, to 5 cm wide, appressed; lobes narrow, 0.5-2 mm wide; lower surface whitish to pale brown, with simple rhizines; isidia present or not; apothecia brown, with a lecanorine margin; cortex and medulla K+Y, P+ deep Y or O, UV- (atranorin and thamnolic acid); on bark and wood (rarely rock).

Sources include: Meyer (1985).

ID tips: Imshaugia appears like a robust *Physcia*, but is more closely related to *Platismatia* and other Parmeliaceae, though it looks nothing like *Platismatia*. Infrequent with us, but fairly common elsewhere, *Imshaugia* is mostly peripheral to the PNW.

1a Apothecia generally present; isidia lacking. Mainly e US, to c and s RM, to N Dak and e Mont; on bark and wood .... [*I. placorodia*]
1b Apothecia generally lacking; isidia present, laminal, thin and cylindrical. Widespread; on bark and wood (mainly conifers), rarely on rock; uncommon in Wash, Ore, and Cal .... *I. aleurites*

Lasallia—*see* Umbilicaria

Leioderma

One species in N Am.

Leioderma sorediatum

Description: Thallus foliose, small, mostly 0.5-3 cm diam; upper surface gray, with a thick matted tomentum; lower surface whitish, veinless, sparsely rhizinate; lobes 2-4 mm wide; photobiont blue green; soredia grayish, marginal, elongate; apothecia not seen; spot tests negative.

Range: S Ore to BC, always within a few km of the ocean; rare.

Substrate: Most frequent on a thin layer of bryophytes over stems of ericaceous shrubs (*Vaccinium, Rhododendron*); also on conifers.

Habitat: Semi-open habitats on the coast, most often in dune woodlands and deflation plains.

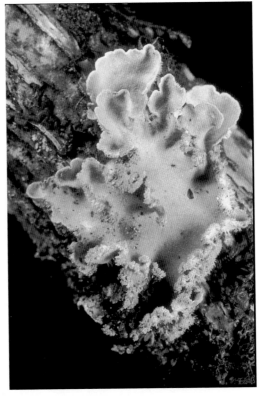

Notes: At a glance one would overlook this species as stunted *Peltigera collina*. But close inspection of *L. sorediatum* shows a lack of veins below and presence of the characteristic matted tomentum of the upper surface. *Erioderma sorediatum* is very similar in appearance to *L. sorediatum* (and occurs with that species), but has distinctly erect tomentum on the upper surface and a P+O medulla.

Sources include: Galloway and Jørgensen (1987), Glavich et al. (2005a), McCune et al. (1997).

Lempholemma

Thallus gelatinous (not stratified), polymorphic, ours dwarf fruticose or warty-foliose, black; cortex lacking; apothecia lecanorine, with sunken disks; spores nonseptate, spherical to ellipsoid; spot tests negative; photobiont blue green (*Nostoc*); on ± calcareous soil, mosses, or rock.

Sources include: Henssen (1968), Schultz (2004), Schultz and Büdel (2002).

ID tips: The dwarf fruticose species can be confused with *Leciophysma, Lichinella,* and *Peccania* (see keys in McCune 2017). Foliose species are diffi-cult to distinguish from *Collema* unless fertile.

1a Thallus with a broadly expanded foliose base with knobby fruiting warts. Spores (7)10-18(20) × 7.5-12.5(15) μm; on mosses over rock in sheltered sites; widespread but rare; BC to Cal, w Mont, and Ariz ....*L. polyanthes* (*L. chalazanum* is a similar species with much larger spores (17)20-33 × 9-13 μm, known from sw N Am but not yet known from PNW.)
1b Thallus dwarf fruticose
    2a Lobes straplike and channeled, longitudinally ridged. Lobes 0.2-0.5(0.7) mm diam, with clusters of isidia in the center; widespread but rare; reported from coastal BC and Id .... [*L. radiatum*]
    2b Lobes cylindrical, not longitudinally ridged, but finely striate
        3a Branches often tipped with swellings (hormocystangia) producing vegetative propagules (hormocysts) with a few cyanobacterial cells and fungal hyphae; spores > 15 μm diam. Thallus a sprawling mat or cushion of fingerlike lobes; lobes 0.1-0.2 mm diam and 1-2(5) mm long; apothecia usually few or absent, terminal or lateral, to 0.4 mm diam; disk black; thalline margin thick, smooth, persistent; spores subspherical, 15-20 μm diam; on calcareous rock........... *L. cladodes*
        3b Lobe tips not swollen and not producing hormocystangia; spores < 15 μm diam. Reported from BC and Id .... *L. intricatum*

Leptochidium

Leptochidium albociliatum

Description: Thallus foliose, gelatinous but with a thin whitish middle layer; upper and lower surfaces greenish black to slate gray to brownish black; lobes 1-5 mm wide, with small but conspicuous (use hand lens) hyaline hairs abundant on lobe margins and apothecial margins, occasionally on the upper surface and isidia; isidia granular to squamule-like, occasionally cylindrical; upper and lower cortex present (*Leptogium*-like); photobiont blue green; apothecia lecanorine, fairly common; spores 2-celled; spot tests negative.

Range: Widespread in w N Am, common e Cas, less often w Cas.

Substrate: Moss, soil, or moss over rock.

Habitat: Somewhat shady to open habitats, mostly in outcrop and talus areas, occasional in steppe, low to mid elevations.

Notes: In the past this species has often been confused with hairy species of *Leptogium*. Although the spores provide a positive separation between *Leptogium* and *Leptochidium* (many-celled vs. 2-celled, respectively), they can also be separated even when sterile by breaking a lobe and inspecting the broken edge with a good hand lens. *Leptochidium* has a thin, pale middle layer, while the broken edge will be completely dark in *Leptogium*.

L. saturninum   L. compactum

Group 2

Aquatic or semi-aquatic

9a Thallus < 0.5 cm broad; nearly crustose with lobes small and indistinct; spores 28-40 × 10-16 μm. Thallus thin (< 100 mm), crustose, filmy, partly granular or warty, sometimes the warts elongating into fingerlike lobes, dark olive to dark brown or gray brown; interior parenchyma-like throughout; apothecia mostly globose, brownish, rim ± thalline, to 0.5 mm diam; known from seacoast sandstone and shale in Oregon, but questionable records; to be sought on calcareous rock in or near fresh water .... *Scytinium aquale*

9b Thallus 0.5-8 cm broad, appressed foliose with lobes distinct and usually 0.2-0.7 mm broad; spores 16-33 × 5-9 μm .... ***Scytinium rivale***

Group 3

Thallus interior parenchyma-like or compactly interwoven

10a Thallus interior with elongate compactly interwoven hyphae, many cells with length to width ratio of 3:1 or longer

11a Substrate usually soil or moss over soil or rock; isidia usually lacking. Thallus foliose or foliose with erect lobules; primary lobes flattened, in cross section 80-120 μm thick, usually < 2 mm broad; upper surface rather smooth; main lobes sometimes membranous in appearance; apothecia common, sessile, to 1.5 mm diam; spores 20-35 × 8-12 μm; widespread in N Am; continental to suboceanic climates, lowland to subalpine, including steppe, both sides of the Cont Div
.... *Scytinium intermedium*

11b Substrate bark and wood; isidia present. Thallus foliose; lobes 2-10 mm broad; interior fungal cells ± isodiametric, with some slightly elongate cells in chains connecting the upper and lower cortex; thallus usually < 3 cm diam; sw Ore and n Cal .... *Scytinium siskiyouense*

10b Thallus interior of ± isodiametric cells (length to width mostly to about 2:1), sometimes with isolated longer cells; substrate various; isidia various

12a Thallus fruticose, sometimes minutely so, OR thallus foliose but
   developing elongate, erect, cushion-forming lobes
   13a On bark or wood..*Scytinium teretiusculum*
   13b On soil, mossy soil, rock, or mossy rock
      14a Thallus wholly of minute coralloid branches, lacking expanded
         lobes
         15a Apothecia lacking thalline margin; spores 1-septate; thallus
            minutely shrubby, of terete to subterete, erect to suberect lobes.
            Common on moss over rock, occasional on soil in steppe
            .... *Polychidium muscicola*
         15b Apothecia with thalline exciple; spores muriform; thallus of
            horizontal sprays of linear, subterete to terete lobes that are ±
            free from the substrate, with a delicate, lacy appearance; on
            moss over soil or rock in cool, montane sites; so far known
            from n Id and Cas........ shade form of *Scytinium nanum* (broad
            sense) (see Lead 18a below)
      14b Thallus with expanded lobes that bear coralloid marginal or
         laminal proliferations, sometimes the coralloid structures nearly
         obscuring the broader primary lobes
         16a Thallus mostly 1-2 cm diam, forming a dense cushion of
            coralloid lobes. Lobes flat to terete, < 0.2 mm broad and 65-80
            µm thick, forming a compact cushion on top of broader lobes
            (to 2 mm wide); apothecia common, to 1.5(3) mm diam,
            becoming deeply concave and surrounded by the coralloid
            cushion; thalline margin entire or lobulate; spores 30-40 ×
            12-14 µm; seldom collected, commonly misapplied; in our area
            montane to alpine, Cas and RM ....*Scytinium tenuissimum*
            (Other small, poorly known *Scytinium* species frequent lower
            elevations in the western states.)
         16b Thallus mostly < 2 mm diam (but forming larger colonies),
            with ascending primary lobes that are dissected and isidiate,
            mainly on the margins .... *Scytinium nanum* (Lead 18a)
12b Thallus foliose
   17a On soil or rock; thallus with very narrow (< 0.2 mm) coralloid or
      granular outgrowths from broader (to 2 mm) spreading lobes
      18a Thallus forming minute, appressed rosettes about 0.5-2 mm
         broad, sometimes with granules or isidia; lobes to 0.5 mm broad;
         on soft rock or soil. Thallus blue-gray to brown black, 65-100 µm
         thick; apothecia with a thalline exciple; spores 20-40 × 10-15 µm,
         with 4-5 transverse septa; fairly common but rarely collected, on
         soil and moss, especially on steep, moist roadcuts and trailcuts,
         in oceanic to suboceanic climates, s Cal n to Wash and n RM
         .... *Scytinium nanum*
      18b Thallus with coralloid outgrowths, developing into a minutely
         fruticose thallus; lobes to 2 mm broad; on moss and soil
         ....*Scytinium tenuissimum* (Lead 16a)
   17b On bark or wood; lobes 0.1-0.6 mm broad

19a On bark of hardwoods; apothecia ± flattened, sessile or partially sunken in the thallus; spores (22)27-34(43) × 12-17 μm. Thallus rosettiform to irregular, often > 2 mm diam; lobes 0.1-0.6(1) mm broad; marginal isidia, knobs, or lobules often present; thallus interior sometimes with scattered elongate cells with length to width over 2:1; on bark of hardwoods, including *Populus*, *Acer*, and *Quercus*; coastal Alas to n Cal, inland to s BC and n Id and w Mont.. *Scytinium cellulosum*
19b On rotten wood, stumps, and plant debris, rarely on standing trees; apothecia globose; spores 20-31 × 10-12 μm. Lobes forming small rosettes radiating horizontally from the apothecia, flat to cylindrical but not distinctly knobby, 0.1-0.3 mm wide; Alas to BC, not known with certainty from Ore and Wash .... [*Scytinium subtile*]

Group 4

Thallus interior with elongate, loose hyphae

20a Thallus smooth to occasionally faintly textured, bluish gray to gray; isidia present, fine and cylindrical...***L. cyanescens***
20b Thallus distinctly wrinkled, gray to brown, dark brown, or black; isidia, if present, granular, cylindrical, clavate, or lobulate
  21a Thallus medium to large, of erect lobes generally > 1 cm tall; lobe margins curling inward, especially at the apices, to form rolled tips ....***Scytinium palmatum***
  21b Thallus various, but lobe tips not tubular-inrolled
    22a Substrate calcareous rock or soil; lobes < 2 mm wide; cortical cells jigsaw-like (LM, surface view); upper surface strongly wrinkled
      23a Thallus truly fruticose. Lobes 1-4 mm tall, 0.3-1 mm thick, simple or moderately branched, irregularly thickened or angular in cross section; lobe tips thickened; on calcareous soil or moss over soil in dry habitats; rare, e Cas.................*Scytinium schraderi*
      23b Thallus cushion-forming foliose. Thallus forming rosettes mostly < 5 mm diam; lobes mostly < 2 mm wide, thick, swollen, often folded or twisted; upper surface with laminal granular isidia; apothecia not yet seen in our material, but with a swollen thalline margin; spores 3-septate to muriform, 27-32 × 10-12 μm; Columbia Basin steppe and RM..............*Scytinium plicatile* group (This group includes the similar but smaller *S. turgidum*.)
    22b Substrate various; lobe width various; cortical cells roundish to polygonal; upper surface strongly wrinkled or not
      24a Thallus lacking isidia and lobules; apothecia present. Thallus small (to about 2 cm diam), roundish, not much divided into lobes, or lobes short and broadly rounded; apothecia always numerous, partially sunken in the thallus; spores 4 per ascus .... ***Scytinium polycarpum***
      24b Thallus isidiate, lobulate, or sorediate

25a Thallus sorediate or with soredia-like granules

26a Thallus thick, with coarse granules or isidia, the propagules irregular in shape, mainly laminal; thallus thick, strongly wrinkled, expanding greatly (to about 500 µm thick) and becoming thick and rubbery when moist; strictly coastal .... *Gabura insignis*

26b Thallus thin, with marginal to submarginal clusters of granules; rare, Santa Cruz Co. to Del Norte Co., Cal .... *S. singulare*

25b Thallus with granular to cylindrical isidia, or becoming lobulate

27a Propagules mainly marginal; thallus cushion-forming foliose

28a Lobe tips curled downward, but away from the tips the margins are often inrolled; marginal propagules blunt, stubby lobules to long dichotomously dissected and narrowly strap shaped; upper surface wrinkled but not commonly reticulate-ridged; abundant in the PNW ....*Scytinium californicum*

28b Lobe tips less often downturned; marginal propagules slender, fimbriate to becoming very narrowly dissected; upper surface of expanded lobes developing a network of sharp ridges; occasional in the PNW ....*Scytinium pulvinatum*

27b Propagules mainly laminal or equally marginal and laminal; thallus rather flat to slightly cushion forming

29a Propagules cylindrical isidia, becoming long and slender (to about 1 mm × 0.35-0.1 mm) and coralloid, sometimes so dense as to obscure the foliose part; habitat oceanic forests, mainly on trees and shrubs .... *Scytinium tacomae*

29b Propagules isidia or lobules, but not long and coralloid

30a Propagules fingerlike to clavate isidia, often like a dimpled or deflated balloon, shiny, very dark, laminal and marginal; habitat dry. Lobes weakly wrinkled, 2-3 mm wide, thin (0.10-0.15 mm); lobe margins dentate; on soil or mosses over soil or rock in exposed sites; se BC to n Cal, e to Mont..........*Scytinium subaridum*

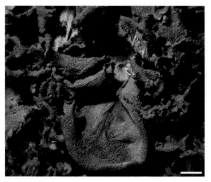

30b Propagules lobules, sparse to dense, laminal and often partly marginal. Thallus thick, (100)150-250(500) µm, thinner at the margins; lobes heavily wrinkled, 1-6 mm wide, the margins frequently downrolled; spores 35-48 × 9-16 µm; BC to Cal, especially common in n Cal ....*Scytinium platynum*

Leptogium cyanescens

Description: Thallus gelatinous (non-stratified), foliose, small to medium sized, to 5 cm diam; lobes mostly 2-4(12) mm wide, broadly rounded at the tips, 35-110 μm thick when moist, bluish gray to dark gray (not brown); lobe surface smooth, without definite wrinkles but sometimes delicately textured; isidia usually abundant, cylindrical to clavate or sometimes flattened and lobulate, laminal; interior of closely spaced interwoven hyphae forming a distinct crisscross pattern in cross section (LM; see photo); apothecia unknown in the PNW; spot tests negative.

Range: Alas to Ore w Cas, usually near the coast; rare in the PNW but widespread in the n hemisphere.

Substrate: On hardwood and conifer bark, rotten logs, and rocks.

Habitat: Moist forests, usually near water (creeks, rivers, or the ocean).

Notes: This species is included here, even though it is rare in the PNW, because it is common elsewhere in the n hemisphere and because the name has been frequently misapplied to other species, including *Scytinium tacomae*. *Leptogium cyanescens* has a thallus cross section that is distinctive among our species of *Leptogium*. In contrast to the loosely tangled hyphae in most of our larger *Leptogium* species, the hyphae of *L. cyanescens* form a crisscross pattern. Even without examining a thin section, *L. cyanescens* should be recognizable by its gray thallus that is not at all brown and its smooth, broadly rounded, isidiate lobes.

5 μm

Leptogium pseudofurfuraceum

Description: Thallus gelatinous (non-stratified), medium sized, mostly 2-4 cm broad; upper surface gray to brown or dark brown, distinctly but finely wrinkled, the wrinkles extending out the lobes then concentric on the lobe tips; lower surface densely white tomentose; lobes mostly 3-8 mm wide with broadly rounded tips; isidia generally abundant, laminal and marginal, clavate to coralloid; apothecia rare; spot tests negative.

Range: Widespread in w N Am, NWT to Mex, inland to S Dak but rather rare except for s Ore to s Cal and Colo to Mex; in our area w Cas and in the Columbia Gorge.

Substrate: Bark, most often hardwood trees and shrubs, occasionally rock.

Habitat: Somewhat dry savannas, woodlands, forests, and outcrop areas.

Notes: Distinguished from most other *Leptogium* by its usually brownish gray coloration, heavily but finely wrinkled upper surface, and presence of isidia, this species appears more like a robust *Scytinium* than a *Leptogium*. *Leptogium pseudofurfuraceum* can be distinguished from *Scytinium*, however, by the dense, fine white tomentum on the lower surface, while *Scytinium* has at most only a few tufts of hairs. The distinct fine wrinkles on the upper surface of *L. pseudofurfuraceum* contrast with the smooth surface of *L. saturninum*. *Leptochidium albociliatum* is easily distinguished by its abundantly but finely hairy lobe margins, which are lacking on *Leptogium pseudofurfuraceum* and *L. saturninum*. *Leptogium pseudofurfuraceum* is morphologically indistinguishable and genetically very close to the European species *L. furfuraceum*.

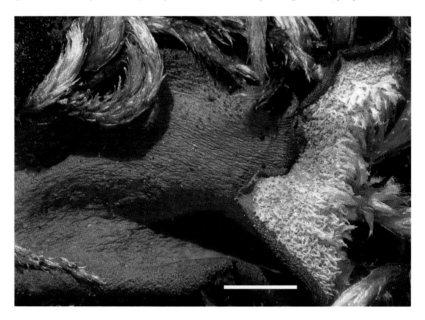

Leptogium saturninum

1 cm

Description: Thallus gelatinous (non-stratified), mostly 2-10(20) cm broad; upper surface blackish gray or black, smooth or slightly wrinkled; lower surface densely white tomentose; isidia generally abundant, granular, becoming globular or compound granular; apothecia rare; spot tests negative.

Range: Widespread.

Substrate: Usually on bark, most commonly deciduous trees and shrubs, especially *Populus trichocarpa,* occasionally on rock or moss over rock.

Habitat: Most frequent in moist riparian forests at low elevations.

Notes: This is the most common of our gelatinous lichens with a whitish-hairy lower surface. *Leptogium pseudofurfuraceum* is less common and has distinct fine wrinkles on the upper surface, in contrast to the smooth surface of *L. saturninum. Leptogium compactum* is lead gray in color and has little space between the hyphae in the thallus interior, while *L. saturninum* is blackish and has spaces between the hyphae. *Leptogium stancookii* is similar in color and size to *L. saturninum,* but has elongate, rather cylindrical, branched isidia. *Leptochidium albociliatum* has abundant hairs on the lobe margins, while the hairs on *L. saturninum, L. stancookii,* and *L. compactum* are usually restricted to the lower surface.

Letharia

Thallus fruticose, brilliant fluorescent yellow green, chartreuse, or yellow; tufted, richly branched, to 15 cm broad; branches solid or with a loosely filled center, usually wrinkled, roundish to irregular in cross section; medulla white; isidia or soredia present or not; apothecia lecanorine, the disk brown or dark brown; dark pycnidia common; spot tests negative.

Sources include: Altermann et al. (2016), Kroken and Taylor (2001), McCune and Altermann (2009), Thomson (1969).

ID tips: This easy genus contains great challenges at the species level. Kroken and Taylor (2001) detected six lineages of *Letharia* in N Am, some of them cryptic—morphologically close to or indistinguishable from one another—but reproductively isolated according to DNA evidence. Putative species not yet formally described are omitted here. The species-level relationships are complex. For example, even though *L. vulpina* and *L. lupina* cannot be distinguished without genetic data, they are not closest relatives, with *L. lupina* closer to one of the fertile lineages (Altermann et al. 2016).

1a Soredia or isidia mixed with soredia present; apothecia rare
.... ***L. vulpina*** (including *L. lupina*)
1b Soredia lacking; isidia sparse or lacking; apothecia usually present
   2a Branches smooth, nearly cylindrical. Apothecia often present, isidia
      sometimes present but sparse; Siskiyou Mts and Sierra Nevada; Cal
      and sw Ore; uncommon or rare...***L. gracilis***
   2b Branches wrinkled, ridged, and with flat spots or depressions between
      the ridges...***L. columbiana***

Letharia columbiana

Description: Thallus fruticose, brilliant fluorescent yellow green or chartreuse, tufted, richly branched, mostly 2-7(15) cm broad; branches solid or with a loosely filled center, wrinkled, irregular in cross section; medulla white; isidia and soredia lacking; apothecia common, the disk brown or dark brown; spot tests negative but disk often containing norstictic acid by TLC.

Range: Widespread throughout the PNW but avoiding the immediate coast.

Substrate: Bark or wood, rarely rock.

Habitat: Common in subalpine forests, high plateaus and ridges to timberline, occasional in low-elevation forests.

Notes: Sometimes occurring with *L. vulpina*, *L. columbiana* is essentially identical in form except that it lacks soredia and isidia. Although recreationists commonly bring home a sample of this stunning lichen on a picturesque branch, indoors the color soon fades to a dull yellow. If placed outside the home, it is also likely to die (usually due to excess moisture). *Letharia* is best enjoyed in place.

Letharia gracilis

Description: Thallus fruticose, brilliant fluorescent yellow green or chartreuse, suberect to drooping or subpendent, to 12 cm long or more, sparsely to moderately branched; branches solid or with a loosely filled center, nearly cylindrical or somewhat flattened or irregular in cross section, nearly smooth to slightly foveolate or ridged and fissured, 0.5–2.5(3.0) mm diameter; medulla white; isidia sparse or lacking; soredia lacking; apothecia common, the disk brown or dark brown; spot tests negative but disk often containing norstictic acid by TLC.

Range: Siskiyou Mts (Cal and Ore) and Sierra Nevada.

Substrate: Bark or wood.

Habitat: Uncommon in montane to subalpine forests.

Notes: Letharia gracilis is distinguished by its relatively sparse branching, absence of soredia and absence or near absence of isidia, and relatively smooth, only slightly foveolate branches, in contrast to the rich branching and irregularly ridged branches of other *Letharia* in N Am.

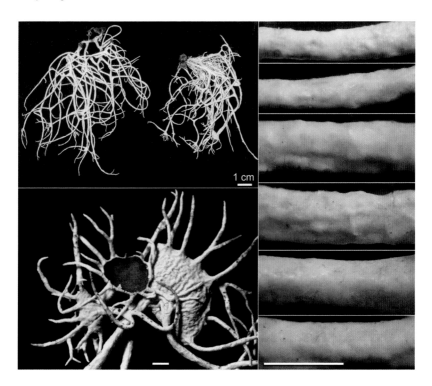

7a Thallus (when dry) pale greenish to olive or brownish (cortex
lacking usnic acid), strongly and coarsely reticulate-ridged;
photobiont green, with cyanobacteria in internal cephalodia;
medulla KC- ....***L. pulmonaria***
7b Thallus yellowish-tinged pale green to olive when dry (cortex
with usnic acid), weakly ridged or undulating; photobiont blue
green; medulla KC+R..***L. scrobiculata***

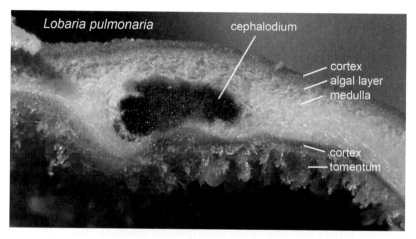

Lobaria pulmonaria cephalodium cortex algal layer medulla cortex tomentum

Lobaria pulmonaria

Lobaria anomala

Description: Thallus foliose, loosely appressed, free at the edges, to 20(40) cm broad; lobes mostly 1-3 cm broad; upper surface light or dark brown, with a network of ridges; lower surface covered with light brown tomentum with scattered white spots (pseudocyphellae); photobiont blue green; soralia roundish to irregular, white, gray, or blue-gray, mainly on the ridges; apothecia brown, uncommon; medulla white; cortex K-; medulla K+Y, P+O (stictic acid group and terpenoids) or spot tests negative (just terpenoids).

Range: Alas to Cal, w Cas, with rare inland disjuncts to w Mont.

Substrate: Bark and wood, most often on deciduous trees and shrubs, occasional on conifers; rarely on rock.

Habitat: Low- to mid-elevation moist forests, including riparian areas, Willamette Valley hardwood forests (including oak savannas and ash swamps), and sporadically in mountain conifer forests.

Notes: Almost identical to *L. anthraspis*, *L. anomala* differs mainly in mode of reproduction and ecology. *Lobaria anomala* is sorediate while *L. anthraspis* lacks soredia and is more frequently fertile. *Lobaria anthraspis* is less common overall and more of a riparian species. Both species were formerly included in *Pseudocyphellaria* because of abundant pseudocyphellae on the lower surface, but molecular data have revealed that they belong to *Lobaria* rather than *Pseudocyphellaria*. Blackish apothecia lacking a thalline margin are commonly seen, but these are produced by a parasitic fungus, *Plectocarpon lichenum*.

Lobaria anthraspis

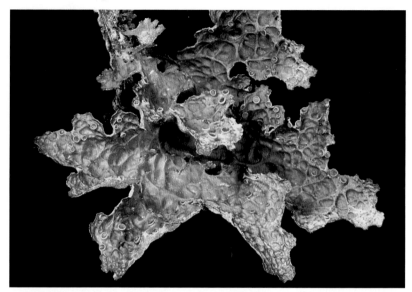

Description: Thallus foliose, loosely appressed, free at the edges, to 20(40) cm broad; lobes mostly 1-3 cm broad; upper surface light or dark brown, with a network of ridges; lower surface covered with light brown tomentum with scattered white spots (pseudocyphellae); photobiont blue green; soredia and isidia lacking; apothecia brown, common; medulla white; cortex K-; medulla K+Y, P+O (stictic acid group and terpenoids) or spot tests negative (just terpenoids).

Range: Alas to Cal, with rare disjuncts inland to n Id.

Substrate: Bark and wood, on conifers, deciduous trees, and shrubs; rarely on rock.

Habitat: Low- to mid-elevation moist forests, especially riparian areas; most frequent near the coast in partially open habitats.

Notes: P- and P+ chemotypes have been reported for both *L. anthraspis* and *L. anomala.* See additional notes under *L. anomala.*

Lobaria hallii

Description: Thallus foliose, mostly
2-12(20) cm wide, single lobed or with
several broadly rounded lobes; lobes to 5
cm broad; upper surface gray to brownish
gray; lower surface pale brown or whitish,
tomentose; upper cortex rough and scabrous, finely but usually sparsely tomentose
(especially toward lobe tips—use hand
lens); photobiont blue green; soredia pres-

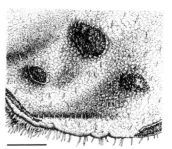

1 mm Detail of hairs at lobe tip

ent in roundish laminal and marginal soralia; apothecia uncommon; spot
tests negative except cortex K+Y, containing unknown substances.

Range: Alas to n Cal, e to near Cont Div in w Mont.

Substrate: Bark and wood, both conifers and hardwoods, e Cas mainly on
Populus trichocarpa.

Habitat: Riparian forests, usually where sheltered and moist.

Notes: This species is regionally rare but locally abundant on *Populus* bark in
moist lowland riparian areas between the Cas and RM crests. W Cas it occurs
sporadically on hardwoods in low- to mid-elevation riparian forests. When
wet, it is difficult to separate from *L. scrobiculata*, which has a distinct yellowish
cast when dry, in contrast to the steel gray color of *L. hallii*. Ambiguous cases
can be resolved by looking for the fine scattered hairs on the lobe tips of *L.
hallii*. In the lab, the negative spot tests of the medulla of *L. hallii* are diagnostic.

Lobaria linita

Description: Thallus foliose, medium to large, mostly 5-15 cm broad; lobes generally 1-3 cm broad; upper surface greenish to brownish or brownish gray, with or without a network of ridges; soredia, isidia, and lobules lacking; photobiont green but with blue-green photobiont in cephalodia; apothecia fairly common; spot tests negative, containing tenuiorin, methyl evernate, and methyl gyrophorate.

Range: Arctic to nw Ore, inland to w Mont.

Substrate: On trees, shrubs, mossy rocks, or alpine sod.

Habitat: Found in two distinct habitats: (1) Montane to alpine, e Cas, usually on alpine sod or mossy rocks. (2) Epiphytic in coastal Alas to w Ore, in and w Cas. Both habitats are moist with coastal influence. *Lobaria linita* greatly increases in abundance north of Washington in BC and se Alas.

Notes: Sun-browned forms on rocks or alpine sod appear very different from the epiphytic forms from coastal forests. The latter are strongly reminiscent of *L. pulmonaria*, and are easily mistaken for that if you don't look for soredia. Occasionally juvenile *L. pulmonaria* will have soredia very sparse or even lacking; these can be distinguished from *L. linita* by the negative K and P tests of the latter. S and e of the Wash Cas *L. linita* is quite rare.

Lobaria oregana

Description: Thallus foliose, large, mostly 5-20(30) cm broad, often becoming pendent when large; lobes generally 1-5 cm broad; upper surface greenish or yellowish green, with a shallow network of ridges; lobules present, mainly marginal; photobiont green but with a blue-green photobiont in internal cephalodia; apothecia occasional; cortex KC+Y (usnic acid), medulla K+Y, P+O, C-, KC- (norstictic and stictic acid group).

Range: Alas to Cal, w Cas; known e Cas from only two sites (BC and n Id).

Substrate: Usually on coniferous trees; sporadically on hardwoods, including *Alnus, Fraxinus*, and shrubs; rarely on mossy rock.

Habitat: Oceanic forests, reaching maximum dominance in mid-elevation old-growth forests (*Pseudotsuga – Tsuga heterophylla*) on w slope of Cas. Occasional in moist low-elevation forests in the foothills.

Notes: This distinctive species is recognized by the yellowish green color, large size, and marginal lobules. In its optimum habitat w Cas it often attains a biomass of more than 1 ton dry weight per hectare. This species is renowned for its contribution of fixed nitrogen to forests w Cas. An uncommon form (the blue-green photomorph sometimes called "*Nephroma silvae-veteris*" or "*Lobaria silvae-veteris*") contains only cyanobacteria in the main thallus but produces green lobules from the upper and lower surfaces. These individuals usually achieve only 1-2 cm diam but have the same chemistry and ITS sequence as normal *L. oregana*. Because the fungal partner of the blue-green and green photomorphs are the same, they have the same name, *L. oregana*.

Lobaria pulmonaria

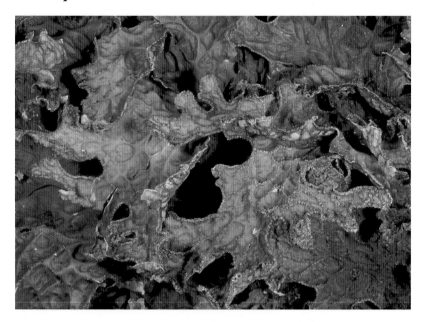

Description: Thallus foliose, medium to large, mostly 5-15 cm broad; lobes generally 1-3 cm broad; upper surface greenish to olive or brownish, with a network of ridges; soredia and/or isidia present, in roundish spots on the ridges and margins, occasionally lobules developing on decadent individuals; photobiont green but with blue-green photobiont in cephalodia; apothecia uncommon; medulla K+Y, P+O, C-, KC- (norstictic and stictic acid group).

Range: Alas to c Cal, inland to w Mont.

Substrate: Trees (conifers and hardwoods), shrubs, and mossy rock.

Habitat: Moist lowland to mid-elevation forests in areas of strong coastal influence; frequent w Cas, occasional in n Id, uncommon to rare elsewhere e Cas.

Notes: The most widespread *Lobaria* in the PNW, *L. pulmonaria* is easily recognized by its greenish color when wet, the presence of soredia, and the distinct network of coarse ridges on the upper surface. Young pre-sorediate specimens may be confused with *L. linita*, which can be separated by its negative K and P tests. *L. pulmonaria* is an old-growth associate e Cas, where it is most often found in riparian areas and in sheltered narrow valleys. W Cas, however, *L. pulmonaria* is more strongly associated with hardwoods than with old conifer forests. *L. retigera*, a rare species from BC to Alaska, is similar in general form but is dark gray (lacking green algae) and has isidia.

Lobaria scrobiculata

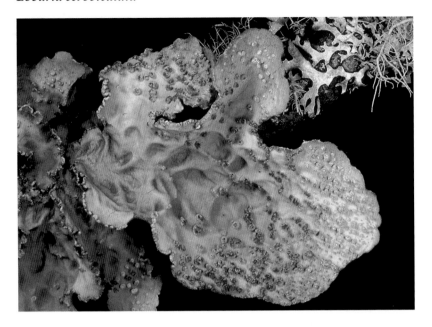

Description: Thallus foliose, mostly 2-10 cm broad, with broadly rounded lobes mostly 1-5 cm wide; upper surface yellowish-tinged pale green to olive gray when dry, dark gray when wet, weakly ridged or undulating; upper cortex often scabrous near the lobe tips but lacking tiny hairs; photobiont blue green; soralia roundish; apothecia rare; cortex KC+Y (usnic acid); medulla K+Y to O, P+O (rarely P-), C-, KC+R (norstictic and stictic acids and scrobiculin); spot tests often weak.

Range: Circumpolar boreal s to Mont and n Cal; rare e Cas where it is known only along the Salmon River and in w Mont.

Substrate: Tree, shrubs, and mossy rocks, rarely on the ground.

Habitat: Most frequent in low-elevation hardwood forests, swamps, and savannas w Cas; also in low- to mid-elevation old-growth conifer forests; e Cas restricted to sheltered mossy outcrop areas, often near lakes or streams.

Notes: Lobaria scrobiculata is occasionally confused with *L. hallii*, which differs in having tiny hairs on the upper surface near the lobe tips and in chemistry. When dry they can be separated by color; *L. scrobiculata* usually has a distinct yellowish tinge while *L. hallii* is gray. Because these species are rare e Cas, they should be collected there only sparingly, if at all.

Massalongia

Massalongia carnosa

Description: Thallus foliose, small, to about 3 cm diam; isidia or lobules present, especially on the margins; upper surface brown to dark brown, lobes generally 2 mm broad or less; lower surface pale with sparse dark rhizines; apothecia with proper margin, brown, uncommon; spores 2-celled; photobiont blue green (*Nostoc*); spot tests negative.

Range: Arctic to Cal and Ariz.

Substrate: Moss over rock, also on humus and soil; rarely epiphytic.

Habitat: Exposed to sheltered outcrop and talus areas, occasionally on shady boulders in the forest, occasionally in steppe; occurring at a wide range of elevations; rare as an epiphyte on moss mats on old-growth conifers.

Notes: Of the small brown lichens containing a blue-green photobiont, this is the most common. Although its form is variable, the narrow isidiate or lobulate lobes forming a compact thallus are characteristic. It is most likely to be confused with *Fuscopannaria* or *Pannaria* species, but those typically have blue-gray or gray tones, and their spores are 1-celled. *Massalongia microphylliza* is a 2-4-celled species, poorly known, from s Cal, with brown squamules to 1.5 mm wide. It has been reported from other western states and provinces based on sterile material.

Sources include: Henssen (1963a); Björk (2009).

Melanelia, Melanelixia, Melanohalea, and Montanelia

Thallus foliose, some shade of brown or olive, small to medium sized; lobes 0.5-5(10) mm broad, short and rounded to elongate, ± flat or slightly convex or concave; upper surface with or without pseudocyphellae (the pseudocyphellae often dark and difficult to see); soredia or isidia often present; lower surface pale to dark brown or black, generally paler marginally (colors in key refer to more central portions), rhizinate; apothecia lecanorine, the disk brown; upper cortex K-, N-; medulla P+R or P-, C+R or C-, KC+R or KC-; on various substrates.

Sources include: Blanco et al. (2004), Divakar et al. (2010, 2012), Esslinger (1977, 1978a).

ID tips: Molecular work has split *Melanelia* into four more natural genera. Because these are similar in appearance, they are keyed together here. *Melanelixia* contains most of our C+R species; *Montanelia* includes most of the narrow-lobed strictly rock-dwelling species; *Melania* retains only a few narrow-lobed rock dwellers; and *Melanohalea* has the remainder, always C- and mostly epiphytes. See also the genus of brown parmelioid species, *Neofuscelia* (= *Xanthoparmelia* subgenus *Neofusca*), which occurs mainly on rock and has an N+ dark blue green cortex and granular isidia or soredia.

Introductory Key

1a Thallus with isidia, soredia, or both
  2a Thallus with isidia, the isidia occasionally pustular and fragmenting
    into soredioid masses .... **Group 1** (Lead 3)
  2b Thallus with soredia or with both soredia and isidia...**Group 2** (Lead 10)
1b Thallus lacking isidia and soredia .... **Group 3** (Lead 14)

Group 1

Isidiate

3a Medulla KC+R, C+R or pink
  4a Isidia fine, 0.1-0.25(0.4) mm long and 0.02-0.06 mm diam, seldom
    branched; soredia also generally present ...... ***Melanelixia subaurifera***
  4b Isidia larger, mostly 0.2-1.0 mm long and 0.05-0.1 mm diam, often
    branched; soredia absent................................. ***Melanelixia glabratula***
3b Medulla KC-, C-
  5a Isidia cylindrical
    6a Isidia arising as small conical to hemispherical papillae with
      pseudocyphellae at the tip (25-40× lens), when mature often with
      small lateral branches ....***Melanohalea elegantula***
    6b Isidia arising as small spherical to hemispherical papillae without
      pseudocyphellae at the tip, when mature simple or isotomically
      branched
      7a On rock; isidia not at all dorsiventral. Thallus to 7(10) cm diam;
        lobes 1-4(6) mm broad; upper surface pale olive-green to

more often dark reddish brown, occasionally blackish brown, frequently whitish pruinose; arctic and boreal N Am s to Colo; common e Cas ....*Melanohalea infumata*

7b On bark or wood; isidia growing into dorsiventral lobules ....***Melanohalea subelegantula***

5b Isidia not cylindrical

8a On rock; isidia soon growing into small overlapping lobules that ± cover the thallus; medulla UV+ white (perlatolic and stenosporic acids). Lobes 0.1-1(1.5) mm broad; widespread, uncommon .... *Montanelia panniformis*

8b On bark or wood, occasionally on rock; isidia bulbous, spatulate, or growing into dorsiventral lobules; medulla UV-

9a Isidia bulbous to spatulate...................... ***Melanohalea exasperatula***

9b Isidia cylindrical to lobulate ....***Melanohalea subelegantula***

Group 2

Sorediate

10a Medulla C+R or pink, UV- (lecanoric or gyrophoric acid); lobes various in width

11a Pseudocyphellae present on upper surface; on rock; lobes mostly 1-3 mm broad; soredia laminal and marginal, brown, occasionally lacking; widespread; common................................... *Montanelia tominii*

11b Pseudocyphellae lacking or obscure on upper surface; on rock, bark, or wood

12a Soralia laminal, arising from flat areas on the cortex; soredia powdery, granular, or isidioid, often with true isidia developing within and between soralia; cortical hairs absent .... ***Melanelixia subaurifera***

12b Soralia laminal and marginal, the laminal ones arising largely from small, ± hemispherical pustules; soredia granular or isidioid, no true isidia present; cortical hairs generally present on some lobe ends. Lobes to 4 mm broad; upper surface some shade of brown, often with a yellowish or reddish cast; most common on rock, also on bark or wood; widespread; occasional in fairly shaded, moist, low-elevation sites..................................... *Melanelixia subargentifera*

10b Medulla C-, UV+ (perlatolic and stenosporic acids); lobes mostly < 1.5 mm broad

13a Soralia laminal and submarginal, punctiform to strongly capitate and stipitate, arising in part from the pseudocyphellae; lobes often shiny, with obscure to distinct submarginal pseudocyphellae. On rock, rarely on bark or wood; widespread; common .... *Montanelia disjuncta*

13b Soralia generally terminal on the primary lobes and/or on small, often ± erect, lateral branches, arising by gradual disintegration of the cortex; lobes generally dull (occasionally shiny on lobe ends), lacking pseudocyphellae. On rock, rarely on bark or wood; widespread; common .... *Montanelia sorediata*

Group 3

Neither isidiate nor sorediate

14a On bark and wood, rarely on rock
  15a Medulla C+R or pink; lobe tips with tiny hyaline cortical hairs.
    Thallus generally thick, relatively large, olive to brown or dark brown;
    apothecia common; abundant in Cal, disjunct in Id; on hardwoods
    .... *Melanelixia californica*
  15b Medulla C-; lobe tips lacking hyaline cortical hairs
    16a Spores 12-32 per ascus (LM); mostly on hardwoods and shrubs in
      oceanic to suboceanic climates ....***Melanohalea multispora***
    16b Spores 8 per ascus; on hardwoods, shrubs, and conifers, abundant
      in suboceanic to continental climates ..... ***Melanohalea subolivacea***
14b On rock, rarely on hard, dry wood
  17a Lobes weakly to distinctly channeled. Pseudocyphellae present;
    medulla with stictic acid group (K+Y, P+O); arctic-alpine s to Ariz
    and Wash; occasional...................................... *Melanelia hepatizon*
  17b Lobes flat to convex
    18a Medulla C+R or pink
      19a Pseudocyphellae lacking or evident
        as linear pale lobe edges. Lobes
        often strongly reticulately ridged
        (especially near the margins) and
        pitted, mostly 2-3 mm broad; thallus
        often much divided and irregularly
        cushion-forming; on rock; Cal, Colo,
        Ore, and s Wash...*Melanelixia glabroides*
      19b Pseudocyphellae present, subtle or distinct
        ....*Montanelia tominii* (Lead 11a)
    18b Medulla C-
      20a Pseudocyphellae absent or indistinct (occasionally
        pseudocyphellae distinct on warts and apothecial rim);
        thallus with numerous small, shingled lobules, occasionally
        transforming into a cushion of lobules
        ....*Montanelia panniformis* (Lead 8a)
      20b Pseudocyphellae distinct, forming small roundish whitish or
        dark spots; on rock; thallus sometimes with masses of narrow
        lobes but not particularly lobulate
        21a Rhizines mainly marginal; lobes ± flat; medulla UV+, KC+R
          (alectoronic acid), P-, K-; mainly Arctic....... [*Melanelia agnata*]
        21b Rhizines randomly distributed, often sparse; lobes distinctly
          convex; medulla UV-, KC-, P- or P+O (± protocetraric
          acid), K- or K+ brownish. Lobes 0.5-2(3) mm broad, flat
          or occasionally almost terete, occasionally forming tangled
          cushions; upper surface dark olive brown, dark brown, or black,
          shiny; circumpolar, s to Wash and Mont .... *Melanelia stygia*

M. hepatizon

1 cm

Melanelixia glabratula

Description: Thallus foliose, appressed, usually < 6 cm broad; lobes mostly 1-3 mm broad; upper surface olive-green, brown with a reddish or yellowish tinge, or dark brown, often shiny; lower surface dark brown or black; isidia thin, cylindrical, mostly 0.2-1.0 mm long and 0.05-0.1 mm diam, often branched, easily broken off and leaving white spots on the thallus; soredia absent; apothecia occasional; medulla K-, P-, C+R, KC+R.

Range: Widespread, throughout the PNW, common w Cas.

Substrate: Bark (mainly hardwood trees and shrubs) and rock.

Habitat: Common as an epiphyte on hardwoods w Cas in low- to mid-elevation forests and more open habitats; e Cas more often in suboceanic forests on rock in shaded habitats in the mountains.

Notes: Melanelixia glabratula is frequently confused with *M. subaurifera*, another C+ species with similar coloration. *Melanelixia subaurifera* has isidia and soredia, while *M. glabratula* has only isidia, but the incipient isidia and spots where isidia have eroded on *M. glabratula* can be similar to soredia and require examination with a good dissecting microscope. These species can, however, be separated by the size and form of the isidia. While both produce cylindrical isidia, they are mostly short (< 0.25 mm) and unbranched in *M. subaurifera*, but longer (to 0.25-1.0 mm) and often branched in *M. glabratula*.

Isidia

1 mm

Melanelixia subaurifera

Description: Thallus foliose, appressed, usually < 10 cm broad; lobes mostly 1-4 mm broad; upper surface olive-green, reddish brown, or dark brown, often shiny; lower surface pale brown, dark brown, or black; isidia thin, cylindrical, mostly < 0.25 mm long and 0.02-0.06 mm diam, seldom branched, easily broken off and leaving white spots on the thallus; soredia present; apothecia rare; medulla K-, P-, C+R, KC+R.

Range: Widespread, most common e Cas.

Substrate: Bark and wood (mainly hardwood trees and shrubs), less often on rock.

Habitat: In a wide variety of forest and shrub habitats at low to middle elevations.

Notes: If you are uncertain about the presence of soredia, see notes under *Melanelixia glabratula*. *Melanelixia glabratula* largely replaces *M. subaurifera* w Cas. Another C+, KC+ species, *M. subargentifera*, commonly occurs on rock, while *M. subaurifera* is usually on bark or wood. However, both can occur on either substrate and should be separated by the form and position of

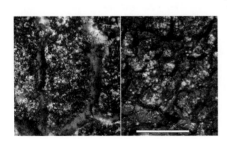

the soralia: marginal and laminal in *M. subargentifera*, but just laminal in *M. subaurifera*. Also, the soralia of *M. subaurifera* usually have small isidia intermixed with the soredia, while isidia are not found in *M. subargentifera*.

Melanohalea elegantula

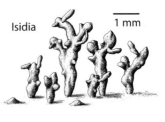

Isidia   1 mm

Description: Thallus foliose, appressed, usually < 6 cm broad; lobes 1-4(7) mm broad; upper surface olive-green or more often brown or dark brown, often whitish pruinose; lower surface pale tan to dark brown or black, often mottled pale and dark; isidia cylindrical, often with small lateral branches; apothecia uncommon; all spot tests negative.

Range: Widespread, throughout w N Am, common e Cas.

Substrate: Bark and wood, less often on rock or mosses over rock.

Habitat: A wide range of forests, savannas, and open habitats, most prominent in the more continental climates of the PNW.

Notes: The key to identifying the epiphytic *Melanohalea* species lies in the form of the isidia. Isidia of *M. elegantula* arise as small conical papillae that elongate into fine, cylindrical isidia. These often branch, many of the branches being produced at right angles to the original isidium. In contrast, isidia in *M. ex-asperatula* are bulbous (pinched at the base and broader above). *Melanohalea subelegantula* has isidia that are initially similar to those of *M. elegantula*, but have a strong tendency to become flattened and lobulate.

Melanohalea exasperatula

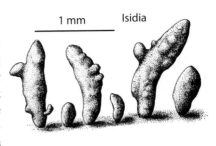

1 mm   Isidia

Description: Thallus foliose, appressed, usually < 6 cm broad; lobes mostly 2-5 mm broad; upper surface olive-green, brown, or dark brown, often shiny; lower surface pale tan to dark brown or black, often mottled; isidia mostly bulbous or compressed club-shaped, sometimes becoming lobulate and flattened; apothecia uncommon; all spot tests negative.

Range: Widespread, throughout w N Am, common on both sides of the Cas.

Substrate: Bark or wood (both conifers and hardwoods), rarely on rock.

Habitat: Frequent in a very broad range of habitats, from low to high elevations, in both continental and oceanic climates, and in deep shade to exposed habitats; very common in montane to subalpine conifer forests.

Notes: This is the most common C-, KC- epiphytic *Melanohalea* w Cas. Its bulbous isidia are diagnostic. When lobulate it can be confused with *M. subelegantula*. These can be separated by the form of the young isidia: cylindrical in *M. subelegantula* and almost spherical in *M. exasperatula*.

Melanohalea multispora

Description: Thallus foliose, appressed, usually < 6 cm broad; lobes mostly 1-3 mm broad; upper surface olive brown, brown, or dark brown; lower surface dark brown or black; isidia and soredia absent, but sometimes with isidia-like papillae; apothecia usually present; spores 12-32 per ascus; all spot tests negative.

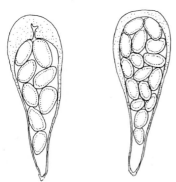

Asci of *M. subolivacea* (left), *M. multispora* (right)

Range: Alas to s Cal, e to Mont.

Substrate: Bark, usually deciduous trees and shrubs.

Habitat: Most common in moist low- to mid-elevation forests; very common w Cas, frequent e Cas in areas of strongest oceanic influence.

Notes: Confident identification of this species requires a thin section of an apothecium viewed under a light microscope. The spores are produced in sacs (asci) containing 12-32 spores per ascus, in contrast to the closely related *Melanohalea subolivacea*, which produces 8 spores per ascus. In most cases, one can make an educated guess: *M. multispora* is typically on shrubs or other hardwoods in more oceanic climates, while *M. subolivacea* occurs on conifers (as well as hardwoods) in more continental and drier climates.

Melanohalea subelegantula

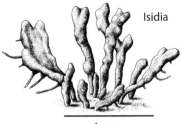

Isidia

1 mm

Description: Thallus foliose, appressed, usually < 6 cm broad; lobes usually 1-3 mm broad; upper surface olive-brown, brown, or dark brown, occasionally lightly pruinose; lower surface dark brown or black; isidia initially slender and cylindrical, 0.1-0.3 mm long and 0.03-0.07 mm diam, lacking pseudocyphellae, in older parts growing into distinctly dorsiventral lobules that may become isidiate and rhizinate; apothecia uncommon; all spot tests negative.

Range: Temperate w N Am; in the PNW most frequent w Cas and more oceanic areas inland to w Mont.

Substrate: Bark or wood.

Habitat: Moist low- to mid-elevation forests.

Notes: Melanohalea subelegantula is distinguished from other epiphytic *Melanohalea* species by its initially cylindrical isidia that develop into lobules. Lobules are also frequently produced by *M. exasperatula*, but in that species the isidia begin with a bulbous form, pinched at the base and distinctly roundish. Juvenile forms of *M. subelegantula* might be confused with *M. elegantula*; look carefully for the tendency of the isidia to become lobulate, indicating the former species.

Melanohalea subolivacea

Description: Thallus foliose, appressed, usually < 6 cm broad; lobes mostly 1-6 mm broad; upper surface olive brown, brown, or dark brown; lower surface dark brown or black; isidia and soredia absent, but sometimes with isidia-like papillae (especially in exposed sites in dry climates); apothecia usually present; spores 8 per ascus; all spot tests negative.

Range: Widespread in w N Am, frequent throughout the PNW.

Substrate: Bark and wood, both conifers and hardwoods.

Habitat: A broad range of open and forested habitats, including dry forests and savannas (e.g., *Pinus ponderosa*), steppe (on *Artemisia* and *Cercocarpus*, etc.), and moist forests; at all forested elevations.

Notes: In much of the West this is the most common *Melanohalea*, with abundant apothecia and no isidia or soredia (exception: hardwoods w Cas where *M. multispora* is most common). *Melanohalea subolivacea* is very similar to *M. multispora*; see notes under that species. The warty forms of *M. subolivacea* may lead to misidentification as one of the isidiate species. However, the abundant production of apothecia and the poorly formed "isidia" are clues that the material is *M. subolivacea*. The occasional reports of *M. olivacea* from w N Am are mostly typos or misspellings, the exception being from coastal se Alas. Two other Alaskan and boreal species will key here and can be distinguished by spot tests: *M. septentrionalis* has medulla K+ brownish, P+O (fumarprotocetraric acid), and *M. trabeculata* has medulla K+Y to O, P+Y to O (norstictic and salacinic acids).

Menegazzia

Thallus foliose, small to medium sized (mostly 2-6 cm broad), appressed; lobes narrow (mostly 1-2 mm broad), hollow, contiguous; upper surface pale gray, cream, or greenish gray, with scattered roundish perforations; lower surface black, brown at the edges; rhizines lacking; soredia present, on short erect lobes; apothecia rare; spores typically 2 per ascus (ours), large and thick walled, the spore wall layered and about 5-6 µm thick; cortex K+Y (atranorin); medulla K+Y, KC-, P+O (stictic acid group), C-; on trees and shrubs.

Sources include: Bjerke (2003), Santesson (1943).

ID tips: This distinctive genus is likely to be confused only with *Hypogymnia*. Both have hollow lobes, but *Menegazzia* develops perforations in the upper surface, while in our *Hypogymnia* species, perforations, if present, are restricted to the lobe tips and axils. Furthermore, *Menegazzia* always has chemistry of the stictic acid group, while *Hypogymnia* has the physodic acid group and other substances, but never stictic acid.

1a Soralia initially collar-like, with granular soredia, the collars becoming lacerate-divided, occasionally much dissected ....**M. subsimilis**
1b Soralia forming a smooth powdery cap, collar, or lip on laminal outgrowths from the thallus, well-developed soralia not or little divided. On trees (mainly hardwoods) and shrubs; widespread, occasional in PNW .... *M. terebrata*

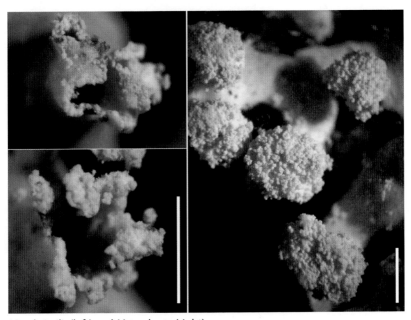

M. subsimilis (left) and *M. terebrata* (right)

Menegazzia subsimilis

Description: Thallus foliose, small to medium sized (mostly 2-6 cm broad), appressed; lobes narrow (mostly 1-2 mm), hollow, contiguous; upper surface pale gray, cream, or pale greenish gray, with scattered roundish perforations; lower surface black, brown at the edges; rhizines lacking; soredia present, on short erect lobes; apothecia rare; cortex K+Y; medulla K+Y, KC-, P+O, C-.

Range: Alas to Cal, w Cas.

Substrate: On trees (mainly hardwoods) and shrubs; especially frequent on *Alnus rubra.*

Habitat: Moist oceanic forests, often in riparian areas, especially common in the Coast Ranges, more scattered on w slope of Cas.

Notes: Once the perforate upper surface is recognized, this genus will seldom be mistaken. Small individuals with few or no perforations or soredia may be confused with *Hypogymnia.* However, the appressed species of *Hypogymnia* have a KC+R medulla, while *Menegazzia* is KC-. Our material was previously all lumped under *M. terebrata*, but is now separated by the form of the soralia. *Menegazzia subsimilis* is much more common in the PNW than *M. terebrata.*

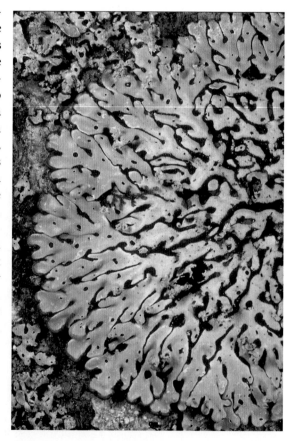

Moelleropsis—*see* Fuscopannaria

Neofuscelia

Thallus foliose, light to dark brown, dull or shiny near the margins, to 10 cm diam or more, appressed; lobes mostly < 3(5) mm wide; pseudocyphellae lacking; lower surface cream, brown, or black; rhizines present, simple, sparse to abundant; isidia or sorediose isidia present, the isidia often bursting or degrading into soredia; apothecia uncommon (to unknown in some species), lecanorine, the disk brown; spot tests various but upper cortex N+ blue green or violet, medulla with divaricatic, glomellic, gyrophoric, hypostictic, perlatolic, or stenosporic acids (spot tests various); on rock or moss over rock, rarely on bark or wood in dry habitats.

Sources include: Blanco et al. (2004) Esslinger (1977, 1978a), McCune et al. (2014b). Although Blanco et al. (2004) placed this genus in *Xanthoparmelia*, McCune et al. (2014b) concluded that their evidence could also be taken as support for its retention as *Neofuscelia*.

ID tips: The N+ blue-green spot test on the cortex is diagnostic for *Neofuscelia*, separating it from other brown parmelioids (*Melanelia, Melanelixia, Melanohalea, Montanelia*), while the others are N-. Many of you, however, will not have access to nitric acid. With practice *Neofuscelia* can be recognized even in the field by the coarse, globular, sometimes sorediose isidia.

1a Medulla K+Y to O, P+ pale O (hypostictic, hyposalacinic, and hypoconstictic acids), UV-. Isidia sparse to dense, pustular, 0.05-0.3 mm diam; lobes mostly 1-2(3) mm wide; upper surface reddish brown to dark olive brown; lower surface dark brown to black; apothecia unknown; on rock; BC to Mont s to Colo and Cal; uncommon; steppe, especially on the basalt plateaus .... *N. subhosseana*
1b Medulla K-, P-, often UV+
  2a Lobe tips and isidia pale brownish to yellowish brown and with a subtle mosaic of light markings.. ***N. loxodes***
  2b Lobe tips and isidia generally brown or dark brown, lacking a mosaic of light markings. Several chemotypes (usually with divaricatic acid, rarely with stenosporic acid): (1) C-, KC-; (2) C-, KC+ pink fading fast; and (3) C+R, KC+R; isidia pustular, 0.1-0.2 mm diam, sparse to dense, resembling soredia when abraded; lobes mostly 1-2(3) mm broad; apothecia rare; on rock, rarely on wood; Wash to S Dak s to Baja Cal, most common in Cal and w Cas .... *N. verruculifera*

Neofuscelia loxodes

Description: Thallus foliose, yellow brown to dark brown, medium sized (mostly < 7 cm diam), appressed; lobes 1-3(5) mm broad; lobe tips and isidia pale brownish to yellowish and with a subtle mosaic of light markings, sometimes the cortex cracking along those marks; isidia pustular, 0.1-0.5 mm diam, soft, often hollow, sparse to dense, often heaped, easily abraded and then mistaken for soredia; lower surface dark brown to black; apothecia uncommon; cortex N+ blue green; medulla K-, P-, C- or C+R, KC+R (glomelliferic, glomellic, and perlatolic acids, rarely with gyrophoric acid).

Range: Wash e to N Dak, s to Cal and Colo.

Substrate: Rock or mosses over rock, very rarely on wood or bark.

Habitat: Fairly common in exposed dry sites, especially on basalt on the Columbia Plateau and Snake River Plain, sporadically into w Mont.

Notes: This and other *Neofuscelia* are common on dry outcrops, especially basalt. Three *Neofuscelia* occur in the PNW. *Neofuscelia loxodes* is most common e Cas and has subtle but distinctive pale angular markings near the lobe tips (use a good hand lens). *Neofuscelia verruculifera* is more common w Cas and s into Cal and is K-. Another species without the pale markings, *N. subhosseana*, is K+Y to O and occurs sporadically throughout the drier parts of the PNW.

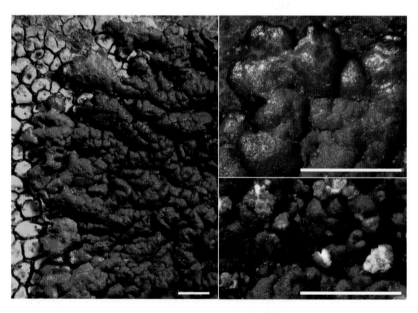

Nephroma

Thallus foliose, brown, gray brown, or yellowish green, appressed or loosely attached; lobes generally 2-20 mm wide; lower surface glabrous, pubescent, or tomentose, but rhizines sparse or lacking, tan to brown or black in part; medulla white or yellow-orange; soredia, isidia, or lobules often present; apothecia on lower surface of lobe tips; disks light to dark brown; spores 3-septate, light brown; primary photobiont blue green (*Nostoc* in cephalodia or throughout the thallus) or green (in spp with cephalodia); spot tests mostly negative, all species containing terpenoids; on rocks, bark, mosses, or tundra sod.

Sources include: James and White (1987), Morris and Stone (2022), Timdal et al. (2020, 2021), Wetmore (1960, 1980), White and James (1988).

ID tips: Brown foliose species with apothecia on the lower surface of the lobes are certainly *Nephroma*. When you find one epiphytic *Nephroma* species you are likely to find other *Nephroma* species on the same branch. With a little practice, all of our *Nephroma* species can be identified in the field.

1a Thallus yellowish green, yellowish brown, or greenish brown; green or bluish gray when wet; photobiont green or blue green; cephalodia present or not
  2a On bark and wood; photobiont blue green; thallus dark bluish gray when wet ....**N. occultum**
  2b On mossy rock or soil; photobiont green, blue-green photobiont restricted to cephalodia; thallus green when wet
    3a Thallus yellowish green; cephalodia internal but visible on upper surface as convex darker spots; upper surface smooth. Thallus to 15 cm broad; lower surface brown to black at the center, tomentose; Alas to se BC.. [*N. arcticum*]
    3b Thallus dull green or brownish; cephalodia internal, not generally visible from above; upper surface smooth, pruinose, or slightly pubescent. Thallus to 10 cm broad; Alas to se BC and sw Alta
    .... [*N. expallidum*]
1b Thallus brown to grayish brown; photobiont blue green; cephalodia absent
  4a Lower surface with whitish papillae surrounded by pale brown tomentum (papillae occasionally lacking on juveniles); upper surface dull, thinly pubescent ....**N. resupinatum**
  4b Lower surface lacking papillae, tomentose or not; upper surface usually shiny, not pubescent
    5a Medulla yellowish, K+R (rarely white and K-) .... **N. laevigatum**
    5b Medulla white, K- or K+ pale yellow
      6a Thallus lacking asexual propagules................................... **N. bellum**
      6b Thallus producing asexual propagules, reliably present except in small individuals

7a Propagules are isidia and/or lobules, clearly completely corticate except where damaged
    8a Propagules are isidia, becoming cylindrical to branched; se Alas and coastal BC.. [*N. isidiosum*]
    8b Propagules are primarily marginal lobules, isidia may be present... ***N. tropicum***
7b Propagules are soredia or isidiose soredia, at least partially decorticate, often coarsely granular to somewhat elongate; widespread
    9a Propagules initially mainly marginal in long, continuous soralia, becoming partially laminal; upper surface undulate or flat, without definite reticulate ridges; upper surface pale brown to gray or darkening .... ***N. parile***
    9b Propagules initially mainly laminal in rounded clumps on ridges; upper surface often with a network of sharp ridges and depressions, but sometimes smooth or undulate; propagules more consistently corticate (isidiose), but soredia are often present; upper surface medium brown to dark brown or partly olivaceous brown. N Alas s to ne Ore and Colo, uncommon... *Nephroma orvoi*

Nephroma orvoi

Nephroma bellum

Description: Thallus foliose, loosely appressed to ascending, to 8 cm diam; upper surface light to dark brown; lower surface tan to brown, lacking tomentum in the PNW; isidia, soredia, and marginal teeth lacking; medulla white; photobiont blue green; apothecia common, on the underside of the lobe tips; spot tests negative.

Range: Widespread; throughout the PNW, but infrequent; primarily boreal, rare w Cas; more common e Cas and northward.

Substrate: Trees, shrubs, and mossy rocks.

Habitat: Moist forests; often riparian hardwoods and shrubs.

Notes: Nephroma bellum is recognized by the absence of characteristics that distinguish the other *Nephroma* species: no isidia, soredia, marginal teeth, or papillae. Other species are frequently misidentified as *N. bellum* if the critical characters are not noted. For example, *N. laevigatum* is often similar in external appearance but has a yellow medulla, easily noted if you cut or break a lobe. Similarly, if you don't spot the papillae of *N. resupinatum* (see illustration), you might mistake it as *N. bellum.* Juvenile specimens of *N. tropicum* with few lobules have also commonly been misidentified as *N. bellum.* These and other species can be separated from *N. bellum* if there is tomentum on the lower surface.

Nephroma laevigatum

Description: Thallus foliose, appressed to ascending, to 8 cm diam; upper surface light to dark brown; lower surface yellowish brown, light brown, or dark brown, bare and smooth; marginal isidia or lobules sometimes present; medulla pale to deep yellow or orange; photobiont blue green; apothecia common, on the underside of the lobe tips; spot tests negative except medulla K+R, lw UV+ (anthraquinones).

Range: Alas to Cal, w Cas, common.

Substrate: Trees (mainly hardwoods) and shrubs, less often on rock.

Habitat: Moist low- to mid-elevation forests, often in riparian areas and on understory shrubs.

Notes: Identification of *N. laevigatum* is clinched by the yellowish medulla and apothecia on the underside of the lobe tips. Specimens with low concentrations of the yellow pigment may be mistaken for *N. bellum* or *N. tropicum*. Snails frequently graze off the upper cortex, leaving the bright yellow (apparently distasteful) medulla behind. This is a conspicuous example of the importance of secondary chemicals in discouraging herbivory of lichens.

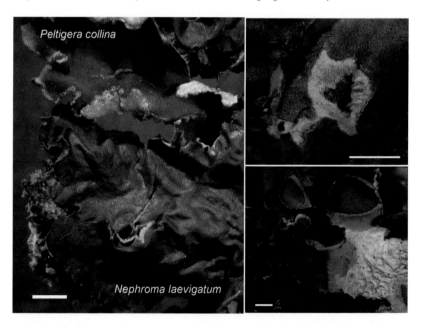

Nephroma occultum

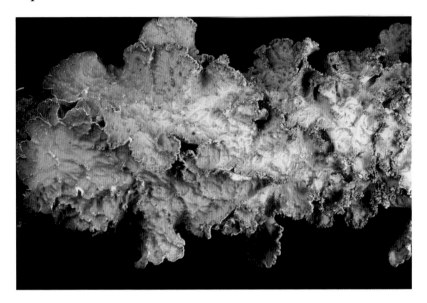

Description: Thallus foliose, appressed, mostly 2-6 cm broad; lobes mostly < 1 cm broad; upper surface yellowish green, yellowish tan, or greenish when dry, often with ± tan edges, blue gray when wet, with a weak or distinct network of ridges; lower surface cream to tan, smooth, lacking tomentum; coarse granular soredia usually present, laminal and marginal; photobiont blue green; apothecia unknown; spot tests negative or cortex KC+Y.

Range: se Alas, BC to Ore, w Cas except for inland disjuncts in BC.

Substrate: Bark and wood of conifers.

Habitat: Almost all known locations are from old-growth forests on w slope of Cas, dominated by *Pseudotsuga – Tsuga heterophylla*, stands tending to have abundant populations of *Lobaria oregana*; on fallen branches or recently windthrown trees; near the forest floor only where the forest is somewhat more open; most frequent in the mid to upper canopy.

Notes: Nephroma occultum is likely to be overlooked because its color when dry is similar to *Lobaria oregana*, which often occurs in tremendous quantities in the stands where *N. occultum* is found. A reasonable search strategy in dry weather is to look for small, appressed individuals similar in color to *L. oregana* or *L. scrobiculata*. In wet weather the color of *N. occultum* changes to blue-gray while *L. oregana* becomes greener. Although *N. occultum* is one of the species that is most restricted to old-growth forests, it has been found occasionally in younger forests.

Nodobryoria

Thallus fruticose, erect, tufted, prostrate, or pendent, generally some shade of reddish brown; branches roundish, uneven, or ridged; short spine-like side branches usually present; soredia and isidia absent; pseudocyphellae absent; thallus cortical cells with knobby jigsaw puzzle-like outlines (LM, see below); apothecia lateral or subterminal, the margin ciliate; disk reddish brown; all spot tests negative; mostly on conifers, rarely rock, and one species on tundra sod.

Sources include: Brodo and Hawksworth (1977), Common and Brodo (1995).

ID tips: The reddish brown color of *Nodobryoria* species is a very useful field characteristic, separating them from most *Bryoria* species. *Nodobryoria* also has short, sharp side branches and negative spot tests. While some *Bryoria* species also have those characteristics, none have the most diagnostic characteristic of *Nodobryoria*, the consistent presence of jigsaw-puzzle-like cortical cells.

1a On rock, soil, or alpine sod; thallus prostrate. Uncommon, BC to Ore, e to
    w Mont ....*N. subdivergens*
1b On bark or wood (rarely rock); thallus tufted or pendent
  2a Thallus tufted, erect or suberect; apothecia almost always present
    ....***N. abbreviata***
  2b Thallus pendent; apothecia occasionally present ....***N. oregana***

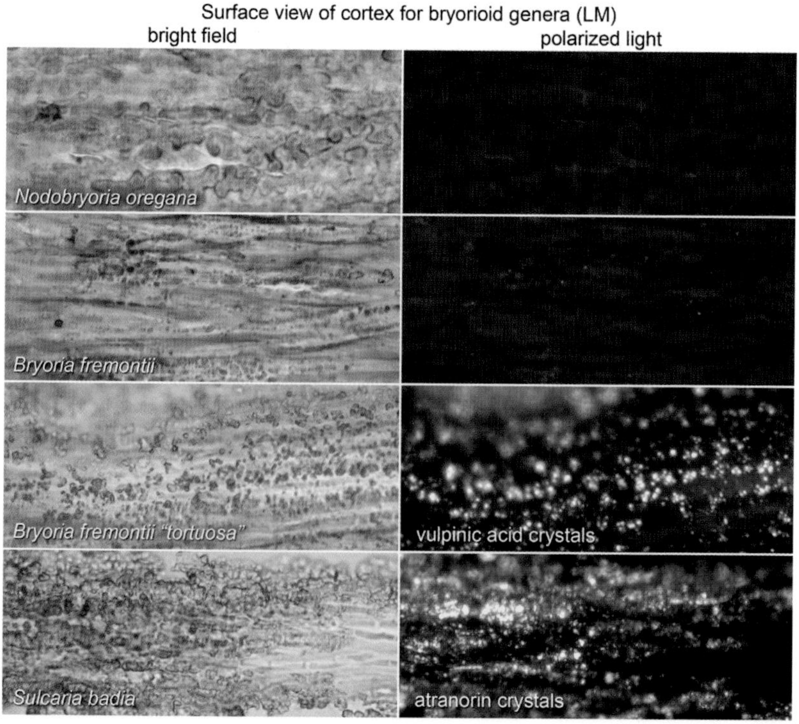

Surface view of cortex for bryorioid genera (LM)
bright field   polarized light

Nodobryoria oregana

Bryoria fremontii

Bryoria fremontii "tortuosa"   vulpinic acid crystals

Sulcaria badia   atranorin crystals

Nodobryoria abbreviata

Description: Thallus reddish brown, tufted, erect or suberect, to 3 cm long; branches roundish or ± angular or uneven in cross section, often appearing spinulose from short, pointed lateral branches; pseudocyphellae absent; soredia and isidia absent; apothecia almost always present, the margin often ciliate; all spot tests negative.

Range: Common from w Mont and c Id to BC, s to Baja Cal; most frequent e Cas.

Substrate: Conifer bark and wood.

1 mm

Habitat: Valleys to near timberline, especially common in open *Pinus* and *Larix* forests e Cas, also in treetops of moist closed forests.

Notes: Commonly confused with only two other species, *Cetraria californica* and *Nodobryoria oregana*. *Cetraria californica* is strictly coastal and more olive brown. *Nodobryoria abbreviata* does not occur on the immediate coast and is always reddish brown. *Nodobryoria oregana*, although colored the same as *N. abbreviata*, is a more elongate, pendent species with slender branches, and it usually lacks apothecia.

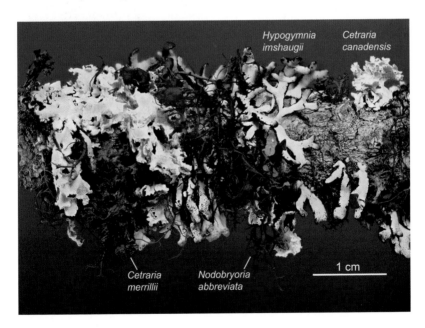

Hypogymnia imshaugii
Cetraria canadensis
Cetraria merrillii
Nodobryoria abbreviata
1 cm

Nodobryoria oregana

Description: Thallus fruticose, pendent, reddish brown, to 20 cm long; branches uneven in diameter, with scattered, short spinulose side branches, 0.1-0.4 mm diam; soredia and isidia lacking; apothecia occasionally present, sometimes ciliate-margined; spot tests negative.

Range: BC to s Cal, e to sw Alta and nw Mont.

Substrate: Usually on conifers, rarely on hardwoods or shrubs, often intricately intermixed with *Bryoria* and *Alectoria* species; sometimes on rock when present as an epiphyte in adjacent forests.

Habitat: Common in moist low-elevation to subalpine forests with relatively strong coastal influence; most abundant in mid-elevation to subalpine forests in the Cascade Range.

Notes: Initially this species can be confused with *Bryoria* species, but the reddish brown color of *N. oregana* makes field identifications easy. When wet, however, the reddish brown color changes to a greenish color. For field identifications of wet material one must rely on the sprinkling of short spinulose side branches

(use hand lens), a feature only rarely found in the other common *Bryoria* species. In ambiguous cases, make a thin slice of the cortex parallel to the long axis of the branch and look for the characteristic jigsaw pattern seen in all *Nodobryoria* species (comparison photos p. 266). *Bryoria americana* from near the coast is often reddish brown, similar to *N. oregana*, but this *Bryoria* is not spinulose and is P+; all parts of *N. oregana* are P-.

Pannaria

Thallus small-foliose, to about 7 cm diam but typically < 3 cm in our species, often forming rosettes with distinct marginal lobes, gray, buff, brown, blue gray, or less often blackish, sometimes pruinose or scabrid; attached to substrate by well-developed blue to black hypothallus or by a tomentum of dark rhizohyphae; lobes generally < 3(5) mm wide; lower surface lacking a cortex; isidia or soredia present or not; apothecia sessile, with thalloid margin; disk reddish brown, dark brown, or black; hymenium I+ blue only around the asci; hypothecium pale or hyaline; spores simple, hyaline, smooth or slightly uneven, the outer spore wall (epispore) sometimes thickened, slightly warty, sometimes apiculate; cortex and/or medulla P+O (pannarin) in our species; photobiont blue green (*Nostoc*); on rocks and trees, mostly in sheltered humid habitats.

Sources include: Jørgensen (1978, 1991, 1994b, 2000b, 2005), McCune et al. (2022), Weber (1965), Vondrak et al. (2013).

ID tips: If you are lucky enough to find a true *Pannaria* in w N Am, you should have little trouble identifying it to species. The distinction between *P. conoplea* and *P. tavaresii* can, however, be tricky in some cases.

Following Jørgensen's revisions, most of the species formerly considered in *Pannaria* are keyed under *Fuscopannaria*. See also *Moelleropsis, Parmeliella, Protopannaria*, and *Psoroma*.

1a Soredia or isidia present
  2a Soredia present, the soredia with a rough, "woolly" appearance. On
      bark or mossy rock in damp, shaded habitats; ne US and Canada, RM
      in Colo and New Mex, also in Alas; not yet known from PNW
      .... [*P. conoplea*]
  2b Soredia absent, but with smooth, coralloid isidia. Black Hills to c and s
      RM; also in se US and c Am .... [*P. tavaresii*]
1b Soredia and isidia lacking
  3a Disk brown to black; habitat arctic tundra. Thallus pale gray to brown,
      of closely contiguous warts and squamules, with whitish markings
      along the edges of the warts and squamules; apothecia with a thick,
      crenate to coarsely warty thalline margin; spores 11-15 × 8-11 µm;
      thallus P+ weak orange or P- (trace of pannarin); on rock or organic
      matter over rock; Arctic ....[*P. hookeri*]
  3b Disk reddish brown; habitat temperate .... *P. oregonensis*

Pannaria oregonensis

Description: Thallus foliose, with well-developed, radiating lobes, forming rosettes mostly < 4 cm broad; marginal lobes to 7-8 mm long and usually 1-2 mm wide; upper surface light blue gray, light gray, to brownish, the margins not white felted-tomentose; lower surface ecorticate, felty with hyaline to blue-black rhizohyphae and branched rhizines that are blue black to whitish, to about 0.7 mm long; isidia and soredia lacking; apothecia common, reddish brown, with an even to slightly beaded, persistent thalline rim; spores 8 per ascus, 11–16(19) × 7–9 μm, with a distinct but thin epispore that is nearly smooth to distinctly warty, often with a short apical extension at one or both ends; cortex P+O, medulla P- or P+O (pannarin).

Range: Coastal se Alas to BC and Ore, rarely inland to foothills w Cas, to be expected in coastal n Cal.

Substrate: Bark and wood of both conifers and hardwoods.

Habitat: Moist lowlands; the largest populations in Ore and Wash are in coastal thickets of old shrubs on wet deflation plains.

Notes: Until recognized as its own species in 2022, *P. oregonensis* was included with either *P. rubiginella* or *P. rubiginosa. Pannaria rubiginella* is a s Am species, while *P. rubiginosa* has a very wide reported range, but is not known from the PNW. *Pannaria rubiginosa* has larger spores (commonly > 16 μm long, mostly 14–20 μm) than *P. oregonensis* (see above) and often has more prominent apical extensions of the epispore. In our area, *P. oregonensis* is most likely to be confused with *Fuscopannaria leucostictoides*, which is P- throughout and has felted-tomentose lobe margins.

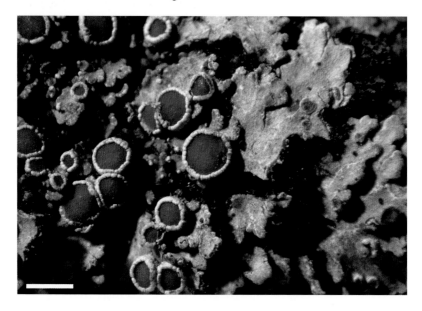

Parmelia

Thallus foliose, medium sized, ± appressed; upper surface gray, greenish gray, pale blue gray, white, or brownish when exposed to sun, occasionally darker; pseudocyphellae laminal or marginal, mostly linear or angular; lower surface black, abundantly rhizinate; isidia or soredia often present, the isidia cylindrical and often branched; apothecia laminal, lecanorine, the disk brown; cortex K+Y (atranorin); medulla K+Y or K-, P+R or O; substrate various.

Sources include: Goward and Ahti (1983), Hale (1987), Hodkinson et al. (2010), McCune (2007), Molina et al. (2017), Skult (1987).

ID tips: One or more species of *Parmelia* can be easily found in most forests and woodlands of the PNW. Identifying them to species will force you to come to grips with several characters: isidia vs. soredia (and the intermediate form in *P. hygrophila*) and squarrose rhizines vs. simple or forked rhizines. Spot tests are not so important here, except in separating *P. pseudosulcata* from other isidiate species.

1a Isidia present; medulla K+ or K-
  2a Isidia short, granular to sparingly branched, often sorediose (use hand lens), dull or weakly shiny
    3a On rock; propagules mostly marginal.............. *P. fraudans* (Lead 11b)
    3b On wood and bark, occasionally on rock; propagules laminal or marginal
      4a Lobes 2-5 mm broad; propagules mostly laminal ...... **P. hygrophila**
      4b Lobes mostly 2-3 mm broad; propagules mostly marginal. BC and n Id .... *P. sulymae*
  2b Isidia becoming long-cylindrical, occasionally richly branched, shiny
    5a On rock, moss, or bark; medulla K+Y to R, P+O or R; rhizines simple .... **P. saxatilis**
    5b On bark (rarely rock); spot tests and rhizines various
      6a Medulla K- or brownish, P+O to R; rhizines simple to apically dichotomous .... **P. pseudosulcata**
      6b Medulla K+Y to R, P+O; rhizines squarrosely branched .... **P. squarrosa**
1b Isidia lacking; medulla K+Y to R
  7a Soredia lacking; on alpine or subalpine sod, rock, and mosses over rock
    8a Laminal pseudocyphellae absent or nearly so; thallus containing norstictic acid in addition to salacinic acid (TLC); upper surface usually grayish pruinose. Oceanic arctic of Alas, with rare disjuncts (Bitterroot Range, Mont) .... *P. skultii*
    (*Parmelia imbricaria* is similar to *P. skultii*, but *P. imbricaria* lacks norstictic acid, has numerous secondary lobules, and occurs on rock in BC and Yukon. In contrast, *P. skultii* has norstictic acid and lacks secondary lobules.)

8b Laminal pseudocyphellae sparse to abundant (reticulate whitish marks, especially towards the margins); thallus lacking norstictic acid, containing salacinic acid; upper surface pruinose or not. Upper surface grayish to brown, often very dark in exposed habitats; medulla white, often aging to orangish in old thalli; widespread arctic-alpine to subalpine; uncommon in PNW .... *P. omphalodes*

7b Soredia present; habitat various

9a Soredia powdery

10a Rhizines simple then becoming squarrosely branched (some dichotomous branching will also be found); soredia remaining heaped in the soralia ....***P. sulcata***

10b Rhizines simple; soredia soon shed from soralia, exposing the medulla and becoming excavate.................................. ***P. barrenoae***

9b Soredia coarsely granular to isidiate; rhizines simple or dichotomous

11a Soredia/isidia mainly laminal; generally on bark or wood .... ***P. hygrophila***

11b Soredia/isidia mostly marginal; on rock. Soredia yellowish; widespread, rare in PNW ....*P. fraudans*

Parmelia barrenoae

Description: Thallus foliose, medium sized, mostly < 6(10) cm diam, loosely appressed; lobes mostly 1-4 mm broad; upper surface whitish gray to greenish or blue gray, with pale angular markings (pseudocyphellae) toward the lobe tips developing into angular sorediate cracks and ridges; isidia lacking, soredia present, marginal and laminal; lower surface black, brown towards the margin; rhizines simple to forked; apothecia rare; cortex K+Y (atranorin), medulla K+Y to R, P+O or R (salacinic acid), C-, KC-.

Range: Mont w to Wash, s through Ore to s Cal, particularly common in the Columbia Basin, e Cas, and from Siskiyou Mts of s Ore s through the Sierra Nevada to San Bernardino Mts.

Substrate: Bark and wood.

Habitat: Dry forests with Mediterranean to suboceanic climates. The climatic restriction yields a much narrower range than that of *P. sulcata*, with *P. barrenoae* being confined to suboceanic dry forests, while *P. sulcata* is much more widespread, ranging from forests in oceanic to continental climates.

Notes: This species is a recent segregate of *Parmelia sulcata*, distinguishable both morphologically and by ITS sequence. Soredia eroding from angular cracks in the upper surface in combination with simple to forked rhizines are characteristic of *P. barrenoae*. In contrast, the similar *P. sulcata* tends to retain soredia in the cracks, developing mounded soralia, and while its rhizines begin simple, they quickly become squarrosely branched. *Parmelia hygrophila* also has simple to forked rhizines, but develops coarsely granular propagules in clusters that do not become conspicuously excavate.

Parmelia hygrophila

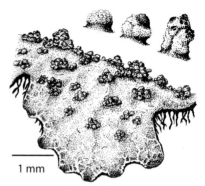

Description: Thallus foliose, medium sized, mostly < 12 cm diam, loosely appressed; lobes mostly 2-5 mm broad; upper surface whitish gray to greenish or blue gray, with pale angular markings toward the lobe tips; isidia dull, not corticate, short granular to sparingly branched, occasionally soredioid, mostly laminal; lower surface black, brown towards the margin; rhizines simple or apically branched, occasionally with a few short lateral branches; apothecia occasional; cortex K+Y (atranorin), medulla K+Y to R, P+O or R (salacinic acid), C-, KC-.

1 mm

Isidia (top); lobe tip (bottom)

Range: Alas to Cal, from the coast inland to central Mont, rare e Cont Div.

Substrate: Bark and wood, rarely rock.

Habitat: Common in moist low- to mid-elevation forests, also in urban and agricultural habitats w Cas.

Notes: Very similar in general appearance to *P. sulcata*, *P. hygrophila* is distinguished by the presence of isidia instead of powdery soredia. The isidia of *P. hygrophila* have the appearance of very coarse granular soredia, since they lack a continuous cortex, unlike the isidia of *P. pseudosulcata*, *P. saxatilis*, and *P. squarrosa*. Although lobe width is highly variable, on average *P. hygrophila* has slightly wider lobes than *P. sulcata*.

Parmelia pseudosulcata

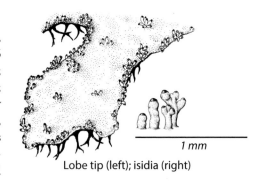

Description: Thallus foliose, medium sized, mostly < 6 cm diam, loosely appressed; lobes mostly 1-3 mm broad; upper surface whitish gray to greenish or blue gray, with pale angular markings toward the lobe tips; isidia shiny, corticate, laminal and marginal; lower surface black, brown towards the margin; rhizines simple or apically branched; apothecia rare; cortex K+Y (atranorin), medulla K-, P+O or R (protocetraric acid), C-, KC-.

Lobe tip (left); isidia (right)

Range: Alas s to Cal, w Cas, rarely inland in BC.

Substrate: Bark and wood, rarely rock, mostly on conifers.

Habitat: Low-elevation moist forests.

Notes: This is the only local *Parmelia* species with a K- medulla. In the field it can often be distinguished morphologically from other isidiate *Parmelia* species by its narrower, more elongate lobes and the tendency for isidia to be concentrated along the lobe margins.

Parmelia saxatilis

Description: Thallus foliose, medium sized, mostly < 12 cm diam, loosely appressed; lobes mostly 1-4 mm broad; upper surface whitish gray to greenish or blue gray, with pale angular markings toward the lobe tips; isidia shiny, corticate, laminal and marginal; lower surface black, brown towards the margin; rhizines simple or apically branched, occasionally with a few short lateral branches; apothecia occasional; cortex K+Y (atranorin), medulla K+Y to R, P+O or R (salacinic ± lobaric acid), C-, KC-.

Range: Widespread in N Am, common throughout the PNW.

Substrate: Noncalcareous rock and soil or mosses over rock, frequently on bark w Cas.

Habitat: The rock-dwelling form grows on both sunny and shaded rock outcrops; the bark-dwelling form lives in moist oceanic conifer forests, in partial to fairly deep shade.

Notes: The rock-dwelling form is readily recognized as the only common isidiate *Parmelia* on rock in the PNW. The bark-dwelling form must be carefully distinguished from several other isidiate *Parmelia* species. The spot tests and isidia of *P. saxatilis* are the same as *P. squarrosa*, but the latter has squarrose-branched rhizines. The bark-dwelling form of *P. saxatilis* is also confused with *P. hygrophila*, which differs in having dull, sorediose isidia in contrast to the shiny corticate isidia of *P. saxatilis* (use hand lens).

Parmelia squarrosa

Description: Thallus foliose, medium sized, mostly < 6 cm diam, loosely appressed; lobes mostly 1-3 mm broad; upper surface whitish gray to greenish or blue gray, with pale angular markings toward the lobe tips; isidia shiny, corticate, laminal and marginal; lower surface black, brown towards the margin; rhizines squarrosely branched (simple towards the margins); apothecia rare; cortex K+Y (atranorin), medulla K+Y to R, P+O or R (salacinic acid), C-, KC-.

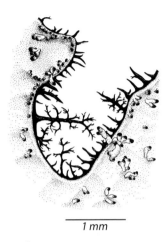

1 mm

Range: Alas to Cal, Coast Range and on the immediate coast.

Substrate: Bark and wood, mainly on conifers.

Habitat: Wet coastal forests, most often within sight or sound of the ocean.

Notes: The combination of isidia and squarrose-branched rhizines are diagnostic. *P. sulcata* also has squarrose-branched rhizines but is sorediate, never isidiate. If the squarrose rhizines are overlooked, *P. squarrosa* would be misidentified as the bark-dwelling form of *P. saxatilis.*

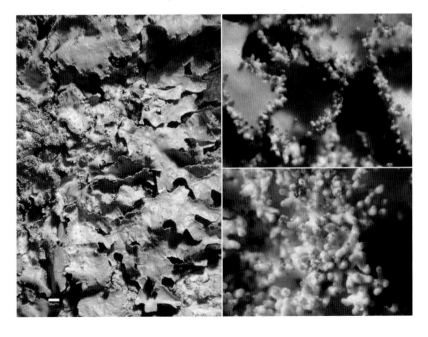

Parmelia sulcata

Description: Thallus foliose, medium sized, mostly < 6(10) cm diam, loosely appressed; lobes mostly 1-4 mm broad; upper surface whitish gray to greenish or blue gray, with pale angular markings toward the lobe tips; isidia lacking, soredia present, marginal and laminal, developing on the angular cracks and ridges; lower surface black, brown towards the margin; rhizines squarrosely branched (simple towards the margins); apothecia rare; cortex K+Y (atranorin), medulla K+Y to R, P+O or R (salacinic acid), C-, KC-.

Range: Widespread, common throughout the PNW.

Substrate: Bark and wood, less often or rock or mossy rock.

Habitat: Occurring in a wide range of habitats, in both continental and oceanic climates, ranging from urban areas to the mountains but dropping out at the highest elevations.

Notes: The angular cracks developing powdery soredia are diagnostic for *P. sulcata* and *P. barrenoae*; see that species for differences. *P. hygrophila* can be similar in appearance but has coarser isidia-like propagules. Young *Parmelia* specimens can be difficult to identify if they lack isidia and soredia. The squarrose rhizines of *P. sulcata*, however, distinguish it from the other local *Parmelia* except for *P. squarrosa*, which is restricted to near the coast.

Parmeliella—*see* key to Fuscopannaria

Parmelina

Parmelina coleae

Description: Thallus foliose, medium sized, mostly < 6 (10) cm diam, rather tightly appressed; lobes mostly 2-3 mm broad; lobe axils often with short cilia; upper surface whitish gray; soredia and isidia lacking, lower surface black, brown towards the margin; rhizines dense, short, simple to sparingly branched; apothecia common, often abundant; disk brown; cortex K+Y, medulla K-, P-, C+R, KC+R (atranorin and lecanoric acid).

Range: S Ore to Cal, Coast Range, Siskiyou Mts, and Sierra Nevada; in Ore most frequent at low elevations in the Rogue River watershed from Merlin to Medford.

Substrate: Bark and wood, mainly on hardwoods and shrubs.

Habitat: Oak forests, savannas, and riparian areas.

Notes: Easily recognized in the field, *P. coleae* is our only whitish gray parmelioid species on bark that lacks soredia and isidia and reliably produces apothecia. It was formerly known as *P. quercina*, but recent work has determined that the w N Am populations are a different species.

Sources include: Argüello et al. (2007).

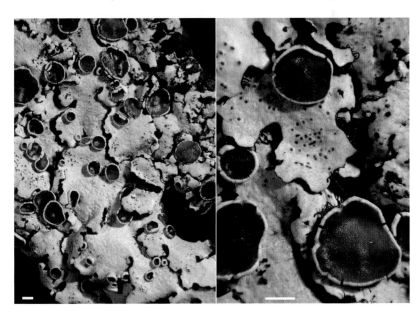

Parmeliopsis

Thallus foliose, to 6 cm broad; lobes narrow, generally contiguous, about 1 mm wide, closely appressed; upper surface whitish gray or pale yellowish green; lower surface brown to brown black with rhizines; soredia present in round laminal soralia; apothecia lecanorine, the disk brown; cortex K+Y (atranorin) or K-, KC+Y (usnic acid); medulla K-, P-, C-, UV+ white (divaricatic acid); most common on conifer bark, also on wood and deciduous trees and shrubs.

Sources include: Meyer (1982).

ID tips: Parmeliopsis is a great genus for learning to see the presence of usnic acid by the yellowish tint in the cortex. In conifer forests in the mountains, both species are quite common, and are identical in form, and can frequently be found growing next to each other.

1a Thallus whitish gray to gray (lacking usnic acid, containing atranorin); upper cortex K+Y .... *P. hyperopta*
1b Thallus pale yellowish green (containing usnic acid); upper cortex not much changed in K ....*P. ambigua*

Parmeliopsis ambigua (left), with yellowish tint from usnic acid, and *P. hyperopta* (right), lacking usnic acid.

Parmeliopsis ambigua

Description: Thallus foliose, to 6 cm broad; lobes narrow, generally contiguous, about 1 mm wide, closely appressed; upper surface pale yellowish green; lower surface brown to brown black; soralia round, laminal; apothecia occasional; cortex K-, KC+Y (usnic acid); medulla K-, P-, C-, UV+ white (divaricatic acid).

Range: Widespread throughout the PNW but most common in more continental climates, increasingly rare toward the coast.

Substrate: Most common on conifer bark, also on wood, deciduous trees, and shrubs.

Habitat: Moist to dry conifer forests, low elevations to subalpine, tolerant of snow burial so often found on tree bases in the mountains.

Notes: Easily recognized in the field by the small, closely appressed greenish or yellowish green lobes, usually with abundant roundish soralia. *Parmeliopsis hyperopta* is essentially identical in form but is pale gray instead of yellowish green.

Parmotrema arnoldii

Description: Thallus foliose, mostly < 12(20) cm broad, loosely adnate, the margins usually free from the substrate; lobes mostly (2) 4-12 mm broad, with marginal black cilia; upper surface whitish, gray, or greenish gray; lower surface black with a broad brown margin; soralia rounded to crescent-shaped, terminal on short side lobes; apothecia rare; cortex K+Y (atranorin); medulla K-, C-, KC+R, P-, UV+ blue-white (alectoronic and alpha-collatolic acids).

Range: Coastal BC to Mex; w Cas.

Substrate: Bark and wood, rarely rock.

Habitat: Coastal conifer forests, hardwood forests, and scrub; infrequent in Willamette Valley and Cas.

Notes: Parmotrema arnoldii commonly occurs with the superficially similar *P. perlatum*. When well developed, *P. arnoldii* has broader lobes than *P. perlatum*, but size is so variable that medium-sized or smaller material cannot be reliably separated on that criterion. Instead, one must rely on the K-, UV+ brilliant blue-white medulla of *P. arnoldii*, as opposed to the K+Y, UV- medulla of *P. perlatum*. *Cetrelia cetrarioides* is similarly broad lobed and UV+, but has minute pseudocyphellae on the upper surface and elongate marginal soralia.

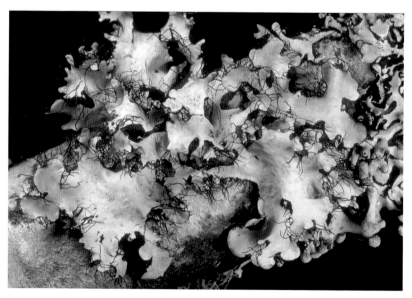

Parmotrema crinitum

Description: Thallus foliose, mostly
< 7(15) cm broad, loosely adnate,
the margins usually free from the
substrate; lobes mostly 2-6(10)
mm broad, with marginal black cilia; upper surface whitish, gray, or
greenish gray; lower surface black
with a brown margin; soredia absent; isidia present, cylindrical or
barrel shaped, often tipped with a
single black cilium; apothecia rare;

cortex K+Y (atranorin); medulla K+Y, C-, KC-, P+O, UV- (stictic acid group).

Range: BC to Mex on the coast.

Substrate: Bark and wood, occasionally on rock.

Habitat: Sporadic along the immediate coast, most abundant in fairly open
forests as on the edge of headland forests and boulders, fences, and trees by
parking and picnic areas.

Notes: The distinctive isidia that often become tipped with cilia are unique in
the lichen flora of the PNW. The isidia are typically well formed and abundant. Young individuals with few or no cilia on the isidia might be mistaken
for *Parmelia saxatilis*, but *Parmelia* lacks marginal cilia on the lobes and angular whitish pseudocyphellae on the upper surface.

Parmotrema perlatum

Description: Thallus foliose, mostly < 7(15) cm broad, loosely adnate, the margins usually free from the substrate; lobes mostly 2-6(10) mm broad, with usually sparse marginal black cilia; upper surface whitish, gray, or greenish gray; lower surface black with a broad brown margin; soralia rounded to crescent shaped, terminal to subterminal, often on short side lobes; apothecia rare; cortex K+Y (atranorin); medulla K+Y, P+O, KC-, C-, UV- (stictic acid group).

Range: BC to Mex on the coast, sporadic in the Coast Range and elsewhere w Cas.

Substrate: Bark and wood, occasionally on rock.

Habitat: Coastal conifer forests, hardwood forests, and scrub, sometimes in more open habitats near the coast; also penetrating inland in low-elevation (usually riparian) forests.

Notes: Parmotrema perlatum is the most common *Parmotrema* in the PNW, followed by *P. arnoldii*. In addition to the distinguishing features noted under *P. arnoldii, P. perlatum* is less restricted to the immediate coast. Because the medulla of *P. arnoldii* fluoresces strongly while *P. perlatum* does not, nighttime exploration with a battery-operated UV lamp provides an interesting way of detecting the two species.

Peltigera

Thallus foliose, small to very large; stratified (gelatinous in one species) upper surface a drab gray, brown, yellowish brown, or greenish; lower surface lacking a cortex, veined (sometimes faint or coalescing); rhizines present, separate, in tufts, or confluent into rows or mats; isidia, soredia, or lobules present in some spp; apothecia lecanorine, often on the tips of upright lobes; disk brown, orange brown, or black; spores multiseptate; spot tests mostly negative; primary photobiont blue green (*Nostoc*) or green; if the latter then with a blue-green photobiont in cephalodia; on the ground, rock, or on trees. One species is aquatic.

Sources include: Goffinet and Hastings (1994), Goffinet et al. (2003), Goward et al. (1995), Goward and Goffinet (2000), Lendemer and O'Brien (2011), Magain et al. (2016), McCune (1984), McCune and Stone (2022), Miadlikowska and Lutzoni (2000, 2004), Miadlikowska et al. (2003, 2011, 2014), Thomson (1950), Vitikainen (1985).

ID tips: The pattern and form of the venation and rhizines are important for distinguishing species. See the photos of lower surfaces of many species in the *P. canina* and *P. polydactylon* groups, pp. 292–294. Additional species known from Alaska and Canada but not in the Lower 48 are not keyed here. Two of these are *P. latiloba* and *P. scabrosella* (Spribille et al. 2010).

Introductory Key

1a Habitat aquatic, growing submerged most of the year; thallus gelatinous, not stratified ....*P. hydrothyria*
1b Habitat terrestrial, never chronically submerged; thallus stratified
  2a Primary photobiont green; thallus bright green when moist
  ....**Group 1** (Lead 5)
  2b Primary photobiont blue green; thallus gray, greenish, blue green, or black when moist
    3a Soredia, isidia, or lobules present, the lobules along the margins or cracks in the thallus ....**Group 2** (Lead 9)
    3b Soredia, isidia, and lobules absent
      4a Upper surface glabrous, lacking tomentum and not scabrous
      .... **Group 3** (Lead 19)
      4b Upper surface tomentose and/or scabrous, particularly toward the lobe tips (scabrosity is a subtle almost crystalline texture in the upper cortex, as opposed to whiter, more superficial pruinosity)
      .... **Group 4** (Lead 24)

Group 1
Primary photobiont green

5a Thallus small (to 2 cm broad), fan-shaped; veins dark, raised, distinct; apothecia horizontal..*P. venosa*

5b Thallus generally larger; veins various; apothecia ± erect
  6a Veins distinct, even toward the center of the thallus, gradually
    darkening inwards
    7a Lower surfaces of apothecia continuously corticate; lobes few; lobe
      margins even or weakly ruffled; cephalodia in central portions of
      the thallus up to 2 mm diam. Thallus large, to 20(30) cm broad;
      lobes to 4.0(5.5) cm broad; upper surface glabrous or with minute
      glassy hairs near the lobe tips; cephalodia appressed throughout;
      rhizines dark brown throughout or occasionally pale near the
      margins; apothecia common, on erect, narrow lobes; on moss,
      mossy rock, and logs; usually at higher elevations where the snow
      lies late, also along the coast; coastal se Alas s to n Wash, inland to se
      BC and w Alta; to be sought in nw Mont and n Id....... *P. chionophila*
    7b Lower surfaces of apothecia patchily corticate to noncorticate; lobes
      many; lobe margins strongly ruffled; cephalodia seldom > 1 mm
      ....***P. leucophlebia***
  6b Veins broad and indistinct, with an abrupt (generally < 1 cm wide)
    transition zone from the pale margins to a gray or black center;
    greenish cortex continuous beneath apothecia
    8a Cephalodia raised, the edges free, easily flicked off with a fingernail,
      occasionally developing into grayish lobules ....***P. britannica***
    8b Cephalodia sessile to sunken, less readily removed, not becoming
      lobulate ....***P. aphthosa***

Group 2

Soredia, isidia, or lobules present

9a Soredia present, coarsely granular, sometimes isidia-like
  10a Soralia marginal, linear.. *P. collina*
  10b Soralia laminal, roundish, occasionally disappearing with age (then
    with roundish scars remaining from the soralia)
    11a Upper cortex glabrous throughout
      ....***P. didactyla*** (young thalli; see below)
    11b Upper cortex tomentose at least along the margins
      12a Thallus of a single rounded lobe, soon strongly concave and
        pouch-like, but often with shallow secondary lobes; outermost
        rhizines mostly tapered to a point................................*P. didactyla*
      12b Thallus usually developing multiple lobes, weakly to strongly
        concave, but not usually pouch-like; outermost rhizines splayed
        apart at the tips or woolly
        13a Upper surface smooth, ± shiny except for the tomentose lobe
          tips, chestnut brown (more gray where sheltered); lower
          surface darkening towards the center of the thallus; rhizines
          often tufted and confluent into rows toward the lobe tips;
          pycnidia frequent on the lobe margins. Thallus to 8 cm wide;
          lobes stiff, fragile, to 1.5 cm wide, roundish, strongly concave
          or plane, lobe tips rounded and upturned; soredia granular,
          brown; lower surface densely veined; veins white, grading

inward to brown or black; apothecia unknown; on moss mats over rock, openings in boreal and montane forests to alpine; se Alas, Yukon to Banff NP and BC.............................. [*P. castanea*]

13b Upper surface dull, tomentose in the lobe tips, then minutely roughened and scabrous toward the thallus center, sometimes nearly glabrous in the center, gray to light purplish brown; lower surface mostly pale (white to pale tan); rhizines ± wooly throughout (p. 293); pycnidia rare. Apothecia rare, on the lobe margins; on mosses over soil, rock, wood, rarely directly on soil; widespread, fairly common.............................. *P. extenuata*

9b Soredia absent; isidia or lobules present

14a Thallus with true laminal isidia that are not restricted to cracks in the thallus; isidia globular to flattened. Thallus small, to 2 cm, often ear shaped, brownish to grayish; alpine to subalpine, occasionally at lower elevations; circumpolar s to Ore and Colo .... *P. lepidophora*

14b Thallus with lobules that are generally most abundant marginally and along cracks in the upper cortex

15a Upper surface tomentose toward the lobe tips

16a Veins narrowly raised; rhizines lightly tomentose.....***P. praetextata***

16b Veins intermediate between low and narrowly raised; rhizines lacking tomentum. Rhizines long, simple to somewhat branched, cream to brown; veins white to brown; upper surface scabrid and/or tomentose; Arctic s to Ore e Cas .... *P. monticola*

15b Upper surface glabrous

17a Apothecia horizontal; outer rhizines in concentric rows; veins indistinct and diffuse ....***P. elisabethae***

17b Apothecia ± vertical on upright lobes; outer rhizines not in concentric rows; veins distinct

18a Veins darkening inward, broad and low; marginal lobules generally well developed...***P. pacifica***

18b Veins generally whitish to pale brown, raised, distinct; marginal lobules often poorly developed.................. *P. degenii* (Lead 22a)

Group 3

Glabrous; no vegetative propagules

19a Apothecia horizontal; disk flat; outer rhizines in concentric rows. Lower surface generally with dark, broadly confluent to ± distinct veins; on soil, rotten wood, and rock; subarctic s to NM, uncommon in PNW .... *P. horizontalis*

19b Apothecia ± vertical on upright lobes; outer rhizines more randomly scattered

20a Rhizines long (to 7-10 mm), slender, rarely branched; thallus large; lobes 2-3 cm broad...***P. neopolydactyla***

20b Rhizines short (mostly < 5 mm); thallus mid-sized; lobes < 2 cm wide

21a Apothecia dark brown to black... ***P. neckeri***

21b Apothecia light brown to orangish brown or brown

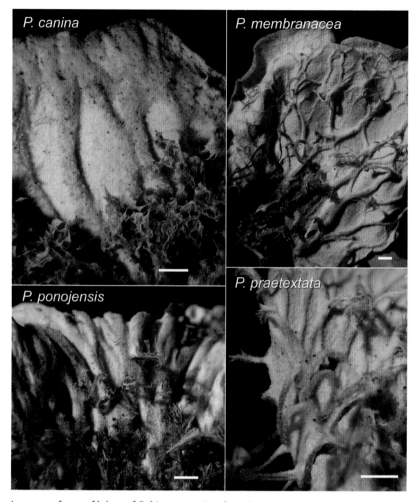

Lower surfaces of lobes of *Peltigera* species, *P. canina* group

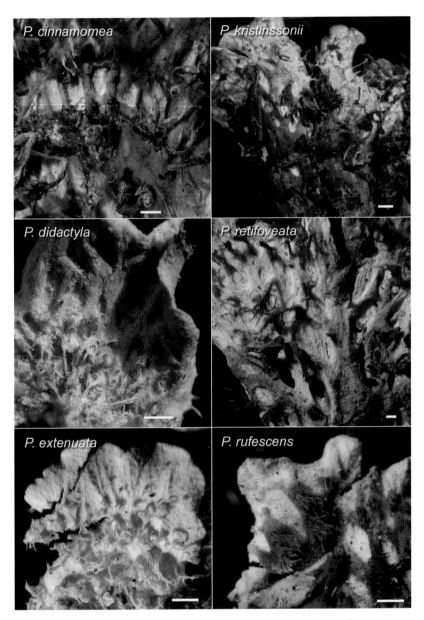

Lower surfaces of lobes of *Peltigera* species, *P. canina* group continued

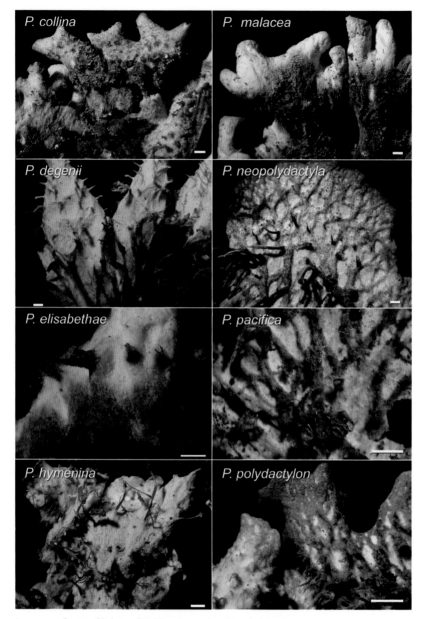

Lower surfaces of lobes of *Peltigera* species, *P. polydactylon* group

Peltigera aphthosa

Description: Thallus foliose, loosely ap
pressed, often large (to 30 cm diam or
more); lobes broadly rounded, mostly 2-4
cm broad; upper surface light greenish or
pale greenish tan when dry, bright green
when wet; veins broad and indistinct,
with an abrupt (generally < 1 cm wide)
transition zone from the pale margins to
a gray or black center; cephalodia sessile
to slightly sunken, not readily removed;
primary photobiont green; soredia,
isidia, and lobules lacking; apothecia
on raised narrow lobes; greenish cortex
present beneath apothecia; spot tests
negative.

Range: Widespread, throughout forested
areas of the PNW.

Substrate: Soil, forest floor, and mossy
rocks.

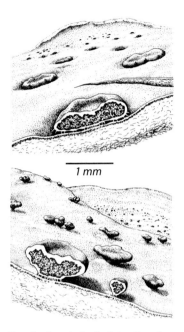

Detail of cephalodia from *P. aptho-
sa* (above) and *P. britannica* (below)

Habitat: Common in more continental climates e Cas but largely replaced by
P. britannica in low- to mid-elevation moist forests w Cas; frequent on roadcuts and in open outcrop and talus areas.

Notes: This common species of cool, moist forests in continental climates is
similar to *P. britannica*. Specimens with tightly appressed cephalodia are assigned to *P. aphthosa*, while those with cephalodia that are slightly raised and
free at the edges should be named *P. britannica*. Peltigera aphthosa also resembles *P. leucophlebia*. Check the backs of the apothecia on fertile specimens.
If the greenish cortex is largely or wholly lacking then it is *P. leucophlebia*. On
sterile specimens one must rely on the lower surface. *Peltigera leucophlebia*
has veins that remain distinct even toward the center of the thallus, while
P. aphthosa and *P. britannica* have indistinct, confluent veins. Also, the lower surface of *P. leucophlebia* gradually changes from white near the edges to
black near the center, while in *P. britannica* and *P. aphthosa* the color change
is relatively abrupt, such that the lower surface shows a pale outer zone and a
dark inner zone.

Peltigera britannica

Description: Thallus foliose, loosely appressed, large (to 30 cm diam or more); lobes broadly rounded, mostly 2-4 cm broad; upper surface light greenish or pale greenish tan when dry, bright green when wet; veins broad and indistinct, with an abrupt (generally < 1 cm wide) transition zone from the pale margins to a gray or black center; cephalodia slightly raised, with free margins (see illustration on preceding page), easily removed with a flick of a fingernail; primary photobiont green; soredia and isidia lacking; apothecia on raised narrow lobes; greenish cortex present below apothecia; spot tests negative.

Range: Alas to n Cal, inland to w Mont.

Substrate: Soil, forest floor, and mossy rocks; occasionally epiphytic in wet old-growth forests.

Habitat: Roadcuts, trailsides, outcrop areas, talus slopes, old-growth canopies, and other partly open habitats in areas of moist oceanic forests.

Notes: This is the most common green *Peltigera* w Cas. A curious form lacking the green photobiont can often be found by searching habitats with abundant *P. britannica*. The lower surface is similar to normal *P. britannica*, but the upper surface is gray, usually with subtle whitish mottling. One can often find this blue-green photomorph reverting to the normal form: look for greenish lobes emerging from the gray thalli.

Peltigera canina

Description: Thallus foliose, loosely appressed, often large (to 50 cm diam or more); lobes broadly rounded, mostly 2-4 cm broad; upper surface grayish or brownish, tomentose, not scabrous; lobe margins undulating, downturned; veins raised and distinct, lacking tomentum, brown but pale towards margins; rhizines branched into bushy tufts (like shaving brushes, see below); primary photobiont blue green; soredia, isidia, and lobules lacking; apothecia on raised narrow lobes; spot tests negative.

Range: Widespread, found throughout the PNW but most common e Cas.

Substrate: On soil, mosses, forest floor, rotten logs, or soil or moss over rock.

Habitat: Forests and relatively open areas (outcrops, talus) or disturbed areas (roadcuts, clearcuts); low elevations to subalpine, mainly in forested regions with a continental climate.

Notes: Formerly used in a very broad sense, the concept of *P. canina* has narrowed in recent years. In most cases one can separate *P. canina* from its relatives without even using a hand lens. The bushy tufts of rhizines distinguish it from *P. cinnamomea, P. membranacea,* and *P. praetextata,* which have discrete slender rhizines. *Peltigera rufescens* and *P. ponojensis* both have lobe edges that are strongly upturned and the lobes are generally smaller (often < 1 cm).

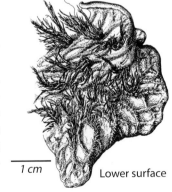

1 cm   Lower surface

Peltigera cinnamomea

Description: Thallus foliose, closely to loosely appressed, often large (to 30 cm diam or more); lobes broadly rounded, mostly 1.5-3 cm broad; upper surface pale grayish or pale brownish, usually tomentose, at least toward the margins; lobe margins undulating, nearly plane to downturned; veins narrow, sharply raised and distinct, threadlike near the margins when dry, lacking tomentum (use lens), cream to pale tan to rusty or cinnamon brown or darker toward the center; rhizines slender, becoming long (to 1 cm), mostly unbranched but often split at the tips, occasionally split to the base into two or three divisions, lacking erect tomentum; primary photobiont blue green; soredia, isidia, and lobules lacking; apothecia common, on raised narrow lobes; spot tests negative.

Range: Alas to Alta, s in RM through Mont and Id to Colo, also in ne Ore, but in Ore and Wash apparently restricted to e Cas.

Substrate: On soil, mosses, forest floor, rotten logs, or soil or moss over rock.

Habitat: Occasional in forests in suboceanic to continental climates at low to mid elevations, becoming rare in the southern part of its range.

Notes: Peltigera cinnamomea is a relatively rare species in Ore and Wash, but becomes more common in the RM of BC, Id, and Mont. In the field look for a large, flat-lying *Peltigera* on the forest floor—it lies flatter than most *Peltigera* species, including the similar-looking *P. membranacea*. Confirmation of *P. cinnamomea* requires observing the absence of tomentum on the veins and rhizines, absence of lobules, and presence of light brown to cinnamon-colored veins toward the center of the thallus. It is most likely to be confused with *P. membranacea*, which has fine erect tomentum on the veins and rhizines.

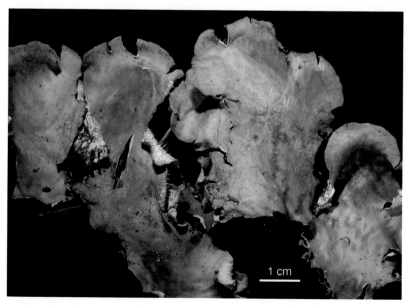

1 cm

Peltigera collina

Description: Thallus foliose, loosely appressed, medium to large (to 15 cm diam or more); lobes broadly rounded or rather elongate, mostly 0.5-1.5 cm broad; upper surface grayish or brownish, smooth or slightly scabrous near the margin, sometimes with a whitish or grayish necrotic surface layer in exposed habitats; lobe margins plane to upturned; veins brown, low, occasionally almost lacking (p. 294); rhizines simple to bushy, sometimes confluent; primary photobiont blue green; soredia usually present on the lobe margins; apothecia black, on raised narrow lobes; spot tests negative.

Range: Widespread, common; throughout the PNW.

Substrate: Mossy bark and rock.

Habitat: Moist habitats from low elevations to subalpine.

Notes: This is the only lichen in the PNW with veins below and sorediate margins. The only other sorediate *Peltigera* spp in the PNW are in the *P. didactyla* group. These are much smaller (usually < 2 cm diam) and have roundish laminal soralia. *Peltigera collina* is a common epiphyte on mossy branches (mainly hardwoods) w Cas. E Cas it occurs mainly on mossy rock. Specimens lacking soredia are often confused with *P. neckeri* and are common w Cas. *Peltigera neckeri* is almost never epiphytic, however, and has a shinier thallus and less distinct veins.

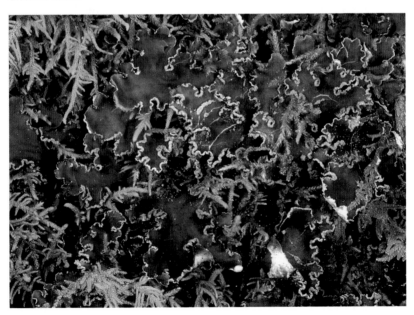

Peltigera didactyla

Description: Thallus foliose, roundish, appressed or edges raised, usually 0.5-2 cm diam; lobes broadly rounded, mostly 0.5-1.5 cm broad; upper surface grayish or brown, tomentose; lobe margins plane to upturned; veins brown, low; rhizines short, white, in the central part of the thallus (p. 293); primary photobiont blue green; soredia present in roundish laminal soralia, sometimes disappearing and leaving roundish scars; apothecia dark brown, on raised narrow lobes; spot tests negative.

Range: Widespread, occasional in forested areas throughout the PNW; occasional in steppe or alpine tundra.

Substrate: On soil or soil over rock.

Habitat: Usually in recently disturbed habitats such as roadcuts, clearcuts, trailsides, and recently burned forests; low to high elevations.

Notes: Easily overlooked, *P. didactyla* is characterized by the small roundish thalli with laminal soralia, usually directly on soil. *Peltigera lepidophora* is similar in its small roundish thallus but it has laminal isidia rather than soralia. Small, shiny specimens of *P. didactyla* can be difficult to distinguish from *P. castanea*, but *P. didactyla* has a more purplish brown color in contrast to the dark chestnut brown color of *P. castanea*. *Peltigera extenuata*, a close relative of *P. didactyla*, has a broader, grayer, less pouch-like thallus, and bushier rhizines than *P. didactyla*.

Peltigera elisabethae

Description: Thallus foliose, often large (to 20 cm diam or more); lobes broadly rounded, mostly 1-3 cm broad, the edges upturned; upper surface grayish or brownish, glossy or slightly pruinose near the margins; veins dark, indistinct and strongly confluent or appearing veinless; rhizines short, tufted, arranged in concentric rows (p. 294); primary photobiont blue green; soredia and isidia lacking; lobules present on the lobe margins and cracks; apothecia brown, on lobes on the same plane as the rest of the thallus; spot tests negative.

Range: Widespread, arctic to Ore and Colo; in the PNW, mainly e Cas.

Substrate: Soil, moss, and mossy rock; occasionally over ± calcareous substrates.

Habitat: Most frequent in cool, moist, rocky habitats, such as outcrop areas and north-facing talus slopes.

Notes: Closely related to *P. horizontalis*, *P. elisabethae* is distinguished by the presence of lobules on the margins and along cracks in the thallus. Also, *P. horizontalis* is never pruinose, but *P. elisabethae* often has pruina on the lobe tips. These are the only two *Peltigera* in the PNW that have horizontal apothecia (not on erect slender lobes) and rhizines arranged in concentric rows. They also share a distinctly glossy upper surface. *Peltigera pacifica* also has a smooth, glossy upper surface and lobules but has distinct veins below.

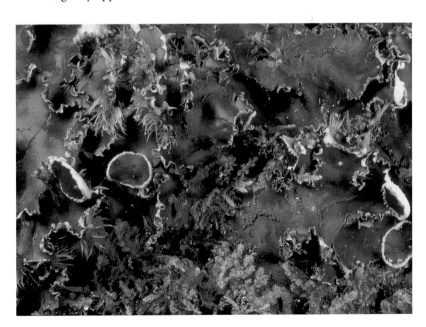

Peltigera hydrothyria

Description: Thallus foliose, gelatinous (non-stratified), medium sized, dark bluish gray to dark gray or black, loosely appressed in ruffles; lobes to about 1 cm wide; lower surface dark, distinctly veined; apothecia common, the disk brownish; spores 3-septate; photobiont *Nostoc*; spot tests negative.

Range: Known from each of the major mountain chains in the US and s Canada, but more common in Cas and Sierra Nevada than in the other ranges.

Substrate: Rocks and gravel, rarely wood.

Habitat: Aquatic, in mountain streams and springs, especially those without marked seasonal fluctuations.

Notes: Previously known as *Hydrothyria venosa*, this distinctive N Am endemic is the only gelatinous lichen with distinct veins and the only gelatinous *Peltigera*. It is also the largest strictly aquatic lichen in the PNW. Look for dark ruffled masses in small spring-fed streams. While searching for *P. hydrothyria* you will encounter ear-like lobes of non-lichenized *Nostoc* attached to submerged rocks. These are more jelly-like than *P. hydrothyria* and are never veined. Recognition of recent segregates of *P. hydrothyria* requires DNA sequencing. For practicality we treat *P. gowardii* and *P. aquatica* as varieties of *P. hydrothyria*. Recognition of the genetic structure at the level of varieties, rather than species, restores practical utility of the species concept in this group, but doesn't prevent others from using the genetic variants at the species level. See more explanation in McCune and Stone (2022).

Peltigera leucophlebia

Description: Thallus foliose, loosely appressed, often large (to 40 cm diam or more); lobes broadly rounded, mostly 2-3 cm broad; upper surface light greenish or pale greenish tan when dry, bright green when wet; veins distinct, with a gradual transition from the whitish veins near the margins to black veins at the center; cephalodia sessile to slightly sunken, not readily removed; primary photobiont green; soredia, isidia, and lobules lacking; apothecia on raised narrow lobes; greenish cortex discontinuous or lacking beneath apothecia; spot tests negative.

Range: Widespread; in forests throughout the PNW.

Substrate: On mossy rock, forest floor, and mossy soil.

Habitat: At all elevations, occasionally in cold or dry habitats but generally in montane conifer forests.

Notes: Peltigera leucophlebia is characterized by its bright green color when wet, distinct veins on the lower surface, and the backs of the apothecia lacking a green cortex. *P. leucophlebia* is very similar to *P. aphthosa* and *P. britannica,* and frequently can be found growing adjacent to

those species. Intermediates are fairly frequent. See discussion of differences under *P. aphthosa.*

Peltigera malacea

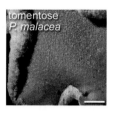

Description: Thallus foliose, medium to large (to 15 cm diam or more); lobes broadly rounded, mostly about 1(3) cm broad, the edges upturned; upper surface dark forest green when wet, grayish or brownish when dry, smooth but always with minute erect tomentum on the lobe tips; veins very indistinct or absent (p. 294); rhizines short, tufted; primary photobiont blue green; soredia, isidia, and lobules lacking; apothecia brown, on upright elongated lobes; spot tests negative.

Range: Widespread, Arctic to Ore and throughout RM; common e Cas, rare w Cas.

Substrate: Usually on mossy rock, less often on litter, occasionally on acidic soils.

Habitat: Low elevations to alpine, usually in semi-open or open areas of rock outcrops or talus; mainly in continental to slightly oceanic climates.

Notes: The nearly veinless lower surface and smooth upper surface with tomentum on the lobe tips are diagnostic. The blue-green photomorph of *P. britannica* can be similar in appearance, but it has subtle white mottling on the upper surface, is dark gray rather than dark green when wet, and often has small lobes reverting to the green photomorph.

Peltigera membranacea

Description: Thallus foliose, loosely appressed, often large (to 40 cm diam or more); lobes broadly rounded, mostly 2-4 cm broad; upper surface grayish or brownish, usually tomentose, at least toward the margins; lobe margins undulating, downturned; veins narrow, sharply raised and distinct, threadlike near the margins when dry, with a fine, erect tomentum (use lens), whitish or dark toward the center (p. 292); rhizines slender, long (often 1 cm), mostly un-

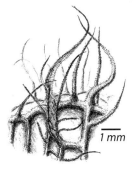

1 mm

branched, covered with fine erect tomentum; primary photobiont blue green; soredia, isidia, and lobules lacking; apothecia on raised narrow lobes; spot tests negative.

Range: Circumpolar, boreal to temperate, with oceanic affinities; Alas to Cal e to Alta and w Mont.

Substrate: On soil, mosses, forest floor, rotten logs, or soil or moss over rock.

Habitat: Common in mesic oceanic forests at low to mid elevations.

Notes: Peltigera membranacea is the most common *Peltigera* w Cas. Its large lobes with narrow, distinct veins and long slender rhizines are characteristic. Confirmation requires observing the fine erect tomentum on the veins and rhizines. *Peltigera praetextata* can be very similar, particularly if lacking regeneration lobules, but can be distinguished by veins with sparse to no tomentum on them. *Peltigera canina* has bushy clumps of rhizines and veins without tomentum. The development of the tomentum on the upper surface of *P. membranacea* varies with climate, the upper surface often being completely glabrous near the ocean. At the most continental localities, the upper surface can be completely covered with tomentum.

Peltigera neckeri

Description: Thallus foliose, loosely appressed, medium to large (to 20 cm diam); lobes broadly rounded, mostly 1-1.5(2) cm broad; upper surface gray or brownish gray, generally glossy, often with ± pruinose margins; lobe margins plane to raised; lower surface with wide, low, dark veins contrasting with white spaces, occasionally the veins few; rhizines mostly dark brown, to 5 mm long, often sparse, simple or more often in bunches (fasciculate); primary photobiont blue green; soredia, isidia, and lobules lacking; apothecia small (up to 6 mm long), dark brown to black, on raised narrow lobes; spot tests negative.

Range: Widespread, uncommon w Cas, more common n and e Cas.

Substrate: Most often on rotten logs, also on humus.

Habitat: Moist conifer forests.

Notes: Sometimes *P. collina* lacks soredia (especially when epiphytic), making it similar to *P. neckeri*. They both have a shiny upper surface and blackish apothecia, in contrast to the usual brown apothecia in *Peltigera*. As compared to *P. collina*, *P. neckeri* has a shinier thallus, less distinct veins, and is almost never epiphytic. *P. collina* is a very common epiphyte w Cas.

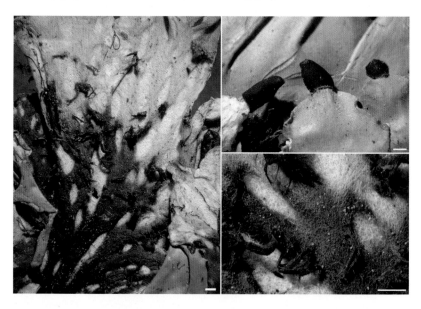

Peltigera neopolydactyla

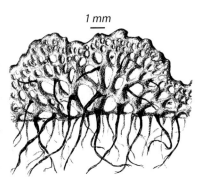

1 mm

Description: Thallus foliose, loosely appressed, often large (to 40 cm diam or more); lobes broadly rounded, mostly 2-4 cm broad; upper surface gray or brownish gray, smooth, lacking tomentum but sometimes faintly pruinose near the margins, rarely slightly scabrous; lobe margins plane or undulating; veins broad, low, pale to dark brown (p. 294); rhizines slender, long (often 1 cm), mostly unbranched; primary photobiont blue green; soredia, isidia, and lobules lacking; apothecia on raised narrow lobes; spot tests negative.

Range: Alas to Cal, inland to Mont.

Substrate: On soil, mosses, forest floor, rotten logs, or soil or moss over rock.

Habitat: Common in coastal forests, increasingly rare eastward where it is restricted to low-elevation forests in areas of strongest coastal influence.

Notes: This is the most common large-lobed, smooth-surfaced *Peltigera* w Cas. The glabrous, pale gray coastal form of *P. membranacea* may be distinguished by its erect-tomentose, narrow, threadlike veins, in contrast to the broad, low, smooth veins of *P. neopolydactyla*. The species is actually a complex of genetically isolated but essentially indistinguishable varieties.

1 cm

Peltigera pacifica

Description: Thallus foliose, to 15 cm diam or more; lobes mostly about 0.5-1 cm broad; upper surface gray or brownish gray, smooth, shiny, lacking tomentum; lobe margins mostly upturned; veins dark, broad and low to somewhat raised, contrasting with the pale interstices (p. 294), paler near the margins; rhizines slender, mostly unbranched; primary photobiont blue green; lobe margins and cracks usually with abundant lobules; soredia and isidia lacking; apothecia on raised narrow lobes; spot tests negative.

Range: Coastal Alas to Ore, mainly w Cas, rarely inland to w Mont.

Substrate: Soil, moss, rocks, logs, and tree bases.

Habitat: Low-elevation moist forests.

Notes: Distinctive in the generally abundant marginal lobules, narrow lobes, and glabrous upper surface. It frequently has a disheveled appearance, with a partially discolored upper surface and margins partially consumed by herbivores. *Peltigera elisabethae* also has a smooth, shiny upper surface and lobules but it has strongly confluent, indistinct veins and has rhizines arranged in concentric rows.

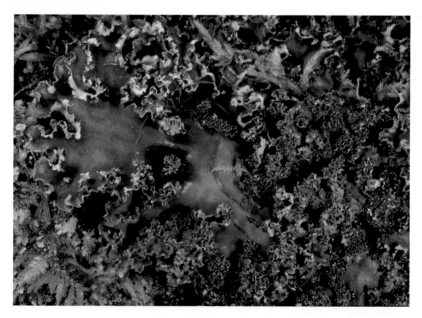

Peltigera ponojensis

Description: Thallus foliose, to 10 cm diam or more; lobes mostly about 1 cm broad; upper surface grayish, brownish, or whitish, tomentose and rarely scabrous; lobe margins upturned; veins raised and distinct, cream to tan throughout (occasionally darker), lacking tomentum, closely set and stout (p. 239); rhizines pale, simple to slightly branched, to 7 mm long; primary photobiont blue green; soredia, isidia, and lobules lacking; apothecia on raised narrow lobes; spot tests negative.

Range: Alas to Cal, inland throughout RM.

Substrate: Soil, moss, and organic debris.

Habitat: Occasional on exposed or partly exposed sites, dry forest fringes to cold semiarid sites; low elevations to alpine.

Notes: The upper surface and general form are very much like *P. rufescens*. However, *P. ponojensis* has pale veins that are exposed, closely set, and ropy in appearance. Furthermore, *P. ponojensis* bears distinct, pale rhizines, while *P. rufescens* has dark veins that are obscured by hedgerows or mats of dark rhizines. Although not formally named, *P. "scotteri"* in the literature is similar to *P. ponojensis* but glabrous above. DNA studies have shown putative *P. scotteri* to be heterogeneous. *Peltigera monticola* has less raised and often somewhat light to medium reddish brown veins, cream to brown rhizines (both simple and branched), and often lobulate margins. It occurs from the Arctic s to BC and Ore.

Peltigera praetextata

Description: Thallus foliose, loosely appressed, often large (to 30 cm diam or more); lobes broadly rounded, mostly 1-3 cm broad; upper surface grayish or brownish, tomentose, occasionally slightly scabrous; lobe margins upturned or downturned in spots; veins raised and distinct, mostly lacking tomentum (p. 292), brown but pale towards margins; rhizines discrete, long, little branched, often with minute erect tomentum; primary photobiont blue green; soredia and isidia lacking; lobules often present on margins and cracks; apothecia on raised narrow lobes; spot tests negative.

1 cm

Range: Widespread, throughout the PNW.

Substrate: On soil, mosses, forest floor, rotten logs, or soil or moss over rock.

Habitat: In a wide range of forests, often where disturbed, such as roadcuts and clearcuts, shade to full sun; low elevations to alpine.

Notes: This very common species is often mistaken for *P. canina*, *P. membranacea*, or *P. rufescens* when the lobules are scarce or lacking. *Peltigera praetextata* differs from *P. canina* and *P. rufescens* in producing slender discrete rhizines, as opposed to the bushy tufts of rhizines in *P. canina* or the mats of dark confluent rhizines in *P. rufescens*. In moist sites, specimens of *P. praetextata* without lobules can be similar to *P. membranacea*. But unlike that species, *P. praetextata* often has upturned margins on the sides of the lobes and little or no tomentum on the veins.

Peltigera rufescens

Description: Thallus foliose, to about 10 cm diam; lobes mostly about 1 cm broad or less; upper surface brownish or whitish, usually strongly pruinose toward the center; tomentose but not scabrous; lobe margins upturned; veins dark, thickly covered with a mat or hedgerows of dark rhizines (p. 293; it is difficult to separate the thallus from the substrate without detaching the rhizine mat from the thallus); primary photobiont blue green; soredia, isidia, and lobules lacking; apothecia on raised narrow lobes; spot tests negative.

Range: Widespread, throughout the PNW but most common e Cas.

Substrate: Soil, mosses, and decaying organic matter.

Habitat: Very common in dry, mostly open sites e Cas, including steppe, dry forest, outcrop areas, and subalpine and alpine ridges.

Notes: Although this is the most common dry-site *Peltigera* e Cas, one should be alert for other species that look similar from above. *P. ponojensis* and *P. praetextata* can be similar to *P. rufescens* but have slender, discrete rhizines. *P. kristinssonii* is scabrous and tomentose in the lobe tips and has discrete clumps of rhizines, while *P. rufescens* is only tomentose and has rhizines confluent into a continuous mat.

Peltigera venosa

Description: Thallus foliose, small (usually 0.5-2 cm diam), often single lobed; lobes broadly fan-shaped, mostly 1-1.5 cm broad; upper surface light greenish or pale greenish tan when dry, bright green when wet; veins distinct, brown, highly contrasting with the white spaces; cephalodia dark bluish gray, on the veins, sometimes growing separately; rhizines dark brown, in submarginal bundles; primary photobiont green; soredia, isidia, and lobules lacking; apothecia borne in the same plane as the thallus, usually present, dark brown to black; spot tests negative.

Range: Widespread; throughout the PNW.

Substrate: Soil.

Habitat: On moist soil, often where recently disturbed but somewhat shaded, especially roadcuts, stream banks, and trailsides.

Notes: Without a doubt the cutest little *Peltigera*. It is distinguished by the small fan-shaped thalli, bright green above, with highly contrasting veins below. If you look closely at the soil around these individuals you might find the free-living blue-green photomorph of this species—tiny dark gray or blackish lobes. This is the fungus associated only with the blue-green photosynthetic partner and lacking green algae.

Peltula

Peltula euploca

Description: Thallus foliose (squamulose or crustose in other species), of small, roundish to somewhat lobed, umbilicate thalli mostly < 1 cm diam; upper surface olive, olive brown, to brownish gray; lower surface pale brownish or grayish, corticate, with an umbilicus; rhizines and tomentum lacking; soredia present on the thallus margins, powdery, blackening, in elongate soralia; photobiont a unicellular cyanobacterium (*Chroococcidiopsis* or *Myxosarcina*); apothecia rare, immersed, < 1 mm diam; spores simple, 6-9 × 3-5 mm; spot tests negative.

Range: Widespread in w N Am but rare n of Cal.

Substrate: Rock.

Habitat: On steep, sheltered, noncalcareous rock faces, including cliffs of the Columbia basalt flows.

Notes: As our only umbilicate foliose cyanolichen, *P. euploca* is a distinctive species. It is the largest member of a mainly squamulose to crustose genus. Two crustose to squamulose species occur in the PNW; see keys and illustrations in McCune (2017), McCune and Rosentreter (2007), and Büdel and Nash (2002).

Sources include: Büdel and Nash (2002), McCune (2017), Wetmore (1970).

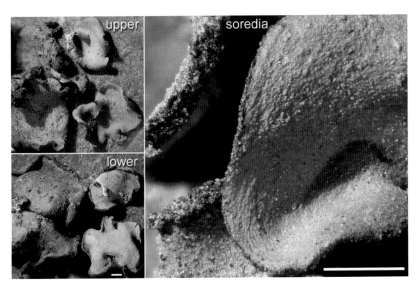

8b Lower surface white to tan. Lobes < 0.3 mm broad, with sparse granular soredia or isidia on some lobe tips or margins. Thallus < 2 cm diam, gray brown; on hardwoods, occasionally on rock; dry areas; s BC, Mont and N Dak s to Colo............................ *P. nigricans*

Phaeophyscia nigricans

Phaeophyscia decolor

Description: Thallus foliose, mostly < 3(8) cm diam; lobes narrow, 0.2-0.5(1) mm broad; upper surface dark gray, gray brown, or dark brown; lower surface brown to black, with dark rhizines; isidia and soredia lacking; apothecia common; spot tests negative.

Range: Widespread in temperate w N Am.

Substrate: Rock or mosses over rock, rarely on hard bark of tree bases.

Habitat: In both forests and steppe, but usually in sheltered microsites. Most frequent on stable boulders and outcrops along streams.

Notes: This small, dark species often grows closely appressed to rock, but most collections will be made where it is overgrowing moss cushions on rock. *Phaeophyscia sciastra* is similar in color and habitat but has sparse to abundant isidioid soredia. *Phaeophyscia nigricans* will sometimes lack soredia and be confused with *P. decolor*, but *P. nigricans* has a pale lower surface and is most frequent on bark. *Phaeophyscia ciliata* is usually broader lobed and on bark. Its upper surface is usually grayer than the brown-gray tone typical of *P. decolor*. The thallus of *P. decolor* contains zeorin (TLC required), while *P. ciliata* does not.

Phaeophyscia orbicularis

Description: Thallus foliose, mostly < 5(10) cm diam; lobes mostly 0.3-1 mm broad; light gray, dark gray, or brownish; lower surface black, rarely pale, with dark rhizines that often project beyond the margins; soredia finely granular, in soralia that are primarily laminal and submarginal, in small roundish points or more irregular; apothecia occasional; spot tests negative.

Range: Widespread in w N Am.

Substrate: Bark, occasionally on rock (including concrete, such as bridges, walls, and tombstones).

Habitat: A wide variety of lowland habitats, often in agricultural, urban, or riparian settings.

Notes: Distinguished by the closely appressed, small, dark thalli, usually with abundant soredia. Check carefully for the pale cortical hairs on the lobe tips that separate *P. hirsuta* from this species. *Phaeophyscia nigricans* is narrower lobed and is pale below. K- material with a pale lower surface but lobes broader than *P. nigricans* should be referred to *Physciella*. *Phaeophyscia sciastra* has coarse, isidia-like soredia in contrast to the fine soredia of *P. orbicularis*.

Phaeophyscia sciastra

Description: Thallus foliose, mostly < 3(8) cm diam; lobes mostly < 0.5 mm broad; upper surface blackish, dark gray, gray, or gray brown, occasionally pruinose; lower surface brown to black, with dark rhizines; soredia dark, isidioid, marginal or laminal, sparse to abundant; apothecia rare; spot tests negative.

Range: Alas to Cal and Ariz; mostly e Cas.

Substrate: Rock or mosses over rock.

Habitat: Common in outcrop and talus areas in dry, continental climates, associated with both siliceous and calcareous rock.

Notes: This common small, dark species of mossy rock outcrops comes in many disguises. The photo below shows a form that is paler than usual; dark gray brown is more typical. In some cases it develops a distinct grayish pruina. In others it is almost black. The characteristic isidioid soredia are often difficult to find, but in other cases they form a continuous crust over much of the upper surface. *Phaeophyscia kairamoi* is a rare species that has similarly coarse isidioid soredia; it is distinguished by abundant pale cortical hairs on the lobe tips and margins.

Physscia

Thallus foliose, appressed or suberect; lobes narrow (generally < 2 mm wide), sometimes with marginal cilia; upper surface white to pale or dark gray, occasionally pruinose; lower surface white to pale brown; rhizines simple or sparingly branched, white to pale brown or with dark tips; soredia or isidia present or not; apothecia lecanorine, the disk brown to black or whitish pruinose; spores brown or dark gray, 1-septate; cortex K+Y (atranorin), other spot tests various; on many substrates.

Sources include: Brodo et al. (2013), Esslinger (1979), Esslinger et al. (2020), Lohtander et al. (2009), Moberg (1977, 1997).

ID tips: See **Table 8** (p. 314) for characters differentiating *Physcia* from other genera. The medullary K test is important in identifying *Physcia* species, but it is a little tricky because the cortex is always K+Y and can bleed the color into the medulla. Therefore, the K test is best made under a dissecting microscope, exposing the medulla with a razor blade so that no upper cortex is touched by the reagent. A dissecting needle can be moistened with K, then touched to the exposed medulla.

1a Lobes with marginal cilia; thallus loosely appressed or ascending
  2a Soredia lacking. Lobes long and narrow (< 1 mm); margins always ciliate; apothecia common, stipitate; on bark; Cal............[*P. rhizinata*]
  2b Soredia present
    3a Soralia on the underside of hood- or helmet-shaped lobe tips
    ....***P. adscendens***
    3b Soralia lip shaped, often on lobes that are flat to reflexed but not helmet shaped... ***P. tenella***
1b Lobes without cilia (occasionally with rhizines projecting from near the margins); thallus loosely or tightly appressed
  4a Soredia present
    5a Soredia just beneath the finely dissected lobe tips. Thallus forming delicate white rosettes of narrow, shiny lobes; medulla K–, but easily misinterpreted as K+Y because of the very thin medulla adjacent to K+ cortices; on rock or moss over rock, usually where semi-sheltered, especially below basalt rimrock; occasional from BC to s Cal, inland to Colo... ***P. occidentalis***
    5b Soredia various laminal and/or terminal but usually easily visible from the upper side
      6a Medulla K+Y; upper surface white to dark gray, white spotted or not (caused by irregular thickness of the upper cortex)
      7a Soralia capitate, laminal and terminal ....***P. caesia***
      7b Soralia lip-shaped, terminal.......................................*P. subalbinea*
      6b Medulla K-; upper surface generally not white spotted

7a Upper surface ± shiny, occasionally weakly pruinose; soralia
 mostly marginal and lip-shaped, less often laminal and
 crateriform; generally on rock or mosses over rock. Lobes
 to 2 mm broad but highly variable in length, width, and
 orientation; widespread; common e Cas .... *P. dubia*
7b Upper surface densely pruinose throughout; soredia marginal,
 linear, coarsely granular, occasionally laminal in older
 portions; generally on bark, occasionally on moss over rock
 ....**P. dimidiata**
4b Soredia absent
 8a On bark
  9a Thallus densely pruinose.. **P. biziana**
  9b Thallus not or weakly pruinose
   10a Medulla K-, otherwise nearly identical to next species;
    widespread; common (especially e Cas) on hardwoods
    .... *P. stellaris*
   10b Medulla K+Y
    11a Lobes mostly 0.2-0.8 mm wide, rather short; apothecia
     extending to near the lobe tips, small (mostly < 1.5 mm) and
     crowded; spores mostly 17-22 μm long; common in PNW,
     boreal and montane .... **P. alnophila**

11b Lobes mostly (0.5)1.0-2.0 mm wide, often elongate; apothecia more centrally located, often larger and sparse; spores mostly 20-25 µm long; common in temperate e N Am but infrequent in PNW.. *P. aipolia*

8b On rocks

12a Thallus not or weakly pruinose. Upper cortex white spotted; lobes mostly about 2 mm broad; apothecia generally present, < 2 mm diam; medulla K+Y; on siliceous rock and mosses over rock, generally in partial shade; widespread; occasional in low to mid elevation forests e Cas.. *P. phaea*

12b Thallus densely pruinose

13a Medulla K-; spores not ornamented ....**P. biziana**

13b Medulla K+Y or K+ dingy rose or orange, especially in discolored medulla; spores faintly ornamented (LM). Thallus mostly < 6 cm diam; upper surface pruinose throughout; lobes 0.8-3 mm broad; on rock or soil over rock; distribution poorly known, Wash, Ore, and Id s to Colo .... *P. magnussonii*

Physcia adscendens

Description: Thallus foliose, small, mostly 0.5-2 cm broad, suberect; lobes <
0.5 mm wide, with 1-4(7) mm long marginal cilia; upper surface white to pale

or dark gray; lower surface white to pale brown; rhizines sparse, white to pale brown or with dark tips; soredia present beneath helmet-shaped lobe tips; apothecia uncommon; cortex K+Y, other spot tests negative.

Range: Widespread; common throughout PNW.

Substrate: Bark (especially hardwoods and shrubs but also occasionally on conifers), rarely on rock.

Habitat: Low- to mid-elevation forests and open shrubby areas, common in agricultural, urban, and suburban areas; partial shade to full sun.

Notes: The small suberect lobes with marginal cilia and soredia beneath helmet-shaped lobe tips are diagnostic. These characters are identical to those of *P. tenella* except for the appearance of the soralia. In *P. adscendens* the soredia line hemispherical bulges at the lobe tips, while in *P. tenella* they occur either as lip-shaped soralia at the lobe tips or as a few grains at the end of an otherwise undifferentiated tip.

Physcia alnophila

Description: Thallus foliose, small, mostly 0.5-2(4) cm broad, appressed; lobes mostly 0.3-0.8 mm wide, lacking marginal cilia; upper surface white to pale or dark gray, usually white-spotted; lower surface white to pale brown; rhizines sparse to dense, white to pale brown or with dark tips; soredia and isidia lacking; apothecia usually present and densely arrayed, occurring almost to the lobe tips, pruinose or not; spores (14)17-22(26) × (6)7.5-9.5(12) μm; cortex and medulla K+Y (atranorin), also containing zeorin without other terpenoids, other spot tests negative.

Range: Widespread; throughout the PNW.

Substrate: Bark and wood, usually on hardwoods.

Habitat: Very common w Cas, occasional e Cas; often in urban and agricultural settings, but also common on hardwoods in the mountains.

Notes: Previously considered a variety of *P. aipolia*, *P. alnophila* has been shown to be a distinct species. *Physcia alnophila* is much more common in the PNW than *P. aipolia*, but s and e of the PNW *P. aipolia* increases in frequency. The species differ in some subtle size characters: *P. aipolia* has somewhat larger spores (17)20-25(31) × (7.5)8.5-10.5(14) μm (compare with above) and longer and broader lobes (mostly 1.0-1.4 mm wide). *Physcia aipolia* also tends to have two or three other terpenoids in addition to zeorin (TLC). E Cas *P. alnophila* is less common than the outwardly similar *P. stellaris*; the distinction in the literature of *P. alnophila* being white spotted in contrast to *P. stellaris* is not useful in the PNW, being poorly correlated with the medullary K reaction.

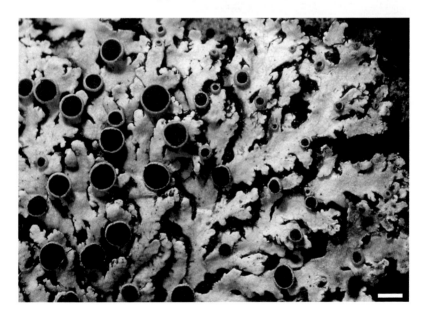

Physcia biziana

Description: Thallus foliose, mostly 1.5-5 cm broad, appressed; lobes mostly 0.5-2(3) mm wide, lacking marginal cilia; upper surface white to grayish, densely white pruinose; lower surface white to pale brown; rhizines sparse, white to pale brown or with dark tips; soredia and isidia lacking; apothecia usually present, often pruinose; cortex K+Y, other spot tests negative.

Range: Widespread, BC to Alta, s to Mex and s RM; throughout the PNW.

Substrate: Bark, wood, or rock.

Habitat: Exposed to partially shaded sites, mostly at lower elevations.

Notes: Distinguished by a thoroughly white pruinose upper surface and a thallus often forming neat rosettes. When heavily pruinose, *Physconia* species may resemble *Physcia biziana*, but *Physconia* is K- and *Physcia* always has a K+Y upper cortex. A very similar species is *Physcia magnussonii*, found on rock between the Cas and RM crests. It has faint pinkish or orange tints in the K reaction of the medulla and has weakly ornamented spores (LM), in contrast to the K- medulla and smooth spores of *P. biziana*. Weakly pruinose *P. biziana* can be mistaken for *P. stellaris*. Moberg (1997) recommended using the browner tinge and more rounded lobes of *P. biziana* in those cases.

Physcia caesia

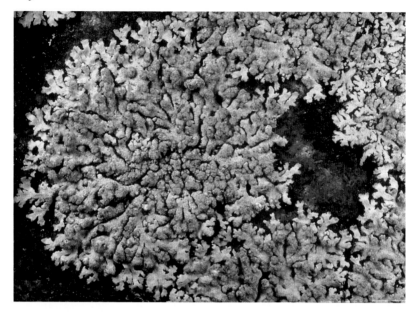

Description: Thallus foliose, mostly 1-5 cm broad, appressed; lobes mostly 0.2-2(3) mm wide, usually lacking marginal cilia; upper surface white to grayish, occasionally dark gray, usually white spotted; lower surface white to pale brown; rhizines sparse, white to pale brown or with dark tips, occasionally projecting like cilia from near the margins; soredia laminal in crater-like to capitate soralia and terminal on short lobes in capitate, irregular, or lip-shaped soralia; apothecia uncommon; cortex and medulla K+Y (atranorin), other spot tests negative.

Range: Widespread; fairly common throughout the PNW.

Substrate: Rock, rarely on bark or wood.

Habitat: Exposed to sheltered rocks in a wide variety of habitats; alpine to sea level, including rocks in the ocean spray zone.

Notes: One of the most common sorediate *Physcia* species on rock, its K+Y medulla distinguishes it from *P. dubia*. The much rarer *P. tribacia* has soredia in a narrow line along the lobe margins and just below them. *Physcia caesia* typically has laminal capitate soredia. A species previously often included under *P. caesia* is *P. subalbinea* (also formerly known as *P. wainioi*). It has terminal lip-shaped soralia and is widespread in w N Am.

Physcia dimidiata

Description: Thallus foliose, mostly 1 4 cm broad, appressed; lobes most-ly 1-2(3) mm wide, lacking marginal cilia; upper surface white to grayish, densely white pruinose; lower surface white to pale brown; rhizines sparse, white to pale brown or with dark tips; soredia marginal, linear, coarsely gran-ular, becoming laminal in old portions; isidia lacking; apothecia uncommon; cortex K+Y, other spot tests negative.

Range: Distribution poorly known, BC to Ore, N Dak, Mont, and Colo.

Substrate: Bark, especially *Juniperus* and *Artemisia*, occasionally on moss over rock.

Habitat: Steppe, open forests, and rock outcrop areas; exposed to somewhat sheltered microsites; low to mid elevations.

Notes: Distinguished by the marginal granular soredia and white-pruinose upper surface. *Physconia isidiigera* can be very similar in appearance but has squarrose-branched rhizines and a K- upper cortex, in contrast to the simple rhizines and K+Y cortex found in all *Physcia* species. *Anaptychia elbursiana* is also similar in appearance, but this has a K- cortex and simple rhizines.

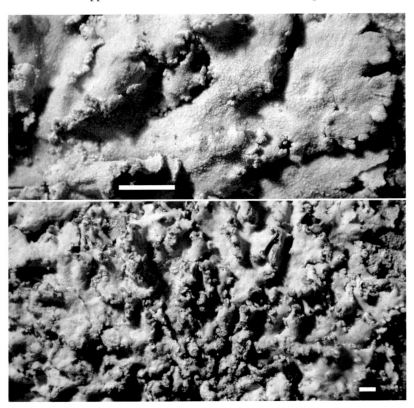

Pilophorus

Thallus of white, gray, greenish, or brownish stalks arising from a crustose primary thallus; stalks simple to slightly branched, cylindrical; cephalodia present on the primary thallus, irregularly shaped, light to dark brown; apothecia terminal, round or elongate, dark brown to black; spores simple, hyaline, 8 per ascus; cortex K+Y (atranorin); medulla K-, C-, KC-, P- (zeorin), or K+Y, P+O (stictic acid), or K-, P- (isousnic acid); on noncalcareous rock (rarely wood or soil), usually in partial shade.

Sources include: Goward (1999), Jahns (1981), Krog (1968).

ID tips: Pilophorus is sometimes confused with *Cladonia* by beginners, but *Cladonia* stalks are invariably hollow and are never associated with cephalodia. *Pilophorus* is a characteristic genus of the cool oceanic climates w Cas, but is rarely seen e or s Cas.

1a Thallus crustose only. Thallus granular to areolate; soredia lacking; cephalodia common; apothecia dark brown to black, strongly convex, sessile; spores 19-24 × 7-8 µm; thallus containing isousnic acid; Arctic s to coastal BC ....*[P. dovrensis]*
1b Thallus with a basal primary thallus (crustose or squamulose) and upright stalks
  2a Stalks sorediate. Stalks short (to about 5 mm tall), covered with granular to powdery soredia; apothecia rare; containing atranorin and zeorin; Alas, BC... *[P. cereolus]*
  2b Stalks lacking soredia
    3a Thallus bright white but stalks with a conspicuously black core; stalks mostly < 4 mm tall.. ***P. nigricaulis***
    3b Thallus gray, brownish, or greenish; stalks usually > 4 mm tall, with an exposed blackish core or not
      4a Apothecia usually lacking or abortive; stalks black, at least at the base and in the interior; thallus P+O (stictic acid present). Stalks simple when young, becoming much branched, to 15 mm tall; basal parts lacking a cortex and exposing the black stalk, upper parts with a granular cortex; rare, Arctic to se Alas and coastal BC ....*[P. vegae]*
      4b Apothecia almost always present, terminal; stalks internally blackish or not but often hidden by the granules or areoles; thallus P- (stictic acid lacking)
        5a Apothecia globose; stalks to 3 cm tall, gray .... ***P. acicularis***
        5b Apothecia cylindrical or club-shaped, longer than broad; stalks to 1 cm tall, olive gray or brownish .... ***P. clavatus***

Pilophorus acicularis

Description: Thallus of mineral gray to bluish gray or greenish gray stalks arising from a granular primary thallus; stalks simple to slightly branched, cylindrical, to 3 cm tall, about 1 mm diam, solid when young, becoming hollow; cephalodia present on the primary thallus, irregularly shaped, light to dark brown; apothecia terminal, ± spherical, black, usually present; cortex K+Y (atranorin); medulla K-, KC-, P-, C- (zeorin).

Range: Coastal Alas to Cal, w Cas and Sierra Nev, rarely disjunct to n Id.

Substrate: Noncalcareous rock, rarely wood.

Habitat: Usually in partial shade in openings in low- to mid-elevation moist forests; also frequent in rocky roadcuts.

Notes: This distinctive species might be mistaken for a *Cladonia*, but the apothecia of *P. acicularis* are black, the stalks are solid, and one can find dark cephalodia associated with the primary thallus. Furthermore, the primary thallus is a gray granular crust in contrast with the greenish squamules of most *Cladonia* species.

Pilophorus clavatus

Description: Thallus of greenish, gray, or brownish stalks arising from a granular primary thallus; stalks simple, cylindrical, mostly < 1 cm tall, to 1 mm diam; cephalodia present on the primary thallus, irregularly shaped, light to dark brown; apothecia terminal, club shaped to cylindrical, black, usually present; cortex K+Y (atranorin); medulla K-, KC-, P-, C- (zeorin).

Range: BC to Ore, w Cas, rare on e slope Cas; very rare inland in n Id and se BC.

Substrate: Noncalcareous rock.

Habitat: Partial to deep shade, usually on rocks in old-growth conifer forests at middle elevations.

Notes: Pilophorus clavatus is characterized by short brownish or greenish stalks tipped with elongate black apothecia and a persistent primary thallus. It is much less common than *P. acicularis*, tends to be found at somewhat higher elevations, is more shade tolerant, and shows a stronger affinity to old-growth forests. *P. clavatus* is usually brownish in relatively exposed sites, greenish in the shade.

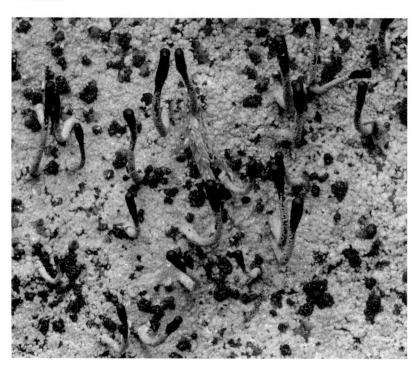

Pilophorus nigricaulis

Description: Thallus of very short stalks arising from a white crustose-are olate primary thallus; areoles convex, roundish, polygonal, or sometimes slightly lobate; stalks simple, cylindrical, with a blackish core and scattered to contiguous white warts or areoles, usually < 5 mm tall, about 1 mm diam; cephalodia present on the primary thallus, roundish to irregularly shaped, light to dark brown; apothecia terminal, roundish to elongate, black, often poorly formed or lacking; cortex K+Y (atranorin); medulla K- or K+Y, P- or P+O, C-, KC- (zeorin ± stictic acid).

Range: Alas to Ore, w Cas.

Substrate: Noncalcareous rock.

Habitat: Cool, moist, rocky slopes, often north facing, usually in the open but sheltered by surrounding topography, such as steep narrow valleys.

Notes: This rare species is detectable at a distance by the bright white crus- tose-appearing thallus growing directly on rock, contrasting with the surrounding darker lichens and bryophytes. Close inspection reveals the stalks. The white granules or warts around a short black axis are diagnostic. Although other Asian and northern species have this characteristic, none of them are known to occur in the PNW.

Platismatia

Thallus foliose, 3-20 cm broad; lobes mostly 1-20 mm broad, usually suberect to erect or drooping, often partly appressed in some spp; upper surface some shade of white, gray, or greenish gray; lower surface black, brown, white, or with patches of these colors; rhizines few; pycnidia marginal, immersed, often absent; soredia or isidia present or not; pseudocyphellae present on the upper surface or not; apothecia lecanorine, the disk brown; cortex K+Y (atranorin), medulla K-, C-, KC-, P- or P+O; on bark or wood, occasionally on rock.

Sources include: Asher et al. (2023), Culberson and Culberson (1968).

ID tips: This genus of large conspicuous lichens is one of the most abundant in the PNW, often contributing huge epiphytic biomass in our forests. Although the multicolored lower surface, with patches of brown, black, and white, is one of the distinguishing features of this genus, this variability can also create confusion if an individual is primarily one color below. Spot tests are of very limited use in this genus.

1a Isidia or soredia present
  2a Lobes narrow, mostly < 3 mm wide; isidia present; soredia lacking
  ....*P. herrei*
  2b Lobes generally broad, mostly > 3 mm wide; isidia and/or soredia
    present
    3a Upper surface with a network of ridges; isidia present; soredia lacking
    ....*P. norvegica*
    3b Upper surface smooth to weakly ridged or wrinkled; isidia and/or
      soredia present
      4a Soralia elaborating into branched isidia or lobules; propagules
        often tipped with a brown spot; thallus with a ragged appearance
        and often dissected margins...*P. glauca*
      4b Soralia persisting, becoming elongate, wavy and marginal;
        propagules not tipped with dark brown spots; thallus with
        general appearance similar to *Cetrelia* because of the broad lobes
        rimmed with soredia...*P. wheeleri*
1b Isidia and soredia lacking
  5a Upper surface with a network of ridges; lobes narrow to broad; medulla
    P+O.. *P. lacunosa*
  5b Upper surface smooth or concave but not ridged; lobes < 5 mm broad;
    medulla P-...*P. stenophylla*

Platismatia glauca

Description: Thallus foliose, to 15 cm broad; lobes generally 5-25 mm broad, suberect or erect; upper surface whitish, grayish, pale greenish gray, or browned in exposed sites; lower surface black, brown, white, or with patches of these colors; rhizines few; soredia or isidia present or not, occasionally proliferating into intricately branched structures bearing isidia and soredia; apothecia rare; cortex K+Y (atranorin), medulla K-, C-, KC-, P- (caperatic acid).

Range: Alas to Cal, inland to Mont and Colo; throughout the PNW.

Substrate: Bark and wood (especially conifers), less often on rock or alpine sod.

Habitat: Moist forests, low elevations to subalpine, rarely alpine, with broad tolerances to shade and moisture.

Notes: One of the most abundant lichens of moist conifer forests, *P. glauca* can be readily recognized with experience as a fluffy, disheveled, large-lobed, grayish or pale gray-green epiphyte. Confirm the species by verifying soredia or mixed isidia and soredia, and the lower surface with patches of at least two of the following colors: black, brown, and white. *Platismatia norvegica* is similarly large lobed but has a network of ridges on the upper surface, usually grows more appressed, and has isidia but no soredia.

1 cm

Platismatia herrei

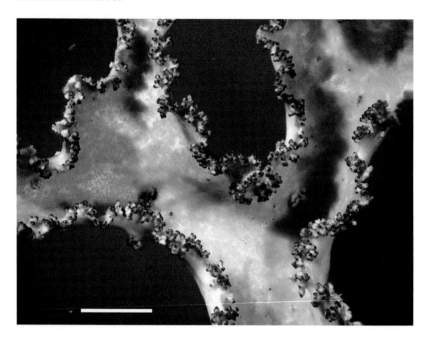

Description: Thallus foliose, suberect to drooping off the substrate, to 10(15) cm long; lobes mostly 1-3(5) mm broad (occasionally broader in basal parts); upper surface whitish, grayish, pale greenish gray, or browned in exposed sites; lower surface black, brown, white, or with patches of these colors; rhizines few; isidia present mainly on the lobe margins; apothecia rare; cortex K+Y (atranorin), medulla K-, C-, KC-, P- (caperatic acid).

Range: Alas to Cal, primarily w Cas but disjunct in n Id.

Substrate: Bark and wood, mainly conifers.

Habitat: Moist conifer forests, from low to mid elevations.

Notes: Large individuals are easily identified by the elongate, narrow lobes with marginal isidia. Some specimens, however, can be similar to some forms of *P. glauca* with narrow lobes, and must be separated by the form of the asexual propagules. *Platismatia glauca* has either soredia or a combination of soredia and isidia, while *P. herrei* has only true isidia (covered with a cortex and therefore shiny).

Platismatia lacunosa

Description: Thallus foliose, suberect or appressed but with edges free, to 10(15) cm broad; lobes mostly 5-10 mm broad; upper surface with a pronounced network of ridges and depressions, whitish, grayish, or browned in exposed sites; lower surface black, brown, white, or with patches of these colors; rhizines few; isidia and soredia lacking; pycnidia marginal, dark, often protruding; apothecia common; cortex K+Y (atranorin), medulla K-, C-, KC-, P+O or R (caperatic and fumarprotocetraric acids).

Range: Alas to Cal, w Cas.

Substrate: On bark and wood (mainly on hardwoods, especially *Alnus*), rarely on mossy rocks.

Habitat: Uncommon in moist riparian forests in the Coast Range and Cas, where it often occurs on upper branches of *Alnus rubra*, often associated with *Menegazzia subsimilis* and *Hypotrachyna sinuosa*. Also present in moist, cool upland sites.

Notes: This rarest of our local *Platismatia* spp is distinctive in its strong network of ridges and depressions, combined with an absence of isidia and soredia. When the pycnidia are protruding to stalked they are sometimes mistaken for isidia. The ridging is always stronger than in *P. norvegica*. The other *Platismatia* spp without isidia or soredia, *P. stenophylla*, has consistently narrower lobes that lack a network of ridges.

Polychidium muscicola

Description: Thallus fruticose, small, mostly < 3 cm diam, but sometimes coalescing to cover larger areas; forming short cushions to about 1 cm high; dark brown or less often grayish brown, dichotomously branched, the branches 60-100 μm diam, cortex of ± isodiametric cells; photobiont blue green (*Nostoc*); apothecia common, disk brown; spores 2-celled.

Range: Arctic to Cal, Colo; throughout the PNW.

Substrate: Moss or soil over rock, rarely on mossy bark of tree bases and roots.

Habitat: Moist rocky sites, usually in outcrop areas or talus slopes; low to high elevations.

Notes: Polychidium muscicola is a common fruticose cyanolichen forming small compact turfs of glossy roundish branches, usually growing on cushion-forming mosses over rock, commonly with brown apothecia nestled among the branches. *Pseudephebe* can be similar in external appearance but is somewhat coarser, has a green photobiont, and tends to grow directly on rock rather than on mosses over rock. *Scytinium* spp usually have lobes that are flattened in at least some portions of the thallus, while lobes of *Polychidium* are never flat. If fertile, a conclusive difference is the multiseptate spores of *Scytinium* in contrast to the 1-septate spores of *Polychidium*.

Pseudephebe

Pseudephebe minuscula

Description: Thallus fruticose but flattened to rock (sometimes bushy and semi-erect toward the center of the thallus), densely branched, dark brown to black; branches mostly < 1 mm diam, round to somewhat flattened in cross section; internodes (0.5)1-3(5) mm; isidia and soredia absent; pseudocyphellae indistinct; apothecia occasional, with a thalline margin; spot tests usually negative, but sometimes with small amounts of norstictic acid and then medulla sometimes K+R, P+O.

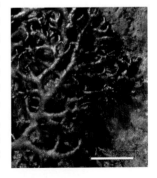

Range: Circumpolar s to Cal and Colo; throughout the PNW.

Substrate: Rock or gravelly soil.

Habitat: Low to high elevations in exposed to somewhat sheltered rocky sites; occasional.

Notes: The species is usually easily identified by the richly branched, hairlike, glossy black mats on rock. Previously, N Am *Pseudephebe* specimens were assigned to two species, *P. minuscula* and *P. pubescens*. Genetic evidence revealed that we have only one species in N Am, *P. minuscula*, except perhaps for a rare unnamed variant in Alaska, while *P. pubescens* is restricted to Europe. Some forms of *Melanelia stygia* appear similar to *Pseudephebe*, but *M. stygia* has distinct pseudocyphellae (use lens) and often a positive P test, while in *Pseudephebe* the pseudocyphellae are indistinct or not apparent and it rarely has a positive P test.

Sources include: Boluda et al. (2016), Brodo and Hawksworth (1977), Garrido-Benavent et al. (2020).

Pseudocyphellaria

Thallus foliose, medium to large; lobes 0.5-3 cm broad; upper surface gray or brown, smooth, weakly ridged, or irregularly wrinkled; lower surface whitish to dark brown, densely tomentose with scattered white or yellow spots (pseudocyphellae), often with scattered long rhizines; primary photobiont blue green or green, if the latter then with internal cephalodia containing blue-green photobiont; soredia or isidia present or not; apothecia brown, with thalline margin; spores elongate, septate; medulla white or yellow; spot tests various; mainly epiphytic, occasionally on rock.

Sources include: Galloway (1992), Imshaug (1950), Lücking et al. (2017), Magnusson (1940), Miadlikowska et al. (2002), Moncada et al. (2013, 2014), Stone et al. (2020), Tønsberg (1999).

ID tips: Our *Pseudocyphellaria* species are readily identifiable without even a hand lens or spot tests. Two of them have recently been transferred to *Lobaria*, where molecular data indicate they belong, despite the presence of conspicuous pseudocyphellae on the lower surface.

1a Pseudocyphellae on lower surface white or gray
  2a Thallus pale greenish gray, often bluish tinged; isidia or lobules present, soredia lacking .... *P. rainierensis*
  2b Thallus brown, grayish brown, or gray; isidia and lobules absent, soredia present or not
    3a Soredia lacking .... *Lobaria anthraspis*
    3b Soredia present.. *Lobaria anomala*
1b Pseudocyphellae on lower surface yellow
  4a Medulla yellow; pseudocyphellae abundant on the lower surface
  .... *P. hawaiiensis*
  4b Medulla white; pseudocyphellae sparse on the lower surface
    5a Upper surface smooth .... *P. citrina*
    5b Upper surface sparsely tomentose, sometimes only tomentose near the margins. .... *P. mallota*

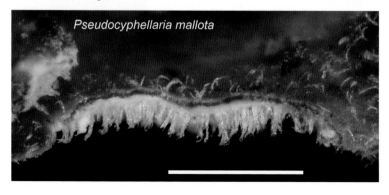
Pseudocyphellaria mallota

Pseudocyphellaria citrina

Description: Thallus foliose, loosely appressed, free at the edges, to 10(15) cm broad; lobes mostly 0.5-2 cm broad; upper surface light or dark brown (grayish in deep shade), sometimes with a shallow network of ridges; lower surface covered with light to dark brown tomentum with scattered yellow spots (pseudocyphellae); photobiont blue green; soralia roundish to irregular, yellow, laminal and marginal; apothecia not seen; medulla white, becoming yellow where exposed; cortex K-; medulla K+Y, P+O (stictic acid group, tenuiorin, terpenoids), C+O, KC+O; soralia UV+ orangish; yellow color from pulvinic acid and calycin.

Range: Alas to Cal, w Cas; very rare inland in se BC.

Substrate: Bark and wood, mainly on hardwood trees and shrubs.

Habitat: Low- to mid-elevation moist forests, usually in valley bottoms and foothills, often in riparian forests, ash swamps, and oak savanna.

Notes: This species is by far the most common of the three species in the PNW with yellow pseudocyphellae. Previously it was lumped under the name *P. crocata*, which does not occur in N Am. *P. citrina* is distinguished from the rare *P. mallota* by the combination of bright yellow soralia and lack of tomentum on the upper surface, while *P. mallota* has tomentum on the upper surface and grayish to brownish soralia that only rarely become yellowish. Near the coast look for *P. hawaiiensis*, which is similar to *P. citrina* but has a yellow medulla, more marginal soralia, and more abundant pseudocyphellae on the lower surface.

Pseudocyphellaria hawaiiensis

Description: Thallus foliose, loosely appressed, free at the edges, to 10 cm broad; lobes mostly 2-10 mm broad; upper surface gray to light or dark brown, glabrous and ± glossy, often with pale angular to reticulate maculae, smooth to weakly reticulately ridged; margins often slightly upturned; lower surface covered with cream-colored to dark brown tomentum with scattered yellow spots (pseudocyphellae); primary photobiont blue green; soralia marginal and laminal, yellowish, roundish to elongate; apothecia not seen; medulla pale to deep yellow; cortex K-; medulla K+ deeper Y, P+O, C+O, KC+O, UV- or dark O; soralia UV+ dark O.

Range: Oceanic areas of w Ore n to Alas (Denali); locally common in a few areas on the immediate coast.

Substrate: Bark and wood, both conifers (especially *Picea sitchensis*) and hardwoods.

Habitat: Coastal conifer forest and ericaceous scrub, mostly within a few km of the ocean; rare in more inland riparian forests.

Notes: This species occurs primarily in old *Picea* forests on the immediate coast of Ore, with only a few more inland locations, but always w Cas crest. In the PNW *P. hawaiiensis* can be confused only with *P. citrina*. Apart from the difference in the color of the medulla (see key), the soralia of *P. hawaiiensis* are both marginal and laminal but often predominantly marginal, as opposed to the predominantly laminal soredia of *P. citrina*. But the easiest differentiating character is that *P. citrina* has very sparse pseudocyphellae on the lower surface, while they are always numerous on *P. hawaiiensis*.

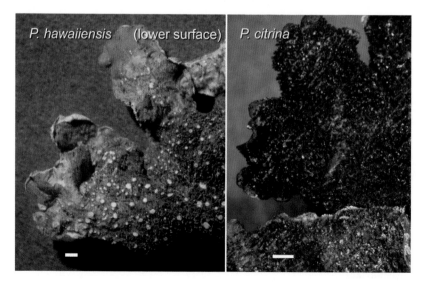

P. hawaiiensis   (lower surface)   P. citrina

Pseudocyphellaria mallota

Description: Thallus foliose, loosely appressed, free at the edges, typically small in the PNW, most often thumbnail size, but up to 4 cm or more; lobes mostly 2-10 mm broad, roundish; upper surface gray to brownish, sparsely tomentose, sometimes only tomentose near the margins; lower surface covered with cream-colored tomentum with scattered yellow spots (pseudocyphellae); photobiont blue green; soralia marginal and laminal, whitish, grayish, or deteriorating to yellow; soredia granular to isidioid; lobules sometimes present; apothecia not seen; medulla white; cortex K-; medulla K-, P-, C-, KC-, UV- (calycin and pulvinic acid in yellow parts).

Range: Alas s to OP and Ore Cas.

Substrate: Bark and wood of conifers, hardwoods, and shrubs.

Habitat: Cyanolichen-rich forests and semi-open areas at low to middle elevations.

Notes: In the field this rare and inconspicuous species looks like a poorly developed scrap of *Lobaria anomala*, from which it is separated by the yellow pseudocyphellae below and the tomentum on the upper surface. The color and tomentum on the upper surface can be similar to *Lobaria hallii*, but *P. mallota* has yellow pseudocyphellae below. In S Am *P. mallota* is more common than in the PNW.

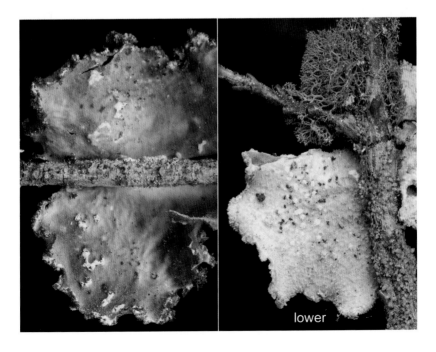

Pseudocyphellaria rainierensis

Description: Thallus foliose, large, loosely appressed to pendulous; lobes 0.5-3 cm broad; upper surface gray or pale greenish gray, often bluish tinged, smooth or irregularly wrinkled; lower surface whitish to light brown, tomentose, with white spots (pseudocyphellae); primary photobiont green, with internal cephalodia containing blue-green photobiont; lobules and isidia present; apothecia rare, reddish brown, with thalline margin; medulla white; cortex K+Y; medulla K- or brownish, P-, C-, KC-.

Range: Southern BC to Ore, w Cas.

Substrate: Bark and wood of conifers, often overgrowing moss mats.

Habitat: Moist old-growth forests at low to mid elevations, usually dominated by *Pseudotsuga* and *Tsuga heterophylla*, usually in the lower to mid canopy; rare.

Notes: The combination of grayish color, pseudocyphellae, and lobules or isidia is diagnostic. For a search image in the field, keep in mind a large-lobed species, grayish to bluish tinged (greenish gray if wet), larger and flatter than *Platismatia glauca*. One of the lichen species most strongly restricted to old-growth forests, it is nevertheless rarely found in younger forests. The patchy distribution, even in prime habitat, suggests strong dispersal limitation.

Punctelia

Thallus foliose, ± appressed, to 15 cm broad; upper surface pale greenish gray or pale gray with brownish lobe tips, with white spots (pseudocyphellae); lobes mostly 0.2-1 cm broad; lower surface cream to light brown, pale pinkish brown, or black; rhizines simple; soredia or isidia present or not, but our species with laminal and marginal soredia; apothecia rare in our species, lecanorine; cortex K+Y (atranorin); medulla C+R (gyrophoric or lecanoric acid), KC+R, P-, K-.

Sources include: Aptroot (2003), Hale (1965a), Krog (1982), Lendemer and Hodkinson (2010).

ID tips: Local specimens of *Punctelia* belong to the *P. subrudecta* group, but DNA sequences indicate that this broadly applied name contains a set of similar-appearing species that are challenging to identify. In the past, specimens from the PNW have been assigned variously to *P. subrudecta*, *P. perreticulata*, and *P. jeckeri*, but these all apply to the same species complex and it appears that none of these is correct; the taxonomy of this group in w N Am has yet to be worked out.

1a Soredia lacking; isidia present. Lower surface pale brown to cream; rhizines pale; medulla P-, K-, KC+R, C+R; widespread and common in e N Am with rare disjuncts e of the Cont Div in Mont; on bark and rock ....[*P. rudecta*]
1b Soredia present; isidia and lobules lacking
  2a Lower surface whitish to pale brown or pinkish brown; medulla containing lecanoric acid .... **Punctelia sp.**
  2b Lower surface usually at least partly black; medulla containing gyrophoric acid
    3a Thallus with at least the margin brown; pseudocyphellae laminal and marginal, irregularly shaped, partly linear. Widely disjunct in central RM, coastal n Cal, and Great Lakes; mainly on rock, less often on bark or wood; coastal rock outcrops in BC and San Juan Islands, Wash ....*P. stictica*
    3b Thallus, except for sometimes a thin dark margin, usually uniformly gray; pseudocyphellae laminal, rounded, varying in size rather than in shape. Frequent in e N Am, infrequent in Cal; reported from BC; not reported from Ore and Wash.....................................[*P. borreri*]

Punctelia sp.

Description: Thallus foliose, ± appressed, to 15 cm broad; upper surface pale greenish gray, with white spots (pseudocyphellae); lobes mostly 0.3-1 cm broad; lower surface light brown, occasionally dark brown; rhizines simple; soredia laminal and often marginal; apothecia lecanorine; cortex K+Y; medulla C+R, KC+R, P-, K-.

Range: Infrequent, BC to w Ore, w Cas.

Substrate: Bark and wood.

Habitat: Low-elevation urban, suburban, agricultural areas, and foothill forests.

Notes: The most frequent local *Punctelia* species is most closely related to *P. caseana*, while other individuals that are similar in appearance have sequenc es similar to *P. jeckeri*. This difficult species complex includes *P. caseana, P. jeckeri*, and *P. ulophylla*. This group as a whole is probably underreported in the PNW, because it blends in with the much more common *Parmelia sulcata* and *P. hygrophila*, two abundant species with similar size and color to *Punctelia*. In its typical form, however, this *Punctelia* is our only gray parmelioid species with a pale lower surface and C+R medulla. Uncommonly the species may have a dark brown lower surface. *Punctelia stictica* and *P. borreri* are similar to our local *Punctelia* in form and spot tests, but are black below and are rare in the PNW.

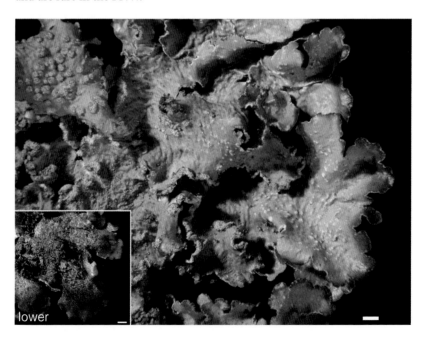

Ramalina

Thallus fruticose, erect or pendulous, pale greenish or tinged yellowish from usnic acid; branches irregular to flattened or filamentous; soredia present or not; apothecia lecanorine; disk concave, pale greenish, yellowish, or tan; spot tests various but cortex mostly KC+Y (usnic acid); mostly on bark.

Sources include: Bowler (1977), Bowler and Rundel (1977, 1978), Gasparyan et al. (2017), LaGreca (1999, 2020), LaGreca et al. (2020), Rundel and Bowler (1976).

ID tips: The taxonomy of *Ramalina* in w N Am is quite tentative; known unresolved taxonomic issues exist for almost all species. As yet molecular data for *Ramalina* in the PNW are almost nonexistent. Meanwhile, the species keyed below are, for the most part, readily distinguished by morphology. Chemical data (spot tests and/or TLC) are quite helpful in some groups, particularly in separating the UV+ evernic-acid containing species (*R. obtusata* and *R. labiosorediata*) from other sorediate species, as well as the often P+ *R. farinacea* from *R. subleptocarpha.*

1a Thallus partially hollow, perforate, small (generally < 2-3 cm long), ± translucent
   2a Soredia lacking .... ***R. dilacerata***
   2b Soredia present

Ramalina obtusata

357

3a Branches slender, delicate, < 1 mm wide; soralia small, rounded,
 terminal on main and lateral branches.............................*R. roesleri*
3b Branches broader, mostly 1-4 mm wide; soredia mainly beneath
 hooded lobe tips, often cup- or lip-shaped. Medulla K-, P-, UV+
 (evernic acid); widespread, uncommon; low-elevation swamps and
 floodplains e Cas (photo previous page)..........................*R. obtusata*
1b Thallus solid or netlike, but not with scattered perforations, often > 3 cm
 long, opaque
 4a Thallus forming delicate nets, the nets either coarse and conspicuous or
 (in coastal forms) tiny and sparse......................................*R. menziesii*
 4b Thallus not forming nets
 5a Branches fine, mostly < 0.5 mm, pendulous, the tips often hooked or
 curled and ending in a minute soralium..........................*R. thrausta*
 5b Branches coarse, mostly 0.5-2.0 mm, pendulous or tufted, the tips
 not regularly hooked or curled
 6a Soredia lacking. Reported for PNW from Linn Co., Ore
 ....*R. americana*
 [If the semi-hollow, perforate branches of *R. dilacerata* are
 missed, it would key here. *R. sinensis* would also key here,
 reported by Goward (1999) from BC. Further study of these
 species in w N Am is needed.]
 6b Soredia present
 7a On rock. Mainly boreal..[*R. intermedia*]
 7b On trees and shrubs
 8a Soralia terminal and lip-shaped, also laminal but rarely
 marginal...*R. labiosorediata*
 8b Soralia marginal, with or without laminal soralia, the rare
 terminal soralium being small and not expanded into a lip
 shape
 9a Main branches narrow (0.5-2 mm); thallus P+R or P-;
 soralia mainly marginal but also laminal; branches
 dichotomous..*R. farinacea*
 9b Main branches broad, mostly 3-8 mm; thallus P-; soralia
 laminal and marginal; main branches ± palmate
 ....*R. subleptocarpha*

Ramalina dilacerata

Description: Thallus fruticose, tufted, to 3 cm long, pale greenish or tinged yellowish; branches mostly 0.5-2 mm broad, irregularly flattened or roundish, with scattered perforations into the ± hollow interior; soredia and isidia absent; apothecia almost always present, pale greenish; cortex KC+Y (usnic acid); medulla K-, C-, KC-, P-, UV+ white (divaricatic acid).

Range: Alas to Cal, w Cas, inland to w Mont.

Substrate: Hardwood trees and shrubs, also frequent on conifers.

Habitat: Most common in riparian forests and shrubs at low elevations; e Cas mainly in areas of strongest oceanic influence.

Notes: This small species is distinctive in the consistent presence of apothecia, translucent appearance, and perforations into the ± hollow branches. Large specimens have been placed in the Australian species, *R. inflata*, but it is not clear that this occurs in N Am. On the other hand, two unpublished studies by different research groups have shown that *R. dilacerata* in w N Am consists of two distinct species. Typical size appears to vary geographically, with most individuals e Cas being small (< 1.5 cm) and many individuals w Cas being over 1.5 cm.

Ramalina farinacea

Description: Thallus fruticose, tufted to pendulous, to 10 cm long, pale greenish or tinged yellowish; branches mostly 0.5-2 mm broad, irregular to flattened to almost filamentous, dichotomous; soredia mainly marginal, occasionally laminal; apothecia lecanorine; disk concave, pale greenish, yellowish, or tan; cortex KC+Y (usnic acid); medulla P+R or P-, K+O or K-, C-, KC- (protocetraric acid or salacinic acid or only usnic acid).

Range: Alas to Cal, common w Cas, uncommon to rare e Cas, inland to w Mont.

Substrate: Bark and wood, both conifers and hardwoods.

Habitat: W Cas found in low- to mid-elevation forests; also on trees in agricultural and urban areas; e Cas mainly in low-elevation riparian habitats.

Notes: A common and usually easily recognized species, *R. farinacea* can be difficult to differentiate from *R. subleptocarpha* in valleys w Cas. E Cas *R. farinacea* sometimes grows with *R. obtusata* and *R. labiosorediata. Ramalina obtusata* has translucent, sparsely perforate branches and both have stronger tendencies to produce expanded terminal soralia than does *R. farinacea*. Stunted, deeply shaded individuals of *R. farinacea* can be nearly esorediate, with only a few grains along the margins and thin, hooked branch tips.

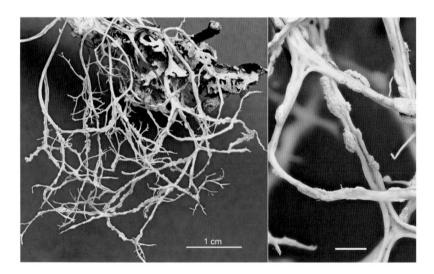

Ramalina labiosorediata

Description: Thallus fruticose, tufted to ± pendulous, to 5 cm long, pale green-ish or tinged yellowish; branches mostly 0.5-3 mm broad, flattened; soredia in roundish to elongate soralia, terminal (and often lip-shaped) and marginal, occasionally laminal; apothecia not seen; cortex KC+Y (usnic acid); medulla K-, C-, KC-, P-, UV± blue white (evernic acid).

Range: Widespread but uncommon; in the PNW mainly between Cas and RM.

Substrate: Bark and wood.

Habitat: Low-elevation swamps, often with *Picea*.

Notes: The name of the local material is in doubt; specimens of "*R. pollinaria*" from e N Am have been shown by DNA sequences to be *R. labiosorediata*, but specimens of "*R. pollinaria*" from the PNW have not been tested. This material is easily confused and often occurring with *R. obtusata*, which is distinguished by scattered perforations into partially hollow branches. Also, the soredia of *R. obtusata* are often beneath hooded lobe tips, which are seldom seen in *R. labiosorediata*. Both of these species contain evernic acid, giving a UV+ bluish white reaction in the medulla; this substance and reaction are absent from the far more common *R. farinacea* and *R. subleptocarpha*.

Ramalina menziesii

Description: Thallus fruticose, pendulous (to 1 m long), pale greenish, forming fine nets but also usually with broad strap-like portions, sometimes predominantly filamentous, especially near the coast; soredia and isidia lacking; apothecia occasional; cortex KC+Y (usnic acid); medulla K-, C-, KC-, P-.

Range: Pacific coast from se Alas to Baja, never e Cas.

Substrate: On broad-leaved trees and shrubs, occasional on conifers.

Habitat: Fog zone along the coast and in the Willamette-Puget trough, especially along rivers and in forested wetlands, often thickly draping whole forests or individual trees.

Notes: No other *Ramalina* in the PNW forms nets or is as long as this species. At a distance it can be confused with *Usnea* or *Alectoria*, but these are readily distinguished in hand. Coastal forms with tiny, fine nets can be confused with *Alectoria sarmentosa*, but that species is KC+R, has elongate, weakly spiraling pseudocyphellae, and does not have distinctly flattened portions.

Ramalina roesleri

Description: Thallus fruticose, tufted, to 3(4) cm long, pale greenish to whitish; branches mostly < 1 mm broad, irregularly flattened or roundish, with intricately divided tips, sparsely perforate into the partially hollow interiors; soralia small, rounded, terminal on main and lateral branches; apothecia not seen; cortex KC+Y (usnic acid); medulla K-, C-, KC-, P-, UV+ white (sekikaic acid).

Range: Circumboreal, s along the coast to Ore.

Substrate: Bark and wood.

Habitat: Coastal *Picea sitchensis* and *Pinus* forests at lower elevations, usually within a few km of the coast; fairly common within this restricted habitat.

Notes: Distinguished by small, fine branches, with sorediate tips. Shaded, stunted forms of *R. farinacea* may be similar, but these lack perforations into partially hollow branches and are less richly branched than *R. roesleri*. *R. obtusata* is perforate but has broader soralia, is less branched, and is mainly found e Cas.

Ramalina subleptocarpha

Description: Thallus fruticose, tufted to pendulous, to 10(25) cm long, pale greenish or tinged yellowish; branches mostly 2-10 mm broad, flattened, palmately branched near the base; soredia both marginal and laminal; apothecia not seen; cortex KC+Y (usnic acid); medulla K-, C-, KC-, P- (zeorin).

Range: BC to Cal, on the coast and in the Willamette-Puget trough with rare disjuncts to n Id.

Substrate: Bark and wood.

Habitat: Valley bottoms, ash swamps, and riparian hardwood forests, occasionally onto foothills, fairly frequent in urban and agricultural areas.

Notes: Distinguished by its broad, strap-shaped lobes with both laminal and marginal soredia, as well as a palmate branching pattern near the base. *R. subleptocarpha* has often been mistaken for *R. farinacea*, which has narrower lobes, mainly marginal soralia, and usually is not palmate branched near the base. Stunted air-pollution morphs of *R. subleptocarpha* can be difficult to distinguish from similarly stunted morphs of *R. farinacea*.

Ramalina thrausta

Description: Thallus fruticose, pendulous, to 30 cm long, pale greenish; branches filamentous, mostly < 0.5(1) mm diam, the tips often hooked or curled and ending in a minute soralium; apothecia not seen; cortex KC+Y (usnic acid); medulla K-, C-, KC-, P- (unknown substance near divaricatic acid).

Range: Boreal N Am s to n Cal and w Mont.

Substrate: On bark and wood, often conifers.

Habitat: Sporadic in (1) low-elevation moist forests, especially riparian *Picea* or *Abies* e Cas; (2) low-elevation old-growth *Pseudotsuga* in and w Cas; and (3) conifer forests on the immediate coast.

Notes: At a glance this species resembles a slightly off-color *Alectoria sarmentosa.* With close inspection the diagnostic hooked, minutely sorediate branch tips are seen. Also, the elongate pseudocyphellae characteristic of *Alectoria* are lacking on *Ramalina thrausta.* The finely filamentous coastal form of *R. menziesii* can also be similar in appearance. Inspecting the branch tips one should find tiny nets and no soralia to separate these forms of *R. menziesii* from *R. thrausta. Ramalina farinacea* can form hooked sorediate tips in deep shade, but these depauperate thalli are always small (< 6 cm long) and usually show some nonfilamentous, flattened portions.

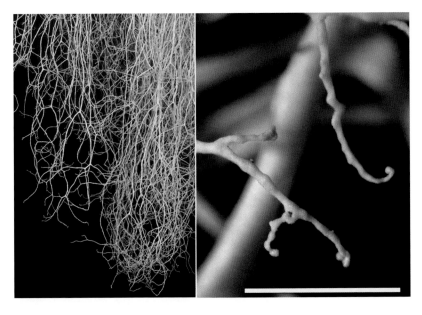

Rhizoplaca

Thallus foliose to crustose, mostly 1-3 cm broad, umbilicate, single-lobed or moderately divided or, when crowded, forming a thick crust of stalked, convex areoles; upper surface greenish, whitish, or yellowish gray; lower surface pale cream, tan, brown, or blue-black, rhizines lacking; isidia and soredia lacking; apothecia common, brown, pink, orange, greenish, or black, with a thalline margin; spot tests various; on rock.

Sources include: Leavitt et al. (2013), McCune (1987). See McCune and Rosentreter (2007) and Jones et al. (2022) for photos of and keys to vagrant species on soil, including *R. haydenii.* None of these is known from Ore and Wash.

ID tips: This fun, distinctive, eye-catching genus is quite common in the PNW, particularly e Cas. Our three species can be readily separated morphologically.

1a Disk brown; lower surface with a network of white cracks; medulla chalky in texture; medulla P+Y or O (pannarin and zeorin). Generally on calcareous rock; widespread in temperate w N Am but only e Cas in PNW .... *R. peltata*
1b Disk brown, olive, black or orange; lower surface brown to black, occasionally radially cracked, but not reticulately cracked; medulla cartilaginous; medulla P+Y or P-; disk various
  2a Disk orange to tan .... ***R. chrysoleuca***
  2b Disk olive black, olive, olive tan, or black .... ***R. melanophthalma***

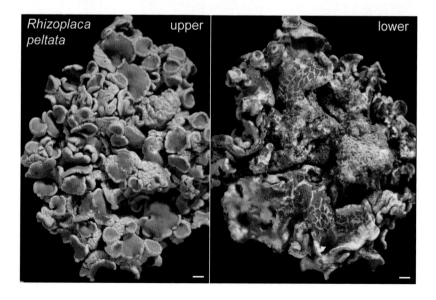

Rhizoplaca peltata / upper / lower

Rhizoplaca chrysoleuca

Description: Thallus foliose to crustose, to 3 cm broad, umbilicate, single-lobed or moderately divided; or, when crowded, forming a thick crust of stalked, convex areoles; upper surface whitish or yellowish gray; lower surface pale to blue-black stained, rhizines lacking; isidia and soredia lacking; apothecia usually present, pink, orange, or tan; medulla K-, P-, rarely P+ deep Y), C-, KC- (usnic acid only or more commonly with placodiolic or pseudoplacodiolic acid, rarely with psoromic acid).

Range: Widespread throughout the PNW but rare w Cas.

Substrate: Noncalcareous rock.

Habitat: Exposed to somewhat sheltered rock outcrops and talus, from low to high elevations in suboceanic to continental climates.

Notes: Readily identified by the pale, umbilicate thallus with orangish apothecia. The concentration of usnic acid in the cortex varies greatly, making the thallus color range from nearly white (with usnic acid barely detectable) to pale yellowish green (with abundant usnic acid). The form with crowded, stalked, bullate areoles is sometimes differentiated as *R. subdiscrepans*, a species mainly in e N Am. *Rhizoplaca peltata*, a much less common species, occurs on calcareous rock, has tan or brown apothecia, and has a P+O medulla.

Rhizoplaca melanophthalma

Description: Thallus foliose to crustose, to 3 cm broad, umbilicate, single-lobed or moderately divided or, when crowded, forming a thick crust of stalked, convex areoles; upper surface greenish or yellowish green; lower surface pale to blue-black stained, rhizines lacking; isidia and soredia lacking; apothecia usually present, black, green, or olive-tan; medulla K-, P- or P+Y, C-, KC- (usnic acid only or more often with psoromic acid, rarely with placodiolic acid).

Range: Widespread, throughout the PNW but uncommon w Cas.

Substrate: Rock, rarely loose on gravelly soil.

Habitat: Exposed to sheltered rock, at all elevations.

Notes: The variable color of the apothecia may cause some initial confusion, but they are never orange like the other common species, *R. chrysoleuca*. *Rhizoplaca melanophthalma* is readily recognized by the greenish to yellow-green umbilicate thallus and black, green, or tan disk. *Rhizoplaca peltata* is similar in thallus color but always has a brown disk, the lower surface has a network of white cracks, and the substrate is typically calcareous. A molecular study (Leavitt et al. 2013) described five cryptic species that all look the same as *R. melanophthalma* and can be separated only by DNA sequences. They are omitted here because they cannot be separated by keys, morphological descriptions, or secondary chemistry.

Ricasolia

Ricasolia amplissima

Description: Thallus minutely fruticose, erect, mostly 3-10(20) mm tall, cream to grayish or olive gray, basal branches becoming dark gray to gray brown at the tips, richly branched, corticate; main branches densely tomentose with fine whitish or hyaline hairs; stalks solid; apothecia unknown; spot tests negative.

Range: Alas to n Cal, w Cas.

Substrate: Bark and wood, on deciduous trees and shrubs, occasional on conifers.

Habitat: Cool, moist mountain forests and oak woodlands.

Notes: Many records of what used to be called *"Dendriscocaulon"* are actually cyanobacterial (blue-green) photomorphs of *Ricasolia amplissima* (or *Lobaria amplissima* if you take a broader view of *Lobaria*); others belong to *Dendriscosticta oroborealis*. We use *"Dendriscocaulon"* in quotes here because it is not a valid taxonomic entity, but has been used for blue-green photomorphs of several species in various genera. These blue-green photomorphs tend to occur in cyanolichen-rich communities. In the PNW these occur in two distinct habitats: low-elevation relatively warm and dry oak woodlands (*R. amplissima* subsp. *amplissima*) and cool, moist conifer forests (*R. amplissima* subsp. *sheiyi*). The photo at right shows an example of subsp. *amplissima* from an oak woodland in s Ore; the photo at left an example of subsp. *sheiyi* from the Coast Range in w Ore.

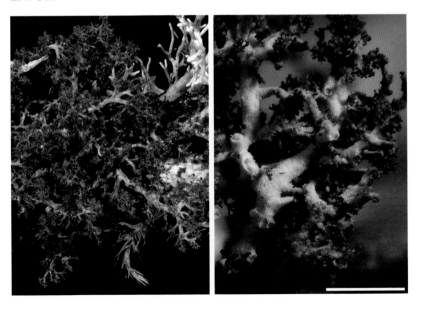

Scytinium

Thallus almost crustose, foliose, squamulose, or fruticose, very small to medium sized; lobes mostly 0.1-4.0 mm wide; gelatinous (non-stratified), gray to brown or dark reddish brown, in a few cases almost black; lower surface bare or with a few scattered tufts of hyphae; cortex a single layer of roundish cells; isidia present or not; soredia present in only one species; interior of thallus with a cellular structure (parenchyma-like) throughout or of loosely interwoven hyphae or of compactly interwoven hyphae; apothecia usually with thalline margin, the disk reddish brown; spores hyaline, ellipsoid, submuriform to muriform, 8 per ascus; spot tests negative; photobiont blue green (*Nostoc*); on rock, bark, moss, and soil, in moist or dry habitats.

Sources include: Jørgensen (1994a), Jørgensen and Goward (1994), Jørgensen and Nash (2004), Jørgensen and Tønsberg (1999), Košuthová et al. (2019), Martin et al. (2002), Otálora, et al. (2008, 2014), Sierk (1964), Stone and Ruchty (2008).

ID tips: The differences in internal anatomy described above are very useful for separating the small species. Seeing the internal anatomy requires a thin section observed under the compound microscope. Section *Scytinium* dry, then transfer the sections to a drop of water on a slide. Sectioning is made much easier by embedding small pieces in PEG 3350, a water-soluble wax sold as a laxative (see Introduction, Sectioning lichens, p. 22). See photos of small soil-dwelling spp in McCune and Rosentreter (2007).

See key to species and comparison with *Leptogium* under *Leptogium*.

Scytinium californicum

Description: Thallus foliose, small, usually cushion forming, with suberect to erect narrow (mostly 0.2-2(5) mm wide) lobes, grayish (in the shade) to brown or dark brown (in the sun); lobe margins irregularly cut or lobulate, isidiate, or finely fringed; upper surface generally with raised wrinkles; lobe tips usually curved, often reflexed, but often incurved on the sides of the lobes; thallus interior of loosely interwoven hyphae; apothecia occasional, 3(5) mm wide; spores (22)25-35(42) × (11)12-14(17) µm.

Range: Widespread, abundant throughout the PNW.

Substrate: Most often on mossy rock or soil or humus over rock, occasionally on tree trunks in moist climates at low elevations (especially w Cas).

Habitat: Most commonly seen in outcrop and talus areas at low to mid elevations, where it is usually intermixed in moss cushions over rock; in both open and sheltered sites.

Notes: Scytinium californicum is highly variable in form but usually has dissected lobes that are lobulate or fringed. *Scytinium pulvinatum* looks similar but often has narrower, sharper lobules and less consistently recurved or incurved lobes. It is much less frequent in the PNW than *S. californicum*. *Scytinium subaridum* can be similar in general appearance, but it has laminal true isidia. The variation in this group is currently under intensive taxonomic study. Specimens in the complex were often identified to *S. lichenoides*, but apparently this species does not occur in the PNW. Such specimens should be referred to *S. californicum* or *S. pulvinatum*.

Scytinium palmatum

Description: Thallus suberect to erect, very dark brown to blackish brown in exposed sites, brownish gray where shaded, generally 1-3(6) cm tall, sometimes forming extensive cushions; lobe margins curling inward, especially at the apices, to form nearly tubular tips; upper surface wrinkled, the wrinkles running lengthwise on the lobes; apothecia common, adnate to sessile; spores 30-56 × 10-20 μm, with 5-9 transverse septa and 1-2 longitudinal cross-walls.

Range: Alas to Cal, mainly w Cas, disjunct to n Id.

Substrate: Soil, gravel, rocks, mosses over soil or rock, rarely mossy bark.

Habitat: Roadcuts, roadsides, poorly revegetated areas in clearcuts, outcrops, and talus areas; low to mid elevations; populations on bark tend to be in hardwood stands at low elevations.

Notes: This species varies a lot in size and growth form, but it consistently produces the diagnostic rolled lobe tips, making it difficult to confuse with any other species. The role of this species as a soil-stabilizing and nitrogen-fixing pioneer on abused soil has not been fully appreciated. Most people do not look carefully at the blackish gelatinous growths so common along gravel roads and roadcuts w Cas. This blackish material is commonly *S. palmatum* and/or non-lichenized *Nostoc*. The latter can be lobed but is relatively amorphous and jelly-like.

Scytinium polycarpum

Description: Thallus small, mostly 0.5-2 cm diam, roundish, brownish or gray; lobes broadly rounded to orbicular; upper surface wrinkled when dry; apothecia abundant, always present, partially sunken; spores 4 per ascus, 25-40 × 12-15 µm, submuriform to muriform, ellipsoid to somewhat pointed at the ends.

Range: se Alas s to n Cal, w Cas.

Substrate: Mossy bark, usually hardwoods (especially *Fraxinus* and *Quercus*) and shrubs, very rarely on mossy rock.

Habitat: Common in oak savannas and ash swamps in the valleys, frequent in foothill conifer and deciduous forests, sporadic at middle elevations.

Notes: One of the most common epiphytic *Scytinium* species in w Ore and Wash. The abundant apothecia are partially sunken in a small roundish gray or brown thallus, making this species easy to recognize in the field. The identification can be confirmed in the lab by observing thin sections of the apothecia under a microscope. Look for the asci containing just 4 spores, a unique feature of *S. polycarpum* as compared to our other local species (which have 8-spored asci).

Scytinium rivale

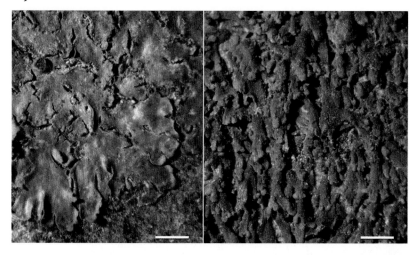

Description: Thallus flat, spreading, tightly appressed with lobe edges narrowly upturned (left photo) or lobes somewhat imbricate (right photo), dark gray, blackish green, dark gray brown, or blackish; lobes elongate, narrow, the rounded apices 0.2-1.0(1.5) mm broad; margins entire to wavy or occasionally irregularly lobulate; interior parenchyma-like throughout; apothecia occasional, adnate to sunken, to 1 mm diam; spores 16-33 × 5-9 µm.

Range: Both sides of the Cont Div in Mont, s to Colo, w to Wash and Cal; seldom collected but frequent in its restricted habitat.

Substrate: Noncalcareous rock, rarely on hard submerged wood.

Habitat: Small to medium-sized streams in the mountains, in or near water, submerged much of the year.

Notes: Most observers in the upright (standing) position will overlook the dark, tightly appressed rosettes formed by *S. rivale*. At a distance it looks similar to the dark brown crustose species (*Staurothele* and *Verrucaria*) with which it associates. Look for it on rocks in and along clear, cold streams in the mountains. *Scytinium rivale* varies considerably in form, suggesting that it may include more than one species or that it shows great environmental plasticity.

Scytinium tacomae

Description: Thallus foliose, gray to brown, to 5 cm diam, nearly flat to some-what cushion forming; lobes mostly 2-4 mm wide, rounded to irregular, the edges flexuose, often upturned, nearly smooth to finely wrinkled, the wrin-kles often concentric on young lobes; isidia laminal and marginal, cylindrical, becoming knobby or branched, usually numerous, mostly 0.1-0.3 mm long; thallus interior of loosely interwoven hyphae; apothecia occasional, the mar-gin often isidiate; spores (28)37-52 × (10)12-14(15) μm, with 5-7 transverse septa; spot tests negative.

Range: Occasional w Cas and Sierra Nevada, apparently a regional endemic, BC to central Cal.

Substrate: Bark and moss over bark, usually hardwoods and shrubs, less often on mossy rock.

Habitat: Low to mid elevations, usually semi-sheltered microsites.

Notes: Scytinium tacomae is easily confused with other smallish *Scytinium* species with wrinkled lobes and isidia, but is usually distinguishable by its dense laminal isidia and rounded, flexuous lobes on bark or mossy bark. *Scytinium siskiyouense,* from sw Ore and n Cal, is similar but internally has compact cells (Group 3 in key, p. 222) rather than loose hyphae, and it often has somewhat broader lobes than *S. tacomae.*

Solorina crocea

Description: Thallus foliose, to 6(10) cm broad, appressed, lobate; upper surface light greenish brown; lower surface bright orange, veined, tomentose; medulla orange; both blue-green and green photobionts present in separate layers; isidia and soredia lacking; apothecia usually present, sunken into the thallus, the disk brown; spores brown, 2-celled; medulla K+ purple (solorinic acid).

Range: Circumpolar s to Cal and NM.

Substrate: Soil (both calcareous and noncalcareous).

Habitat: Subalpine to alpine, often in seepage areas.

Notes: No other brown to green lichen has a bright orange lower surface, although this may not be visible from above when it is growing closely appressed. Most of the other species of *Solorina* are rare or as yet unknown from Ore and Wash, but should be sought on seepy soil over calcareous rock in alpine and subalpine areas.

Sphaerophorus

Thallus fruticose, forming erect tufts or cushions to 10 cm diam; color white, light gray, cream, or orangish tan or blackish when exposed; branches round to slightly flattened or irregular in cross section, solid; cortex shiny; soredia and isidia absent, though slender branchlets may function as propagules; apothecia terminal, spherical, bursting open irregularly, exposing a black spore mass (mazaedium); spores globose, dark, 1-celled; spot tests various but always UV+ blue white.

Sources include: Högnabba and Wedin (2003), Ohlsson (1973b), Tibell (1975), Wedin (1993a, 1993b), Wedin et al. (2009).

ID tips: Beginners sometimes mistake *Sphaerophorus* for *Cladonia*, but the solid stalks of *Sphaerophorus* easily distinguish it from *Cladonia*, which always has hollow stalks. All *Sphaerophorus* species show variation from sun-browned forms to nearly white shade forms, although some species solarize more readily to brown than others.

1a Main branches and secondary branches of similar diameter; medulla I- or pale bluish; usually on rock or moss over rock or soil. Alas to BC
.... [*S. fragilis*]
1b Main branches conspicuously thicker than secondary branches; medulla I+B or purple
2a Substrate terrestrial, arctic-alpine... [*S. globosus*]
2b Substrate epiphytic
3a Thallus with abundant coralloid branches; main branches roundish in cross section, smooth .... **S. tuckermanii**
3b Thallus with few coralloid branches and those mainly confined to near the base; main branches roundish to angular in cross section, with irregular depressions and ridges (foveolate)........**S. venerabilis**

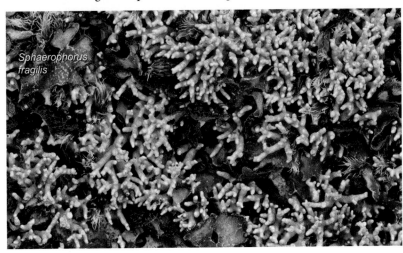

Sphaerophorus fragilis

Sphaerophorus tuckermanii

Description: Thallus fruticose, forming erect tufts to 10 cm diam; color white, light gray, cream, or orangish tan when exposed; branches round in cross section, solid; cortex shiny; apothecia common, terminal, spherical, bursting open irregularly, exposing a spore mass (mazaedium); cortex UV+ blue white (sphaerophorin); medulla K+ deep Y, P+O (thamnolic acid) or K-, P- (squamatic acid).

Range: Alas to Cal, with rare inland disjuncts to n Id and nw Mont.

Substrate: Bark and wood.

Habitat: Moist forests at low to mid elevations.

Notes: The perplexing morphological variation in form of the *S. globosus* complex, to which this species belongs, has created much confusion, including misapplication of *S. tuckermanii* to what is now *S. venerabilis*. Högnabba and Wedin (2003) demonstrated three distinct evolutionary lines: *S. globosus* growing on the ground in arctic-alpine habitats and as an epiphyte in Europe; *S. tuckermanii*, a PNW epiphyte with many coralloid branches; and *S. venerabilis*, a PNW epiphyte with few coralloid branches. Superficially similar to *Sphaerophorus*, *Bunodophoron melanocarpum* occurs from the OP northwards and is distinguished by its flattened branches.

1 cm

Sphaerophorus venerabilis

Description: Thallus fruticose, forming erect tufts to 10 cm diam; color white, light gray, cream, or orangish tan when exposed; branches roundish to angular or irregular in cross section, solid, the surface smooth to irregularly ridged and pitted (foveolate); cortex shiny; apothecia usually present, terminal, spherical, bursting open irregularly, exposing a spore mass (mazaedium); cortex UV+ blue white (sphaerophorin); medulla K+ deep Y, P+O (thamnolic acid) or K-, P- (squamatic acid).

Range: Alas to Cal, with rare inland disjuncts to n Id.

Substrate: Bark and wood.

Habitat: Moist forests at low to mid elevations.

Notes: Sphaerophorus venerabilis and *S. tuckermanii* are similar in distribution and chemical variation, but have recently been separated using molecular tools (see under *S. tuckermanii*). Typical *S. venerabilis* is easily recognized by its relatively coarse and fertile appearance, with relatively few fine branchlets, and its foveolate main branches. Occasional specimens will seem morphologically intermediate with *S. tuckermanii*; in those cases, conclusive identification requires molecular data.

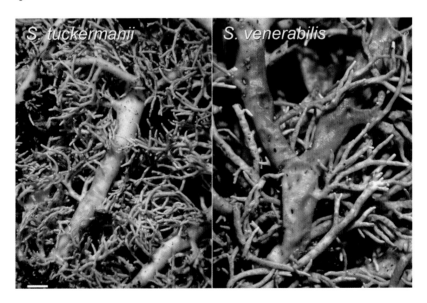

Stereocaulon

Thallus with erect to prostrate stalks (pseudopodetia), with or without a crustose or squamulose primary thallus, but a few species crustose and without stalks; stalks solid, white, cream, or gray, slightly to moderately branched, cylindrical to slightly flattened, bearing projections (phyllocladia) and often covered by tomentum; phyllocladia granular, globular, coralloid, or squamule-like; cephalodia external, conspicuous or not, warty or compound warty, grayish, tan, brown, or blackish, sometimes minutely shrubby when containing *Stigonema*; apothecia lecideine, terminal or lateral; disk red brown to dark brown; spores cigar shaped, but often more tapered at one end, mostly 3-7-septate, often slightly curved, spot tests various but cortex always K+Y and P+ pale Y (atranorin); medulla usually either K-, P-, KC+ reddish or orangish, UV+ blue white (lobaric acid) or K+Y or R, P+O, KC-, UV- (stictic acid group); photobiont green with cyanobacteria (*Nostoc* or *Stigonema*) in cephalodia; on rock, soil, or moss mats.

Sources include: Lamb (1977, 1978), McCune et al. (2019).

ID tips: The genus of cyanobacteria in the cephalodia can be helpful in identification. To determine this, examine thin sections of the cephalodia under the compound microscope. *Stigonema* makes short chains of relatively broad cells when in cephalodia of *Stereocaulon*. Sometimes it escapes into a more recognizable form with long multi-rowed filaments. *Nostoc* makes longer chains of smaller, bead-like cells.

K and P spot tests are quite helpful in many cases, with two caveats. First, some P reactions of stictic acid are slow, so be sure to wait at least 5 minutes before drawing a conclusion. Second, old herbarium specimens naturally discolor to brownish and can look slightly orangish brown when moistened with reagents, regardless of the chemical content. With such specimens you can reduce uncertainty if you make an acetone extract on a slide or white spot plate, remove the lichen fragments, then test on the white residue. Use a white background if extracted on a glass slide.

Miriquidic acid can be detected by the S test. An acetone extract on a glass slide is treated with a drop of 0.5% sulfuric acid, let dry, then heated over a flame (a few seconds) or in an oven (100 °C for 15 minutes). Miriquidic acid gives a pink, violet, or purple reaction. In contrast, lobaric acid tends to react to the S test as a dull orangish or brown or simply charring to gray.

Occasionally *Stereocaulon* species are parasitized by *Catillaria stereocaulorum*, which produces dense arrays of small black apothecia. These contain hyaline 1-septate spores that are ellipsoid, in contrast to the narrow, multiseptate spores of *Stereocaulon*.

Introductory Key

1a Thallus crustose; stalks (pseudopodetia) lacking..........................**Group 1**
1b Thallus fruticose; stalks (pseudopodetia) present
   2a Soredia present...**Group 2**
   2b Soredia lacking, though phyllocladia may be rather small and granular
      3a Phyllocladia (at least some) with darker centers and paler margins or
         sides...**Group 3**
      3b Phyllocladia more uniform in color
         4a Medulla P+O or R (containing stictic acid, with or without
            norstictic acid)...**Group 4**
         4b Medulla P–, containing other substances or with atranorin only
            5a Thallus directly attached to rock AND with little or no
               tomentum AND containing lobaric acid (medulla KC+
               reddish, UV+) or with atranorin only (KC-, UV-)........**Group 5**
            5b Thallus on the ground OR with conspicuous tomentum OR with
               a substance other than lobaric acid (spot tests various)
               6a Thallus containing lobaric acid (medulla KC+ reddish, UV+)
                   AND stalks usually distinctly tomentose ....**Group 6**
               6b Thallus lacking lobaric acid and containing another substance
                  (bourgeanic, porphyrilic, miriquidic, or perlatolic acids;
                  KC-, UV+ or UV-); stalks tomentose or not..............**Group 7**

Group 1

Crustose

7a Soredia lacking; lobaric acid lacking, containing atranorin only (UV–,
   KC–); apothecia usually abundant, lecideine; areoles uniform in color.
   Southeast Alas to Ore Cas ....*S. nivale*
7b Soredia present or absent; lobaric acid present (medulla UV+ blue
   white, KC+ reddish) along with atranorin; areoles often with darker
   or translucent centers and whitish edges similar to phyllocladia in *S.
   vesuvianum*, though the darker centers often small or punctiform;
   apothecia fairly frequent, but also commonly sparse or absent. Common
   in concave or sheltered topography in lava flows and talus slopes, in or
   near where snow accumulates; Ore Cas ....*S. oregonense*

Group 2

Sorediate

8a Pseudopodetia usually < 3 mm tall. Primary phyllocladia persistent;
   apothecia rare; cephalodia usually present, dark brown, botryose or in
   blackish spiky tufts, containing *Stigonema*; thallus containing lobaric
   acid; on moist rock; in N Am mainly ne, relatively rare in PNW and
   Alas.. [*S. pileatum*]
8b Pseudopodetia usually > 3 mm tall

9a Phyllocladia with darker centers and paler margins (cortical windows)
  10a Medulla K+Y, P+O, UV–, containing stictic acid. On rock;
    tomentum very thin to none; range as in *S. vesuvianum*
    .... sorediate morph of **S. vesuvianum**
  10b Medulla K–, P–, UV+ white, containing lobaric acid
    11a Phyllocladia grading into small partly corticate granules, not
      clearly sorediate. Thallus similar to *S. symphycheilum* but stalks
      longer, to 22 mm long; cephalodia dark brown, compound warty;
      apothecia infrequent, terminal; on rock or soil or moss over rock,
      usually firmly attached to rock; fairly common in coastal Alas
      ....[*S. klondikense*]
    11b Phyllocladia small and warty or compound warty, mostly
      0.2–1.0 mm diam, dissolving into granules and soredia, forming
      terminal, capitate, subspherical soralia. Stalks mostly < 10(15)
      mm long, simple or sparingly branched; on mossy rock, soil over
      rock, or soil; widespread in n hemisphere, infrequent in Alas s to
      Wash Cas and ne N Am; also in Fennoscandia; not particularly
      associated with oceanic climates........................ *S. symphycheilum*
9b Phyllocladia more uniform in color
  12a On semi-sheltered or vertical rock; medulla P+ deep Y or O
    (containing stictic acid)..**S. spathuliferum**
  12b On exposed soil or rock; medulla P–, containing lobaric or
    porphyrilic acid
    13a Phyllocladia replaced by squamule-like expansions, with
      sorediate edges; thallus containing lobaric acid (KC+ reddish,
      UV+); on soil. Soredia relatively fine, phyllocladia granular to
      compound warty or somewhat flattened, dissolving into soredia,
      sometimes absent; uncommon in se Alas, widespread but
      infrequent in n hemisphere ....[*S. coniophyllum*]
    13b Phyllocladia granular to warty; thallus containing porphyrilic
      acid (KC- UV-); on rock, less often on sand or gravel. Thallus
      to 2.5(4.5) cm tall; apothecia terminal, solitary or clustered,
      to 3 mm diam; cephalodia containing *Nostoc*; widespread, n
      circumpolar s to BC ....*S. botryosum*

Group 3

Phyllocladia with darker centers; fruticose; esorediate

14a Medulla P+ O or P+R, often slowly so (stictic acid); on rock, firmly
  attached; stalks usually lacking tomentum or nearly so ...**S. vesuvianum**
14b Medulla P– (porphyrilic or lobaric acid) though cortex P+Y (like all
  Stereocaulon, with atranorin)
  15a Thallus containing lobaric acid (KC+ reddish, UV+). Phyllocladia
    grading upwards into small corticate granules; photobiont *Stigonema*;
    se and sw Alas.. [*S. klondikense*]
  15b Thallus containing porphyrilic acid (KC-, UV-). Thallus forming
    compact cushions mostly 1-3 cm high; stalks tomentose toward the

tips but bare below; phyllocladia almost continuously covering the tips of the pseudopodetia, warty to flattened-peltate, at least some with darker centers; cephalodia pale to dark brown or pale grayish, compound warty, with *Nostoc* or *Stigonema*; apothecia rare, terminal and lateral; on soil, mosses, or gravel, rarely on solid rock; widespread in the N Am Arctic, s to se Alas...[*S. arenarium*]

Group 4

Thallus P+O or R; fruticose; esorediate; phyllocladia uniform in color

16a Thallus directly attached to rock; pseudopodetia with little or no tomentum; thallus cushion forming, to about 3 cm tall; phyllocladia warty to compound-warty, becoming granular at the tips of the stalks; cephalodia usually containing *Stigonema*
....granular morph of **S. spathuliferum**

16b Thallus on soil, humus, or moss, or loosely attached in rock crevices or on humus over rock; pseudopodetia heavily tomentose or not; thallus usually 3-12 cm tall; phyllocladia various but not becoming granular at the tips of the stalks; cephalodia usually containing *Nostoc*

17a Medulla UV+, lobaric acid present in addition to stictic acid and satellite compound..........................P+ chemotype of **S. tomentosoides**

17b Medulla UV-, lobaric acid lacking, containing stictic acid and satellite compounds. Apothecia usually numerous, mostly 0.3–1.0 mm diam; phyllocladia granular to warty or compound warty, becoming flattened lobulate, sometimes contiguous, conspicuously flattened when grown in shadier habitats; abundant in Alas and boreal N Am, scattered s in RM..*S. tomentosum*

Group 5

On rock; little or no tomentum; P-, KC+R, UV+, containing lobaric acid

18a Thallus developing into loose, tall (3–8 cm) cushions; apothecia common, large (often > 3 mm diam), terminal; stalks minutely tomentose in part..........................**S. intermedium** (see under **S. sterile**)

18b Thallus forming short, dense cushions, usually to 1–3 cm tall

19a Habitat oceanic; thallus forming small cushions 2–5 cm diam and 1–2 cm high, but becoming much taller when fertile; phyllocladia initially granular, sometimes flattened and lobate, becoming coralloid
.... **S. sterile**

19b Habitat boreal to arctic; thallus to about 1.5 cm high, forming dense cushions; phyllocladia coralloid (isidia–like), not flattened or squamulose, but sometimes dividing in a plane and digitate; stalks often discoloring to orangish tan; containing lobaric acid or atranorin only; photobiont *Stigonema* or *Nostoc*; tightly attached on rock; Alas
....[*S. subcoralloides*]

Stereocaulon spathuliferum

Description: Thallus forming small dense cushions to 2(3) cm tall and 4 cm diam or coalescing into broader cushions, white, cream, or grayish cream; stalks moderately branched, glabrous or nearly so, sometimes flattened in cross section; phyllocladia small, granular, fragile, grading into soredia; cephalodia often sparse but rather conspicuous, brown to gray or whitish gray, warty to compound warty, often developing into cauliflower-like clumps to 2 mm diam or more, containing *Stigonema*, less often *Nostoc*; apothecia not usually seen, but sometimes fertile in coastal Alas; medulla K+Y or O, P+O (stictic and norstictic acids), KC-, UV-.

Range: Alas to Ore, w Cas, rare.

Substrate: Rock (noncalcareous).

Habitat: Sheltered microsites in cool moist habitats, especially talus slopes and cliffs.

Notes: Stereocaulon spathuliferum is the only sorediate P+O *Stereocaulon* growing on sheltered rock faces in Ore and Wash. *Stereocaulon sterile* is similar in size but is esorediate, P-, and grows on more exposed rock surfaces. Like *S. spathuliferum, S. symphycheilum* is also sorediate, but it is P- and has phyllocladia with darker centers contrasting with pale margins. Other sorediate *Stereocaulon* spp occur in BC and Alas.

Stereocaulon sterile

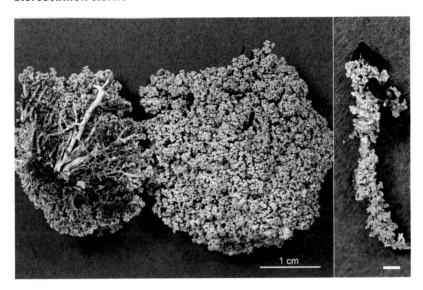

Description: Thallus with stalks forming dense cushions to about 3 cm tall, white, cream, or gray; stalks slightly to moderately branched, glabrous or minutely tomentose in part; phyllocladia tiny, granular to coralloid, fragile; cephalodia external, conspicuous or not, irregularly globose, containing *Stigonema*; apothecia usually lacking, terminal; cortex K+Y; medulla K-, P-, KC+ violet or orangish, UV+ white (lobaric acid).

Range: Alas to Cal, common w Cas in the PNW, rarely inland to suboceanic n Id.

Substrate: Rock (noncalcareous).

Habitat: Sunny to sheltered rock outcrops, talus, and lava flows; low to moderately high elevations.

Notes: One of the most frequent *Stereocaulon* spp in the PNW, *S. sterile* is characterized by the low compact cushions growing directly on rock and the stalks that have little or no tomentum. A rare species, *S. spathuliferum*, also makes small dense cushions on rock, but it is sorediate and P+O. The name *S. intermedium* has been applied to multiple species, and in N Am it has been applied to a developmental modification of *S. sterile*. The DNA sequences of N Am *S. intermedium* and *S. sterile* are nearly identical to each other. It appears that when *S. sterile* becomes fertile (photo at right), the stalks elongate and separate, becoming the form referred to as *S. intermedium*.

Sticta, with Dendriscosticta

Thallus foliose, mostly 2-5(10) cm broad, often with an odor of shrimp or fish; lobes broadly rounded or somewhat elongate; upper surface black or brown; lower surface with light to dark brown tomentum and sparse craters (cyphellae); photobiont blue green (*Nostoc*); isidia or soredia often present; apothecia rare; spores narrow, septate; spot tests negative; on bark, wood, or mossy rock.

Sources include: McCune et al. (1997), McDonald et al. (2003), Moncada and Lücking (2012), Moncada et al. (2013), Moncada et al. (2021), Simon et al. (2018), Spribille et al. (2020), Tønsberg and Goward (2001).

ID tips: Although rather few species of *Sticta* occur in the PNW, phylogenetic studies have shown that our species concepts have been too broad. The species previously known as *S. weigelii* and *S. fuliginosa* are both species complexes, only some of which have been described at this writing. Meanwhile the genus is easily recognized, as it is the only one that produces true cyphellae, crater-like structures in the lower surface that are surrounded by a differentiated rim. This is the edge of the "basal membrane" that lines the cyphellae. Thin sections of that membrane under LM are increasingly being used to differentiate species.

1a Lobes green when fresh, containing green algae; associated with a
　　fruticose, bluish gray growth form containing cyanobacteria that may
　　be free living or attached to the green lobes. Coastal BC and perhaps
　　southwards, but the green form is known with certainty only from the
　　Kispiox area of BC .... [*Dendriscosticta oroborealis*]
1b Lobes gray to brown, lacking green algae but containing cyanobacteria
　　(*Nostoc*)
　2a Isidia lacking (soredia or lobules may be present)
　　3a Soredia present; on bark or wood.. **S. limbata**
　　3b Soredia lacking; on mossy rock ....**S. arctica**
　2b Isidia present, soredia lacking
　　4a Isidia laminal; upper surface black or dark gray; lobes rotund
　　... **S. fuliginosa**
　　4b Isidia marginal; upper surface brownish; lobes divided, not rotund
　　　5a Isidia shrubby, becoming richly branched from a basal stalk; cells
　　　　of basal membrane of cyphellae papillose (LM); thallus mostly
　　　　1-2 cm diam ....**S. torii**
　　　5b Isidia cylindrical to coralloid but rarely shrubby and without a
　　　　distinct basal stalk; cells of basal membrane of cyphellae smooth
　　　　(LM); thallus mostly 2-5 cm diam ....**S. rhizinata**

Sticta arctica

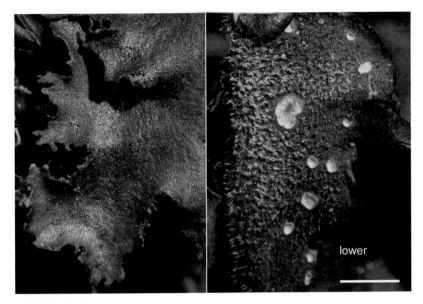

Description: Thallus foliose, brownish, mostly 2-5 cm broad, often with an odor of shrimp or fish; lobes 2-5(12) mm broad; upper surface tan to dark brown; lower surface black centrally to pale grayish brown near the margins, with light to dark brown tomentum and sparse craters (cyphellae); photobiont blue green; lobules present, marginal, mostly 0.1-0.5 mm broad, simple or lobate, sparse or abundant; isidia and soredia lacking; apothecia not seen; spot tests negative.

Range: Arctic s to nw Ore, rare.

Substrate: On tundra sod or mossy rock.

Habitat: Oceanic tundra, coastal bluffs, and exposed subalpine summits near the coast.

Notes: This uncommon species of the Arctic has rarely been found south of Alas; known from mossy outcrops at one site each in Ore (Clatsop Co.) and Wash (Island Co.).

Sticta fuliginosa

Description: Thallus foliose, mostly 2-10 cm broad, often with an odor of shrimp or fish; lobes broadly rounded; upper surface black or dark gray; lower surface with light brown tomentum and sparse craters (cyphellae); photobiont blue green; isidia fine, laminal, the same color as or darker than the thallus; apothecia not seen; spot tests negative.

Range: Alas to Cal, w Cas.

Substrate: Usually on bark or wood of hardwoods; occasionally on conifers, rarely on rock.

Habitat: Fairly common in warm, moist, low-elevation forests, especially valley and foothill hardwoods such as ash swamps and oak forests. Sporadic in mountain conifer forests.

Notes: The blackish, isidiate upper surface and the tomentose lower surface with cyphellae are characteristic. Like other *Sticta* species, it has a definite odor of fish or shrimp. Sun forms feel thick and are densely coated with black isidia. Shade forms are thinner, browner, and have more sparse isidia. The cyphellae are typically fewer than on our other *Sticta* species and may be absent on small specimens. The name *S. fuliginosa* has been applied broadly; recent research resulted in the segregation of *S. torii*; current research will result in further revisions and narrower species concepts.

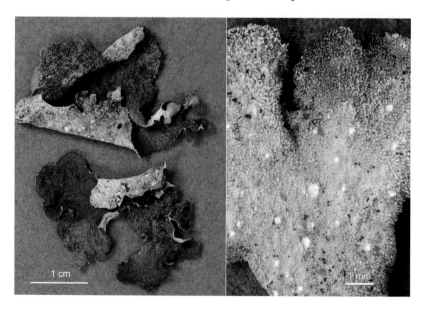

Sticta limbata

Description: Thallus foliose, mostly 1-6 cm broad, often with an odor of shrimp or fish; lobes broadly rounded; upper surface brown; lower surface with light brown tomentum and sparse craters (cyphellae); photobiont blue green; soredia laminal and submarginal to marginal, in roundish to irregular or elongate soralia; apothecia not seen; spot tests negative.

Range: Alas to Cal, w Cas, with rare inland disjuncts in BC.

Substrate: Bark and wood, rarely rock.

Habitat: Fairly common in warm, moist, low-elevation forests, especially valley and foothill hardwoods such as ash swamps and oak forests. Sporadic in mountain conifer forests.

Notes: Sticta limbata is distinguished by its smallish, rotund, brown thallus with soredia above and cyphellae and tomentum below. At low elevations its habitats are very similar to *S. fuliginosa* and it frequently occurs with that species. However, *S. limbata* is infrequent in mid-elevation montane conifer forests where *S. fuliginosa* is frequent.

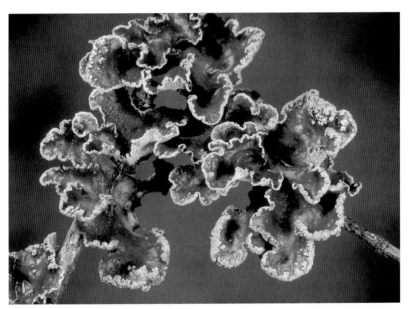

Sticta rhizinata

Description: Thallus foliose, mostly 2-8 cm broad, often with an odor of shrimp or fish; lobes somewhat elongate; upper surface brownish or blackening, grayish in the shade; lower surface with light to dark brown tomentum and sparse craters (cyphellae), often with rhizines; cells of basal membrane of cyphellae smooth (LM); photobiont blue green; isidia marginal, cylindrical, often branching; apothecia not seen; spot tests negative.

Range: Alas to Cal, w Cas.

Substrate: On bark, wood, and moss mats on trees and shrubs, frequently on conifers; rarely on mossy rock.

Habitat: Cool, moist, old-growth conifer forests at middle elevations; uncommon; rarer and more strongly associated with old-growth than our other *Sticta* species.

Notes: Sticta rhizinata has been called *S. beauvoisii* or *S. weigelii* in the past, but those names apply to species elsewhere in the world. *Sticta rhizinata* is distinguished from other local cyanolichens by its marginal isidia and the tomentose, cratered lower surface. The lobes are more narrowly divided than in our other epiphytic *Sticta* species. Although *S. rhizinata* is distinguished in S Am, where it was described, by its conspicuous rhizines, local material is not necessarily rhizinate. Specimens from mossy rocks should be carefully distinguished from *S. arctica*, an even rarer species that is at the s limit of its range in nw Ore. Although similar in appearance, *S. arctica* lacks cylindrical isidia, instead having flattened marginal lobules. *Sticta arctica* occurs in open rock outcrop areas at low or high elevations near the coast.

Sticta torii

Description: Thallus foliose, to 2 cm diam; lobes one to few, roundish to some-what elongate, to about 7 mm wide, entire to lobulate-lacerate; upper surface gray, brown, or dark brown; lower surface cream to brownish, with similar-ly colored dense tomentum, sparse craters (cyphellae), and rhizines absent; cyphellae 60–135 μm deep, cells of basal membrane ± isodiametric, 6–8 μm wide, papillose (lower right photo); isidia abundant, marginal and laminal, but often continuous along margin, with ascending branches, often forming a dense cluster on a stalk, darker than the thallus, usually dark brown, shiny; apothecia unknown; spot tests negative.

Range: Apparently rare from coastal se Alaska to Ore on the immediate coast.

Substrate: Bark and wood of hardwood trees and shrubs and conifers.

Habitat: Cyanolichen-rich habitats in oceanic forests, especially on twigs.

Notes: This recently described species was formerly lumped under *S. fuligi-nosa*. They differ in the form and placement of the isidia, which are mainly marginal and densely branched at the tip of a stalk in *S. torii*, vs. mainly sim-ple laminal isidia in *S. fuliginosa*. *Sticta rhizinata* also has primarily marginal isidia but has more elongate lobes and lacks papillae on the cells of the basal membrane of the cyphellae.

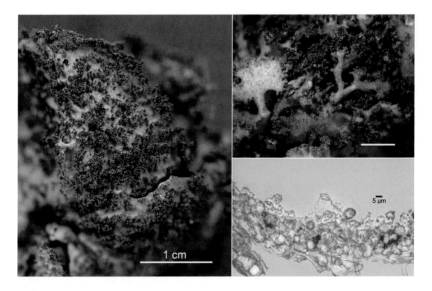

Sulcaria

Our three taxa (*S. badia*, *S. spiralifera* var. *spiralifera*, and *S. spiralifera* var. *pseudocapillaris*) are keyed under *Bryoria*. See photos of cortical structure under *Nodobryoria*. Myllys et al. (2014) demonstrated that *Bryoria pseudocapillaris* and *B. spiralifera* belong in *Sulcaria* and synonymized them. McCune and Stone (2022) treated them at the varietal level under *S. spiralifera* to recognize the chemical and subtle morphological differences between them.

Sulcaria badia

Description: Thallus pendulous, filamentous, to 50 cm long; branches to 0.4(1.0) mm diam, somewhat flattened, twisted, with conspicuous grooves and broadening considerably at the branch points, usually pale brown with a purplish cast or dull chestnut brown; pseudocyphellae, white, elongate, developing into deep furrows; isidia and soredia absent; apothecia unknown; cortex K+Y (atranorin), C-, KC+Y, P+Y or brownish; medulla with spot tests negative.

Range: Wash (extirpated) to Cal, w Cas; mainly in sw Ore and n Cal.

Substrate: Bark and wood, mainly hardwoods, rarely conifers.

Habitat: Well-lit savannas to partly shaded forests at low elevations. One site is known from a coastal *Pinus* forest.

Notes: This species is rather rare, is known from only a handful of locations in w Ore, and is apparently extirpated from Wash. The type specimen is from an old orchard near Philomath, Ore, collected decades ago. The species is distinguished from *Bryoria* by the elongate, twisted pseudocyphellae which develop into grooves, along with a K+Y, P+Y cortex but other spot tests negative. In the field, look for *Bryoria*-like material but with a pale purplish brown cast. Because its habitat is typically on the valley fringes, land development and air pollution are increasing threats to its existence.

Sources include: Brodo and Hawksworth (1977), Glavich (2003), Myllys et al. (2014), Peterson et al. (1998).

Seirophora

Seirophora contortuplicata

Description: Thallus fruticose, yellow to orange, small to medium sized, forming low tufts or prostrate, in sparse to dense, tangled mats or cushions usually < 2 cm diam, but sometimes fusing into larger colonies; occasionally greenish, beige, or whitish when shaded; branches dorsiventrally flattened, 0.3-1.0 mm wide, irregular, warty, often ciliate; rhizines generally lacking, attached to the substrate by rhizohyphae or hapters; photobiont green (*Trebouxia*); apothecia rare, lecanorine, marginal or laminal; spores septate, polarilocular, 8 per ascus, 10-12 × 6.0-7.5 µm; cortex K+ purple (anthraquinones); on rock.

Range: Alas to s Cal and s RM, e Cas, e to w S Dak.

Substrate: Calcareous rock and moss or humus over rock.

Habitat: Sheltered microsites on calcareous rock outcrops in continental climates, especially in crevices and beneath overhangs; forested mountains, cold steppe, and tundra.

Notes: This species is an often overlooked but fairly frequent occupant of sheltered sites on calcareous rock (limestone, dolomite, and calcareous sandstone) throughout much of w US and Canada.

Sources include: Arup et al. (2013), Fródén and Lassen (2004), Rosentreter and McCune (1996).

Seirophora contortuplicata

Teloschistes

1a Thallus mostly > 2 cm broad or long; on trees and shrubs; coastal
.... ***T. flavicans***
1b Thallus mostly < 2 cm broad; in crevices of calcareous rock; e Cas
.... ***Seirophora contortuplicata***

Teloschistes flavicans

Description: Thallus fruticose, erect or drooping, mostly 2-8 cm long, moderately branched, bright or dull yellow-orange, greenish yellow, or even pale greenish or whitish when grown in shade; branches narrow, mostly < 0.5(1) mm diam, roundish in cross section, with pointed tips and short pointed side branches; attached to the substrate by rhizohyphae or hapters; soredia yellowish, in roundish soralia; apothecia not seen in our area; spores septate, polarilocular; cortex K+ purple-red (anthraquinones); medulla with spot tests negative.

Range: Coastal Ore to Cal and Mex.

Substrate: On bark and wood.

Habitat: Coastal headland forests (usually *Picea sitchensis*), rare.

Notes: This widespread, mainly tropical and subtropical species occurs sporadically up the West Coast to n Ore. Only one substantial population has been found in Ore, along with a small number of minor populations. The bushy, orange, sorediate thalli cannot be mistaken for any other species. The soredia can vary from sparse to frequent. Other species of *Teloschistes* (*T. chrysophthalmus, T. exilis*) occur e and s of the PNW.

Sources include: McCune et al. (1997).

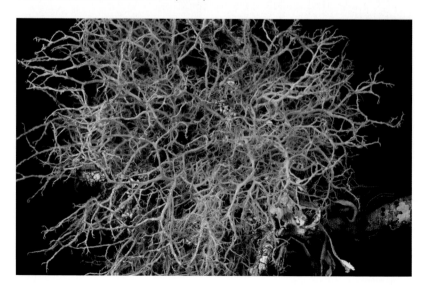

Thamnolia

Thallus fruticose, occasionally forming upright turfs but usually prostrate or sprawling, white to bluish white or cream, roundish or irregular in cross section, tapering to points, simple to slightly branched, hollow, corticate, to 6 cm long and mostly 1-2 mm diam; apothecia unknown; pycnidia rare; spot tests various; arctic-alpine to subalpine.

Sources include: Nelsen and Gargas (2009), Onut-Brännström et al. (2017, 2018), Sato (1962), Sheard (1977).

1a Cortex UV+Y, K- or K+Y. Common e Cas on alpine sod, occasionally on
    windswept subalpine peaks and ridges; arctic s through RM and Wash
    Cas ....*T. subuliformis* squamatic chemotype
1b Cortex UV- (brownish to dark purplish under UV in dark), K+ deep Y to
    O; arctic to Cal, in and w Cas ..... *T. subuliformis* thamnolic chemotype

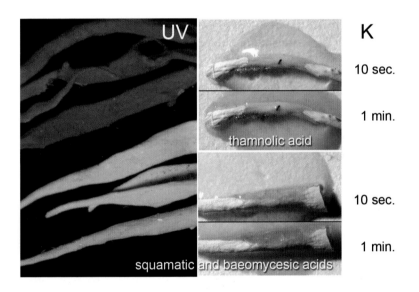

403

Thamnolia subuliformis

Description: Thallus fruticose, occasionally forming upright turfs but usually prostrate or sprawling, white to bluish white or cream, roundish in cross section, tapering to points, simple to slightly branched, hollow, corticate, to 6 cm long and mostly 1-2 mm diam; apothecia unknown; two chemotypes in our area: (1) cortex UV-; medulla P+O to R, K+Y to O (thamnolic acid); (2) cortex UV+Y, P+Y, K- or K+Y (squamatic and baeomycesic acid).

Range: Arctic to Cal and through RM.

Substrate: Moss and *Selaginella* mats, often on thin soil over rock.

Habitat: Exposed subalpine to alpine ridges and peaks, occasionally at low elevations in open rocky sites.

Notes: Formerly segregated into *T. vermicularis* and *T. subuliformis* purely on the basis of chemical content, the two local chemotypes are now thought to belong to the same species, based on DNA sequence data. True *T. vermicularis* is now thought to be geographically restricted to an area in Europe. A third species, *T. tundrae* occurs in Eurasia and the Aleutian Islands and always has squamatic and baeomycesic acid. All are essentially identical in form, but partially differ in chemistry and distribution. In the PNW, the two chemotypes are mostly geographically separate, as given in the key above. Mixed populations are found within the Cas. In addition to spot test and UV differences, specimens with thamnolic acid chemotype become pinkish with storage and stain the adjoining paper brown.

Tholurna

Tholurna dissimilis

Description: Thallus dwarf fruticose, of short erect gray stalks to 2(5) mm tall and 1 mm broad, hollow or nearly so; stalks arising from a squamulose to nearly crustose primary thallus; apothecia terminal, generally present, dissolving into a powdery mass of dark spores; spores 1-septate; spot tests negative except P+ pale Y diffusion onto paper.

Range: Alas and BC to Ore, w Cas.

Substrate: Conifer twigs, rarely on other trees (e.g., birch), rarely on rock.

Habitat: Exposed subalpine ridges and peaks, occasionally at low to mid elevations in cool, moist sites.

Notes: Once seen this lichen can be confused with no other. The stalks form a coarse stubble tipped with powdery black spore masses, usually on twigs of subalpine conifers growing in the open.

Sources include: Otto (1983), Tibell (1975).

Tingiopsidium

Thallus minutely foliose to lobate crustose, olive to olive brown, mostly forming small appressed rosettes < 2 cm diam; lobes narrow and elongate (generally < 0.2 mm wide to 2 mm long) that can become overlapping; upper surface often longitudinally ridged, with radiating lines of photobiont below the upper cortex separated by minute furrows; medulla weakly stratified; lower surface and rhizines pale; hypothallus hardly or not apparent; asexual propagules common, lobules or isidia, laminal and marginal; apothecia common in one species, otherwise rare; apothecial margin lecanorine or lecideine; disk red brown, flat to convex; spores nonseptate or 1-septate, 10-20 × 3-4 μm; photobiont blue green (*Scytonema*-like or *Nostoc*); spot tests negative.

Sources include: Hafellner and Spribille (2016), Henssen (1963b), Spribille and Muggia (2013).

ID tips: Tingiopsidium is a genus of small, olive colored, northern oceanic rock dwellers. Some species in the genus were formerly included in *Vestergrenopsis*. Only one species, *T. sonomense*, is common in Ore and Wash, but the other two species become more frequent northward along the coast into Alas. Sterile specimens might be confused with *Placynthium* (keyed in McCune 2017), which usually has a black hypothallus, sometimes conspicuous, which *Tingiopsidium* lacks.

1a Isidia and lobules lacking; apothecia usually present; spores nonseptate. Arctic s to se Alas, rare in s Wash...*T. elaeinum*
1b Isidia and/or lobules present; apothecia occasionally present; spores nonseptate or 1-septate
  2a Isidia very slender, cylindrical to ± flattened; lobes very thin, elongate, fan-tipped, and tightly appressed; apothecia margins thalline; spores nonseptate. Rare w Cas, more common northward........... *T. isidiatum*
  2b Isidia present or not, but with abundant narrow lobules, the lobules mostly flattened (some cylindrical) and arising from fine divisions of the lobe margins, often becoming densely shingled on the upper surface; lobes thicker, shorter, and less tightly appressed than the preceding species, generally fan-tipped in part; apothecia margins variable, thalline or with a false proper margin; spores 1-septate ....***T. sonomense***

Tingiopsidium sonomense

Description: Thallus minutely foliose to lobate crustose, mostly forming rosettes < 2 cm diam, of narrow elongate lobes (generally < 0.2 mm wide to 2 mm long) that become shingled in older parts of the thallus; lobules narrow but generally ± flattened, occasionally cylindrical, primarily arising from the lobe margins; upper surface olive brown, often longitudinally ridged; lower surface and rhizines pale; apothecia uncommon, the margin variable, with thalline rim or lecideine; disk red brown, flat to convex; spores 1-septate, 10-20 × 3-4 µm, 8 per ascus; photobiont blue green; spot tests negative.

Range: RM and Cas s to Ariz and Cal; seldom collected.

Substrate: Noncalcareous rock.

Habitat: Exposed or shaded outcrops and talus, often in cracks, sheltered areas, and drainage lines.

Notes: Tingiopsidium isidiatum is very similar but much rarer than *T. sonomense* in Ore and Wash, and it occurs in cooler, wetter habitats than *T. sonomense.* Unlike these rock-dwelling species, another tiny cyanolichen, *Koerberia biformis,* occurs on bark. It is known from many locations in Cal, but only one site in Wash and no sites in Ore. The species is distinguished by its minute blackish olive thallus, appressed lobes, dark brown-black apothecia, very slender and twisted spores (LM), and cylindrical isidia.

Umbilicaria

Thallus foliose, umbilicate, roundish and undivided or lobed, to 18 cm diam or larger; upper surface some shade of brown, gray, or black (rarely red), variously textured (smooth, rugose, radiate-ridged, coarsely warty or blistered); lower surface beige, tan, gray, dark brown, or black, papillose or not; rhizines or plate-like structures sometimes present, rhizines and lower surface sometimes sooty black, the color produced by a fine covering of small asexual propagules, "thalloconidia"; isidia or soredia present in a few spp; apothecia lecideine, black, often fissured or with a central sterile button, partially sunken in the thallus to sessile or almost stalked; spores simple or muriform, mostly hyaline, pale brown to brown in a few species, ellipsoid, oval, or irregularly lumpy; medulla C- or C+R, KC- or KC+R, mostly K-, P-; on rock.

Sources include: Allen et al. (2022), Davydov et al. (2010, 2011, 2017), Kofranek and McCune (2008), Llano (1950), Løfall and Timdal (2005), McCune (2018), McCune and Curtis (2012), McCune et al. (2104a), Poelt and Nash (1993).

ID tips: Spore examination has not been routinely used in *Umbilicaria*, but the large variation in spore features among species in the PNW makes it a very useful tool, particularly with problematic specimens. Spot tests are seldom useful in *Umbilicaria* and are therefore mostly omitted from the following species accounts. TLC can, however, be quite useful in some cases, so we include characteristic lichen substances where helpful.

Introductory Key

1a Isidia or soredia present..**Group 1** (Lead 9)
1b Isidia and soredia lacking
  2a Thallus crustose to squamulose ....**Group 2** (Lead 10)
  2b Thallus foliose
    3a Upper surface conspicuously blistered or warty, the warts typically
      1-5 mm diameter and 1 mm or more high; rare disjuncts (subgenus
      Lasallia) ....**Group 3** (Lead 12)
    3b Upper surface variously textured (often rugose) but not
      conspicuously warty; common
      4a Apothecia present but not concentrically fissured, the disk smooth
        or with a central sterile button, sometimes with very fine
        secondary concentric fissures
        5a Apothecia lacking fissures; disk flat, smooth ...**Group 4** (Lead 13)
        5b Apothecia with a central sterile button, otherwise smooth or
          with very fine secondary concentric fissures
          ....**Group 5** (Lead 16)
      4b Apothecia lacking or, if present, with concentric or radiating
        fissures
        6a Thallus almost crustose or squamulose. .... *U. lambii* (Group 2)

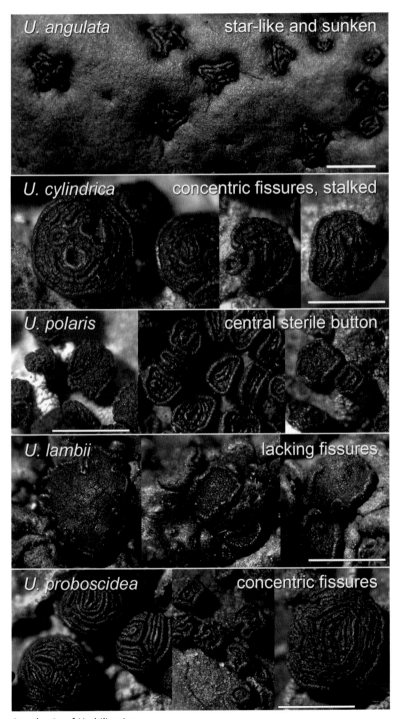

Apothecia of *Umbilicaria.*

Umbilicaria americana

Description: Thallus foliose, umbilicate, roundish to somewhat lobed, to 20 cm diam or more; upper surface whitish gray to gray; lower surface sooty black, granular to areolate; rhizines abundant, slender, sooty black, not ball tipped, sometime stubby and degenerate; isidia and soredia lacking; apothecia very rare, with concentric fissures; spores simple; thallus containing gyrophoric acid (medulla C+R, KC+R).

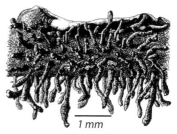

U. americana, rhizines

Range: Widespread, throughout the PNW; also in e N Am.

Substrate: Rock (noncalcareous).

Habitat: Steep rock faces and cliffs, usually in sheltered, cool sites in moist forested areas in the mountains, usually at low to middle elevations.

Notes: Umbilicaria americana grows larger than any other *Umbilicaria* species in w N Am and will commonly form masses of large, floppy thalli on near vertical walls of noncalcareous rock in moist environments. *Umbilicaria vellea* is very similar but usually smaller and arctic-alpine. It is differentiated by its two-part rhizines: sooty black and stubby below with a pale elongate tip. In contrast, *U. americana* has completely sooty, slim and elongate or stubby rhizines. Some *Dermatocarpon* spp are superficially similar but bear spores in perithecia, lack the sooty black coating on the rhizines and lower surface, and are C-, KC-.

Umbilicaria angulata

Description: Thallus umbilicate, to 5(9) cm diam, monophyllous to lobate; upper surface brown to gray brown, sometimes with faint violet tones, matte to slightly shiny, smooth to faintly areolate, not frosty pruinose near the umbo; lower surface dark brown to black, papillose or finely verrucose; rhizines sparse to more often dense and tangled, seldom projecting from the margins, slender to flattened, not ball-tipped, usually brown to black but sometimes whitish; platy structures often present among the rhizines but mostly near the umbilicus; isidia and soredia lacking; apothecia angular to stellate, initially level with the thallus, becoming sessile; spores simple, (13)17-23(25) × 8-13 μm; thallus containing gyrophoric (major) and lecanoric acids and zeorin.

Range: Ore Cas n to coastal Alas.

Substrate: Noncalcareous rock.

Habitat: Outcrops and talus, often on steep faces.

Notes: Although *U. angulata* and *U. semitensis* are similar in general form, they are readily distinguished by their spores and DNA sequences. *Umbilicaria angulata* has simple spores whereas *U. semitensis* develops larger muriform spores. Their ranges overlap somewhat, with *U. angulata* occurring from Alas to n Cal and *U. semitensis* from sw Ore to s Cal; both can be found in a transitional area in sw Ore and n Cal.

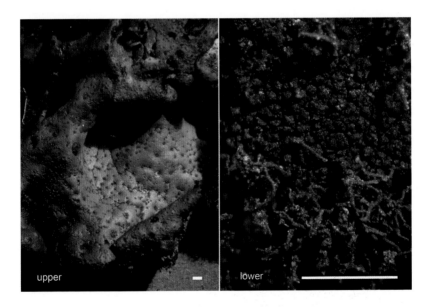

upper   lower

Umbilicaria cylindrica

Description: Thallus foliose, umbilicate, roundish to somewhat lobed, mostly < 4 cm diam; upper surface grayish pruinose; lower surface smooth to finely granulose, brown to pinkish; rhizines light brown to black, occasionally sparse, generally long and usually concentrated toward the margins, usually conspicuously projecting from the margins when viewed from above; isidia and soredia lacking; apothecia almost always present, short stalked, with concentric fissures (photo p. 409); spores 11-15.5 × 6-11 µm; cortex and medulla C-, KC- (with or without norstictic acid).

Range: Alas to Ore, w Cas, Cascades and Coast Range.

Substrate: Rock (noncalcareous).

Habitat: Infrequent, subalpine to alpine, becoming more frequent northward along the coast into se Alas.

Notes: Umbilicaria cylindrica is easily recognized by its grayish color, short-stalked apothecia, and long rhizines projecting from the margins. *Umbilicaria virginis* also can have projecting rhizines but has a nearly smooth disk with a central button, in contrast to the concentric fissures of *U. cylindrica.* Additionally, *U. virginis* usually has a medulla C+R and KC+R (gyrophoric acid), while *U. cylindrica* is unusual in this genus in its C-, KC- reactions (from the absence of gyrophoric and lecanoric acids).

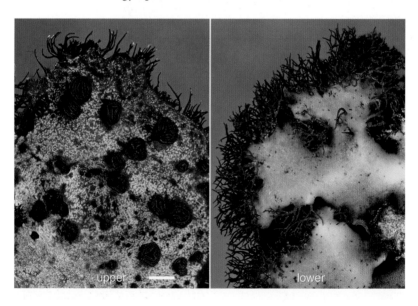

Umbilicaria deusta

Description: Thallus foliose, umbilicate, roundish to somewhat lobed, mostly < 4(9) cm diam, but often in extensive colonies of contiguous to shingled individuals; both upper and lower surfaces dark brown to blackish brown; lobes rather thin and brittle when dry; lower surface smooth to irregularly pitted; rhizines lacking; isidia laminal, fine to coarse, occasionally minutely squamulose, varying from sparse and fine to thickly covering almost the whole upper surface; soredia lacking; apothecia unknown from N Am; thallus containing gyrophoric, lecanoric and ± hiascic acids.

Range: Widespread and common; circumpolar, s to Ariz and Cal; also common in ne N Am.

Substrate: Rock (noncalcareous).

Habitat: Damp rocks in humid microsites, mostly at low elevations to subalpine, occupying an impressive array of macroclimates from temperate to arctic.

Notes: The only isidiate species of *Umbilicaria* in the PNW, *U. deusta* occupies wetter rock surfaces than most members of the genus, often being found on rocks that receive seepage or water trickles during parts of the year. Once seen the species is easily recognized in the field.

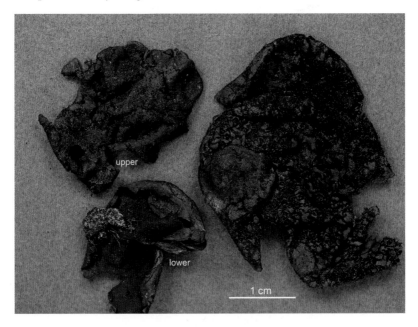

Umbilicaria dura

Description: Thallus umbilicate, monophyllous; upper surface brownish gray, brown, or very dark brown, not discolored near the umbo, matte to shiny, epruinose, to 4 cm diam and 2–10 mm thick; margins entire to irregularly incised, slightly to strongly rolled under; upper surface smooth to areolate (similar to *U. torrefacta*), developing coarse folds, imperforate to moderately perforate with pinholes (hold up to light); lower surface black or dark brown, lacking thalloconidia, warty, usually developing a dense mat of multiple levels of black or dark brown parallel plates and rhizines over part or all, the mat to about 5 mm thick, rarely with patches not covered by plates or rhizines and thus exposing the warty lower cortex; rhizines few to abundant, cylindrical to flattened, dark brown to black, to 3 mm long; isidia and soredia lacking; apothecia subsessile to sessile, black, gyrose, to about 4 mm diam, initially elliptic, becoming angular to stellate, then compound and forming dense aggregations; spores simple, hyaline, 8 per ascus, ellipsoidal, 7–10(13) × 4.5–6(7) μm; thallus containing gyrophoric acid (major) and lecanoric acid (trace).

Range: Alas to Ore Cas; apparently endemic to the PNW of N Am.

Substrate: Noncalcareous rock.

Habitat: Montane to alpine, sheltered to exposed surfaces of rock outcrops, usually on ridges and peaks. The type collection is from an open rocky ridge-top at 1600 m in the central Ore Cas, west of the crest.

Notes: The species is named for the tough, hard thallus. It is similar to *U. torrefacta* but has multilayered dark brown to black plates on the lower surface. *Umbilicaria multistrata* is another species very similar to *U. dura*, but so far *U. multistrata* is known only from coastal Alas. While typical *U. torrefacta* is easily distinguished from *U. dura* and *U. multistrata* by its pale lower surface and thin thallus, *U. torrefacta* can be similar in appearance to the other two species in having a moderately thick thallus and black lower surface. Reliably separating that form requires TLC or ITS sequence.

— upper   — lower

Umbilicaria havaasii

Description: Thallus foliose, umbilicate, roundish to broadly and irregularly lobed, mostly < 10 cm diam, rather fragile when dry; margins often perforate and lacerate; upper surface gray, gray brown, or blackish, partially grayish to white pruinose, especially on reticulate ridges near the umbo, the ridges fading radially, distinct or weak; lower surface light tan near umbilicus, dark gray brown toward the margins, with sooty black patches (from thalloconidia), the gray areas scabrid; sooty black rhizines often present, frequently in tufts, sometimes the tufts protruding upwards through holes in the surface; soredia and isidia lacking; apothecia infrequent, substipitate; thallus containing gyrophoric acid ± umbilicaric acid.

Range: Alas to n Cal (Siskiyou Co.), rarely inland to w Mont.

Substrate: Rock (noncalcareous).

Habitat: Outcrops and talus, often on steep rock faces, montane to alpine, most frequent in areas with oceanic to suboceanic climate.

Notes: Umbilicaria havaasii can form large fluffy colonies on rock, often with a floppy appearance and somewhat dangling when on steep rock faces, in contrast to the more appressed growth form of many *Umbilicaria* species. The degree of development of black sooty thalloconidia on the lower surface is highly variable. Some thalli have a completely pale brown, brown, or gray lower surface, and belong to a clade closely related to *U. proboscidea.* The edges of the thallus often develop perforations around the margin (hold up to the light), which will help identify individuals lacking thalloconidia. In Ore and Wash the species often grows with *U. rigida.*

Umbilicaria hirsuta

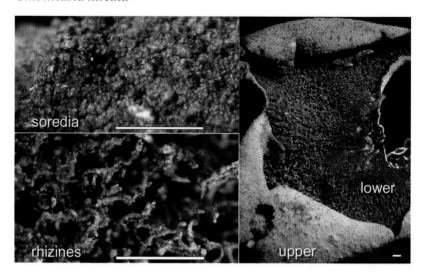

Description: Thallus foliose, umbilicate, roundish to somewhat lobed, mostly < 6 cm diam, margins often turned under; gray to brown gray above; lower surface pale gray, brown, or black, with pale to dark rhizines and ± platy structures; isidia lacking; soredia present, powdery or granular, diffuse, near the margins on the upper surface; apothecia rare (not seen by us); thallus containing gyrophoric and lecanoric acids.

Range: Sporadic from Arctic to Cal and Colo, w S Dak, e N Am and Mex; rare in the PNW (Jackson Co., Ore; n Id, w Mont).

Substrate: Rock (noncalcareous).

Habitat: Sheltered rock faces in forested or semi-open areas.

Notes: This is the only sorediate species of *Umbilicaria* in N Am. The soredia develop from disintegration of the upper cortex close to the margins. The soredia are often sparse and inconspicuous, generally not grouped into discrete soralia, and thus easily overlooked without careful inspection with a hand lens. The species is apparently rare throughout its range in N Am, having a puzzling sporadic distribution. A good search image is a pale brownish gray thallus (more like a *Dermatocarpon* than an *Umbilicaria* in color) and sheltered habitats, similar to *U. americana* and *U. vellea*, rather than the exposed habitats of many *Umbilicaria* species.

Umbilicaria hyperborea

Description: Thallus foliose, umbilicate, roundish to somewhat lobed, mostly < 5 cm diam; upper surface rugose with contiguous, contorted, round-topped ridges, mid to dark brown; lower surface brown or brown black, sometimes somewhat gray in part, sometimes black but lacking sooty thalloconidia, smooth or finely papillose; rhizines lacking; isidia and soredia lacking; apothecia common, with concentric fissures; spores 12-17(20) × (5.5)7-9(11) μm.

Range: Widespread, throughout the PNW except near the coast.

Substrate: Rock (noncalcareous).

Habitat: Exposed rock outcrops; w Cas mainly at higher elevations; e Cas common at all elevations.

Notes: Umbilicaria hyperborea is probably the most abundant and well distributed member of this genus in the PNW, distinguished by the rugose upper surface and the smooth brown or blackish lower surface. *Umbilicaria arctica* is similar but has a partly dove gray lower surface. Shade forms of *U. hyperborea* may also be pale below, but in this case, the upper surface is paler than usual; thalli that are dark brown above but largely gray below are *U. arctica*. *Umbilicaria nylanderiana* is similar but much rarer, distinguished by a black lower surface covered with sooty thalloconidia.

Umbilicaria nodulospora

Description: Thallus umbilicate, monophyllous to more often dividing as it expands into larger polyphyllous colonies, to 2(4) cm diam; upper surface brown to gray brown, faintly to distinctly grayish pruinose, often grayer near the umbo, matte to slightly shiny; upper surface smooth to reticulately cracked, the cracks often deep so that the thallus readily divides into part thalli; submarginal areas occasionally minutely perforate (hold up to light); margins entire to irregularly lacerate or lobulate; lower surface brown to black, lacking thalloconidia, smooth to finely papillose or warty, usually developing a dense mat of parallel or tangled rhizines over part or all of the lower surface, often interspersed with brown to black platy areas, sometimes patches apparent without rhizines or plates; isidia and soredia lacking; apothecia sessile, black, gyrose, to 1.2(2) mm diam, initially angular to stellate, with age protruding, becoming convex and roundish in outline; spores simple, ellipsoidal but usually with 1-2 blunt, shallowly bulging knobs at one end and forming a Y, T, or L shape, (9)10-13(18) × 6-7.5(9) μm; thallus containing gyrophoric acid.

Range: Occasional from c Ore s to n Cal and e to sw Id, endemic to the PNW.

Substrate: Noncalcareous rock, most often on basalt.

Habitat: Lower montane to intermountain valleys, rocky openings in dry forests to semiarid steppe; often in semi-sheltered microsites on steep rock faces.

Notes: This ascospore shape unusual among ascomycetes, lichenized or not. One end of the spore typically has one or two shallow bulges, suggesting a T, L or Y. Even without examination of spores, most *U. nodulospora* can be readily identified. Morphologically the species shares some characters with *U. torrefacta*, such as occasional presence of plates on the lower surface, and perforations in the thallus near the margins. But neither of these characters is as strongly developed in *U. nodulospora* as in *U. torrefacta*.

Umbilicaria phaea

Description: Thallus foliose, umbilicate, roundish to somewhat lobed, mostly < 5 cm diam; upper surface smooth or with a fine network of shallow indentations, brown (rarely red; see var. *coccinea* below); lower surface dark brown to black, covered with fine papillae or warts; rhizines lacking or occasionally a few present; isidia and soredia lacking; apothecia common, with concentric fissures, round or star-shaped, sessile or sunken in the thallus.

Range: Widespread in w N Am, throughout the PNW.

Substrate: Rock (noncalcareous).

Habitat: Exposed to somewhat shaded rock outcrops and talus, common at low to mid elevations, less frequent at high elevations. *Umbilicaria phaea* is more tolerant of high temperatures than any of our other *Umbilicaria* species and thus prominent in hot, dry environments, such as low elevations in most parts of California.

Notes: One of the most frequent species in the PNW, *U. phaea* is recognized by the nearly smooth upper surface and the dark papillose lower surface (use lens). *Umbilicaria phaea* var. *coccinea* (see below) is bright red but otherwise similar. The occasional specimen of *U. phaea* with a few rhizines may key to *U. angulata* or *U. cylindrica*, but those species are more abundantly rhizinate.

Umbilicaria phaea **var.** *coccinea*

Description: Thallus foliose, umbilicate, roundish to somewhat lobed, mostly < 3 cm diam; upper surface orange red, red, or burnt red, smooth or with sparse indentations; lower surface dark brown to black, covered with fine papillae or warts; rhizines lacking; isidia and soredia lacking; apothecia common, with concentric fissures, round or star-shaped, sessile or partially sunken in the thallus; medulla containing gyrophoric and lecanoric acids (medulla C+R) and cortex with an anthraquinone (K+R, lw UV+O) responsible for its red color.

Range: Locally common in and near Siskiyou National Monument, Yreka, and the Shasta River in n Cal, also in adjoining Ore; rarer near Klamath Falls, Ore, and even rarer in c Ore and c Wash. Endemic to the PNW.

Substrate: Basalt and ultramafic rock (noncalcareous).

Habitat: Fully exposed to shaded rock outcrops, boulders, and talus; low to middle elevations.

Notes: Umbilicaria phaea var. *coccinea,* also known as the "lipstick lichen," looks like dabs of red lipstick, usually on dark basalt or peridotite. It grows intermixed with the usual brown *U. phaea* var. *phaea,* but with a much narrower range. One wonders whether it is reproductively isolated—a species in itself—or whether it interbreeds with the associated brown thalli. Recent evidence suggests that gene flow is possible among the color morphs. Rare individuals are intermediate in color.

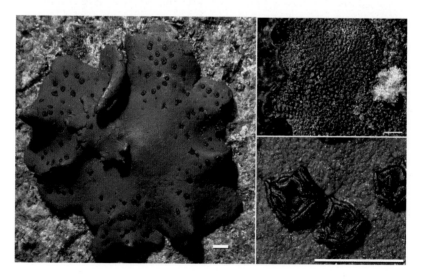

Umbilicaria polaris

Description: Thallus foliose, umbilicate, roundish to somewhat lobed, mostly < 4 cm diam; upper surface with a radially spreading network of frosty ridges; lower surface gray, tan, or brown, smooth; rhizines lacking; isidia and soredia lacking; apothecia usually present, with a central button, sometimes with faint secondary concentric fissures, sessile to almost stalked (photo p. 409); spores 9-14 × 5.7-8 µm; thallus containing gyrophoric (medulla C+R) and umbilicaric acids.

Range: Circumpolar, s to Ariz and Cal.

Substrate: Rock (noncalcareous).

Habitat: Exposed to somewhat sheltered rock, arctic-alpine to subalpine and in cold steppe, very common.

Notes: Of the three species of *Umbilicaria* with apothecia with a central button, *U. polaris* is unique in its pale, smooth lower surface. *Umbilicaria virginis* has rhizines and *U. decussata* has a sooty black lower surface. *Umbilicaria proboscidea* is similar but has disks with concentric fissures and a more northern distribution, to some extent replacing *U. polaris* n of Mont. From Mont s in RM, *U. polaris* is abundant on rocky subalpine and alpine ridges and peaks as well as boulders and cliffs in cold canyons and high valleys.

Umbilicaria polyphylla

Description: Thallus foliose, umbilicate, roundish but generally becoming deeply lobed and divided, mostly < 8 cm diam; upper surface smooth to gently rugose, chocolate brown to dark brown, less often grayish brown; lower surface black, smooth to papillose; rhizines lacking; isidia and soredia lacking; apothecia rare, with concentric fissures.

Range: Widespread, common both w Cas and between Cas and RM crests.

Substrate: Rock (noncalcareous).

Habitat: Exposed to shaded rock outcrops, lava flows, and talus, most common in moist forested areas throughout the PNW.

Notes: The smooth brown upper surface lacking apothecia and contrasting with the smooth black lower surface lacking rhizines are characteristic. Populations with an upper surface more gray brown than chocolate brown are occasional. When dry the curly lobe tips allow you to see the contrasting colors of upper and lower surfaces, unlike many *Umbilicaria* species that are more appressed. *U. polyphylla* also tends to be deeply divided into many lobes ("polyphyllous") rather than forming a single-lobed roundish thallus. Although the upper surface is typically smooth, it sometimes develops a rugose texture similar to but less pronounced than that of *U. hyperborea*.

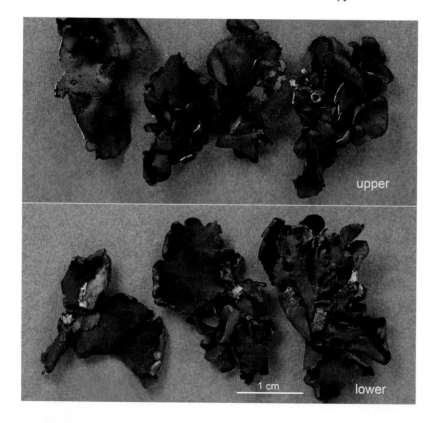

upper

1 cm   lower

Umbilicaria polyrhiza

Description: Thallus single or multi-lobed, to 6(10) cm diam; upper surface dark brown to dark gray brown, smooth, the umbo hardly differentiated; lower surface rhizinate, black, papillate; rhizines ball-tipped at least in part, moderately dense to dense, simple or forked, often quite long and projecting from the margins; soredia and isidia lacking; apothecia uncommon, with radiating fissures, sunken in the thallus; spores hyaline, simple, small, 8.4-9.0 × 3.5-6 μm; thallus containing gyrophoric acid.

Range: Coastal se Alas to n Cal, in Ore and Wash most frequent in the Coast Range and Cas, both sides of the crest, occasionally inland to n Id; reports from farther inland are doubtful and should be re-examined.

Substrate: Noncalcareous rock.

Habitat: Rock outcrop areas, talus, and lava flows, generally in areas surrounded by forests, woodlands, and savanna in oceanic to suboceanic climates, low to mid elevations.

Notes: The combination of numerous ball-tipped rhizines and a brown upper surface is diagnostic for this species. Ball-tipped rhizines also occur in *U. cinereorufescens*, but that species has a gray to brownish gray upper surface and is rare in the PNW (though fairly frequent in Alaska and occasional in central to s RM).

Umbilicaria torrefacta

Description: Thallus foliose, umbilicate, roundish to slightly lobed, mostly < 5 cm diam; upper surface with a network of cracks, light to mid brown; margins finely perforate (hold up to light) and dissected; lower surface light brown or tan, with discontinuous platy structures that become dissected into radiating straps or strands, as well as rhizine-like structures, sometimes rhizines predominating; isidia and soredia lacking; apothecia common, with concentric fissures; thallus chemical content variable but usually with stictic acid (major), occasionally with norstictic acid or gyrophoric acid.

Range: Widespread; very common e Cas, occasional w Cas.

Substrate: Rock (noncalcareous).

Habitat: Open to sheltered rock outcrops and talus, low to high elevations.

Notes: The light brown, platy lower surface combined with the perforate margins are diagnostic. The cracks in the upper surface, like sutures in a skull, are also characteristic of this species. While typical material with those characteristics is easy to identify, forms with a dark brown to black lower surface are more challenging. They are easily confused with *U. dura* or *U. nodulospora*. See additional notes under those species.

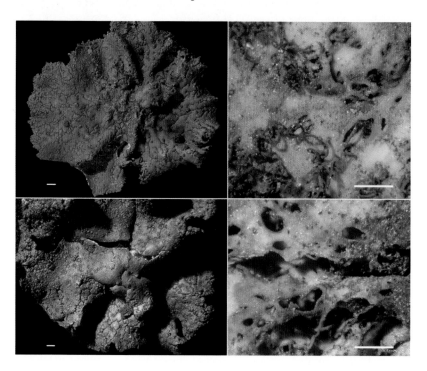

Umbilicaria vellea

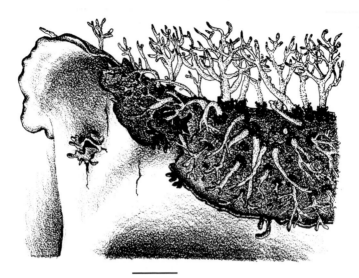

1 mm

Description: Thallus foliose, umbilicate, roundish to somewhat lobed, mostly < 6(10) cm diam; upper surface light gray to gray; lower surface sooty black or often pale between the rhizines; rhizines abundant, with sooty black stumpy bases and narrow, elongate, pale tips; isidia and soredia lacking; apothecia rare, with concentric fissures; thallus most often containing gyrophoric acid (medulla C+R, KC+R), but other chemotypes exist.

Range: Arctic-alpine to subalpine, s in RM to Colo and s in the coastal states to n Cal; also in e N Am; online maps overestimate the range because of confusion with *U. americana*.

Substrate: Rock (noncalcareous).

Habitat: Mostly arctic-alpine to subalpine, occasionally in cool canyons.

Notes: Very similar to *U. americana*, *U. vellea* is somewhat smaller and is more common in northern, alpine, and subalpine areas, while *U. americana* predominates at low to mid elevations. Morphologically the two species were confused for a long time, so many herbarium records of *U. vellea* are actually *U. americana*. They are differentiated by the uniformly sooty black rhizines of *U. americana* in contrast with the two-part rhizines of *U. vellea*: a thick black base with an elongate pale tip. *Umbilicaria cinereorufescens* is superficially similar but typically is smaller, lacks the whitish rhizines, and more often has brownish or cinnamon tints to the upper surface. It is rare in the PNW.

Umbilicaria virginis

Description: Thallus foliose, umbilicate, roundish to somewhat lobed, mostly < 6(10) cm diam; upper surface brownish to gray, sometimes blackening, with or without a network of ridges, but usually either ridged or wrinkled; lower surface gray, tan, or brown, smooth; rhizines brown to buff or pinkish when shaded to black where exposed; isidia and soredia lacking; apothecia common, with a central button (sometimes subtle or lacking when young), sessile to almost stalked; spores 10-15(18) × (5)6.5-9(10) μm; thallus most often containing gyrophoric acid (medulla C+R, KC+R).

Range: Widespread, Arctic s to Mex following the major mountain chains.

Substrate: Rock (noncalcareous).

Habitat: Common on exposed or sheltered outcrops and talus at mid to high elevations; at high elevations it is often the dominant *Umbilicaria* on steep sheltered rock faces.

Notes: Although the upper surface of *U. virginis* is quite variable, the abundant pinkish to tan rhizines and the tan to pinkish center of the lower surface make this species easy to identify. *Umbilicaria virginis, U. polaris,* and *U. hyperborea* are the most frequently collected species of *Umbilicaria* in the alpine areas in and e Cas. Although seemingly distinctive morphologically, DNA evidence suggests that *U. virginis* is actually a species complex rather than a single species.

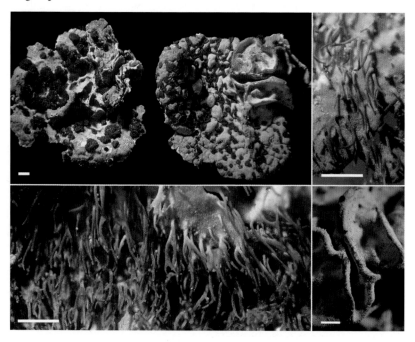

Usnea

Thallus fruticose, tufted or pendulous, pale greenish or yellowish tinged, occasionally reddish brown or blackening; branching various but often with numerous perpendicular short (2-20 mm), lateral branches (fibrils); branches often papillate, round to weakly angular in cross section, with a dense central cord (axis) surrounded by a cottony medulla and a dense cortex; isidia ("isidiomorphs" of Clerc) or soredia usually present, usually in clusters (isidalia and soralia); apothecia rare, lateral or terminal, with a thalloid margin and pale disk; spot tests various; on bark or wood, rarely on rock.

Sources include: Arcadia (2013), Clerc (1984, 1987a, 1987b, 2004, 2008), Elix et al. (2007), Halonen (2000), Halonen et al. (1998), Herrera-Campos and Clerc (1998), Kayes et al. (2008), Lendemer (2006b), Lendemer and Tavares (2003), Lücking et al. (2020), Mark et al. (2016), Miller (2011), Riley et al. (1995), Stone (2018), Tavares (1997), Truong et al. (2013), Wirtz et al. (2006, 2007), Wright (2001).

ID tips: Despite recent work on the taxonomy of *Usnea*, it remains one of the most difficult genera of lichens, apart from a few well-defined species. Thickness of the cortex, medulla, and axis are useful but not infallible characters for separating *Usnea* spp. Measure these relative to the branch diameter. Choose a main branch, but not a basal part, and section it lengthwise with a sharp razor blade. Slice along the midline of the branch. Measure the thickness of the layers with either an ocular micrometer or by taking a digital photograph and measuring the dimensions with image-editing software. Average the two measurements of the cortex and divide by the branch diameter. Repeat for the medulla. Measure the axis and again divide by branch diameter. Express as percentages. These are sometimes abbreviated, for example, CMA = 19/9/50 means cortex 19%, medulla 9%, and axis 50% of the diameter.

All species in the genus contain usnic acid, often with other substances such as barbatic, diffractaic, norstictic, protocetraric, salacinic, squamatic, or stictic acid. Spot tests can be unreliable in *Usnea*, so chemical substances detectable by TLC appear in the key. For example, many specimens with norstictic acid or salacinic acid will yield negative or slow K and P reactions, even though in most lichens these substances yield distinct positive spot test reactions. UV fluorescence is also sometimes challenging, because in most species the medulla and axis are pale in UV light while the usnic acid in the cortex absorbs UV, and is therefore dark. This contrast suggests a UV+ reaction, but it is not the strong fluorescence seen with specimens containing squamatic acid.

Spot tests in *Usnea* are best done on two parts: a main branch that has been cut obliquely or nearly parallel with the long axis, and some branches bearing soredia or isidia. Be sure to check for reactions in the medulla just below the cortex and on soralia near branch tips.

Introductory Key

1a Cortex or central cord distinctly reddish, brown, yellow, or blackening
....**Group 1** (Lead 6)
1b Cortex yellow green to green and central cord white
  2a Thallus pendulous, often long, typically > 12 cm ......**Group 2** (Lead 10)
  2b Thallus tufted, ± erect or drooping, typically < 12 cm
    3a Thallus with sparse to abundant apothecia; soredia and isidia lacking
....**Group 3** (Lead 19)
    3b Thallus with few or no apothecia; soredia and/or isidia usually
      present
      4a Branches pinched at the nodes and slightly to distinctly expanded
        in the internodes, so that individual branch segments are cigar
        shaped.. **Group 4** (Lead 22)
      4b Branches cylindrical
        5a Isidia lacking or present mainly in young soralia; soredia
          present; medulla thicker than the axis.......... **Group 5** (Lead 29)
        5b Isidia usually abundant; soredia present or not; medulla thinner
          than or about equal in thickness with the axis
          .... **Group 6** (Lead 33)

Group 1

Colored axis or cortex (red, yellow, brown, or black)

6a Central cord pale reddish, yellow, or brown
  7a Central axis yellow (photo p. 440); cortex often with red spots
.... ***U. flavocardia***
  (*U. cornuta* s. lat. and *U. fragilescens* var. *mollis* rarely have red spots
  in the cortex and the central axis is occasionally yellowish to orangish
  when breakdown products of salacinic acid are present; the axis of
  these species is white when healthy; see Group 4.)
  7b Central axis reddish, rose, or brown; cortex yellowish green to green
    8a Central axis brown; medulla usually K+O, P+O (constictic, salacinic
      [trace], and diffractaic [trace] acids); cortex not tuberculate;
      isidia and soredia lacking. Thallus pendulous, with cylindrical
      slender branches; medulla compact; cortex matte; annular cracks
      conspicuous, often abundant; papillae absent; fibrils sparse; e N Am,
      known in PNW from a specimen in sub-boreal BC and a specimen
      from near the coast in Curry Co., Ore ....*U. trichodea*
    8b Central axis reddish, pinkish brown, or rose (photo p. 437); medulla
      and axis often C+Y or pale Y, CK+ deep yellow orange (diffractaic
      and barbatic acids and accessory substances); cortex with raised
      tubercles that commonly bear isidia and coarse soredia. Thallus
      pendulous, to 100 cm; base pale to rarely blackened; medulla
      compact; cortex thick and glossy; annular cracks often abundant
      and conspicuous; tubercles sometimes coalescing into ridges; fibrils
      and papillae sparse to abundant; this widely distributed oceanic
      species is rare in PNW, occurring from BC to s Cal.........*U. ceratina*

(Sometimes the rose pigment in the axis is faint. Diffractaic acid is
also found in *U. longissima, U. trichodea,* and rarely in *U. hirta.*)
6b Central cord white
   9a Cortex with a diffuse red pigment or red spots; on bark or wood near
      the coast...***U. rubicunda***
   9b Cortex blackening toward the branch tips; on rock in subalpine to
      alpine ....***U. lambii***

Group 2

Pendulous

10a Main branches very long, rarely dividing, with dense perpendicular
   fibrils; central axis of main branches > 50% of the thickness of branches;
   central axis I+B .... *U. longissima*
10b Main branches often rebranching, fibrils absent to abundant; central axis
   of main branches < 50% of the thickness of the branch; central axis I-
   11a Papillae lacking
      12a Main branches weakly to strongly ridged and pitted; annular cracks
         few or many
         13a Main branches becoming strongly pitted and ridged, bluntly
            or sharply angular in cross section; isidia and soredia lacking;
            papillae lacking..***U. cavernosa***
         13b Main branches weakly ridged or wrinkled in part; isidia usually
            present; papillae present but often sparse or low; annular cracks
            few ....***U. scabrata***
      12b Main branches not or only weakly ridged and pitted; annular cracks
         abundant
         14a Base slightly to distinctly blackened; papillae usually present,
            though often sparse and low; medulla thicker than cortex;
            medulla usually containing salacinic acid (K+Y to R, P+O),
            rarely with usnic acid only (K-, P-). Thallus often 30 cm or more
            long, sometimes resembling *Alectoria;* usually with abundant
            slender branches but never with abundant fibrils; cortex about
            10% of diam; medulla about 17-20% of diam; axis thick (40-
            45% of diam); soralia and isidalia absent to scattered, usually
            minute, borne on small tubercles (warts); usually on conifers;
            uncommon; Coast Ranges and Cas at low to mid elevations; BC
            and se Alas to Ore .... *U. chaetophora*
         14b Base not blackened; papillae absent; medulla as thin or thinner
            than cortex; medulla K+Y or K-, P+R (protocetraric acid ±
            compounds in the stictic acid group) ....***U. subgracilis***
   11b Papillae present
      15a Branches with abundant annular cracks (typically 6-15 cracks per
         cm) that sometimes expose the medulla
         16a Fibrils abundant; cortex thick (15-20% of diam); medulla dense
            and thin (7-15% of diam); base conspicuously blackened;
            papillae usually abundant. Thallus pendent to subpendent, rarely

shrubby, to 25 cm long, with broadly divergent main branches and abundant fibrils; soralia and isidalia often numerous, distinctly raised, closely spaced to confluent, occasionally absent or sparse, typically larger than ½ of the branch diam, isidiate at least when young; medulla K+Y or R, P+Y or O (salacinic acid ± protocetraric and constictic acids, ± barbatic acid (rare)); on conifers, hardwoods, and shrubs; common in oceanic areas from Cal to BC, usually near the coast, sometimes in the Coast Ranges, sporadic in the Willamette-Puget trough .... *U. silesiaca*

16b Fibrils sparse to absent; cortex about 10% of diam; medulla compact, about 17-20% of diam; base distinctly blackened or not; papillae usually sparse and low .... *U. chaetophora* (Lead 14a) (Note: almost any *Usnea* species can develop abundant annular cracks when they are deteriorating or otherwise in poor condition. When in doubt, key both ways.)

15b Branches with few annular cracks (0-6 per cm), those present mainly near the base

17a Main branches with raised wrinkles that may be widely scattered or dense and reticulating with intervening flat or depressed spots; branches uneven in thickness, swollen, often sinuose near the tips; cortex thin (2-9% of diam) .... *U. scabrata*

17b Main branches cylindrical throughout, not ridged; cortex moderately thick (8-17% of diam)

18a Fibrils usually abundant; papillae usually tall, cylindrical, and abundant; isidalia and soralia tuberculate, often arising from scars of detached fibrils; medulla K-, P- or K+Y to O, P+O .... *U. dasopoga*

18b Fibrils absent to sparse; papillae short warty bumps; soralia plane to tuberculate, arising from small tubercles; medulla K-, P+Y .... *U. pacificana* (Lead 36a)

Group 3

Tufted; fertile; no isidia or soredia

This group is poorly studied and generally rare in PNW, except in the Coast Range of n Cal and near Bellingham, Wash. The chemically based dichotomies below are sure to be replaced as our information improves. In general this group is abundantly papillate and fibrillose, the thallus erect to subpendent, with mainly anisotomic branching, a thin, loose medulla, and a thick central axis.

19a Medulla K+Y or K+O

20a Medulla K+ deep Y, P+O, UV-, containing thamnolic acid with accessory alectorialic acid; presence in PNW is questionable ....[*U. florida* (thamnolic acid chemotype)]

20b Medulla K+Y to O, P+O; containing salacinic acid. Papillae narrow, short to intermediate in height; fibrils tapering from base; Cal .... *U. intermedia*

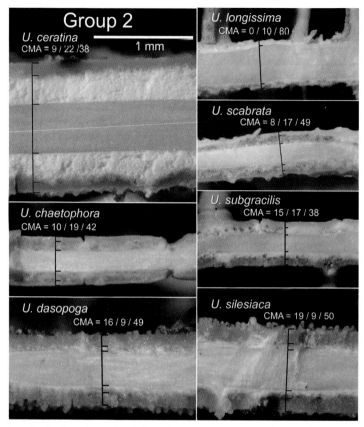

Longitudinal sections of main branches in Usnea. "CMA" refers to relative thick-nesses of the cortex, medulla, and axis, expressed as percentages of the branch diameter (see details under genus description). The CMA values shown are exam-ples only; they will vary among branches and individuals.

19b Medulla K-
    21a Medulla P-, UV+, containing squamatic acid ± alectorialic acid
        ....[*U. florida* (squamatic acid chemotype)]
    21b Medulla K-, P+R, UV-, containing protocetraric acid in high
        quantities; uncommon; BC, n Wash w Cas, perhaps increasing in
        Whatcom and Skagit Cos., rare in sw Ore...................... *U. quasirigida*

Group 4

Tufted; cigar-shaped branch segments

This is the *U. fragilescens* group, characterized by a tufted thallus; glossy, usually thin and fragile cortex; a thick, loose medulla; thin central axis; and cigar-shaped branch segments that are constricted at the nodes.

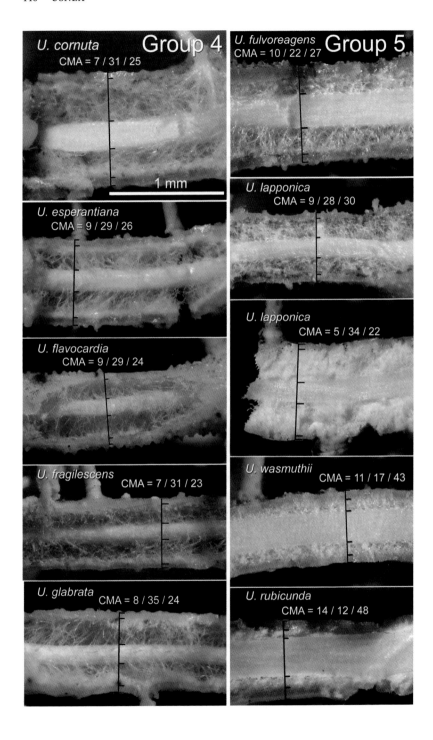

Group 6
Tufted; cylindrical branch segments; isidia abundant

33a Papillae low and indistinct or absent

34a Main branches weakly to strongly foveolate, lacking papillae, generally fibrillose, typically isidiate but sometimes with indistinct soredia; base never blackened... *U. hirta*

34b Main branches never both foveolate and lacking papillae; isidia and/or soredia present; base blackened or not. Cortex translucent, distinctly glossy; fibrils often in fascicles of two to four; papillae low, indistinct to nearly absent; soralia tuberculate when present; branches often segmented by annular cracks; cortex thin (6-12% of diam); medulla loose or dense; medulla either (1) K+Y to O, P+O (norstictic acid ± salacinic, protocetraric, or stictic acids) or (2) K-, P+ deep Y (psoromic acid); on both conifers and hardwoods; apparently rare; BC to Ore on the immediate coast and in the Coast Ranges.... *U. nidulans* (Fascicles of fibrils are occasionally found also in other species; *Usnea nidulans* is also distinguished by other morphological and chemical characters. The combination of norstictic acid and isidia is unique for our *Usnea* species.)

33b Papillae present and conspicuous, at least near the base of main branches

35a Medulla K- (rarely +), P+Y or P-, UV+ blue-white (squamatic and baeomycesic acids ± barbatic acid).

36a Medulla P+Y (squamatic and baeomycesic acids ± barbatic acid). Thallus initially forming robust tufts, becoming pendent, to 20(25) cm long; base slightly to distinctly blackened; annular cracks common near base, often with conspicuous, white, everted medullary rings on main branches; cortex thick (11-19% of diam); medulla dense and thin, 4-20% of diam; axis thick, 36-58% of diam (photo p. 442); fibrils sparse to abundant; soralia punctiform and tuberculate, usually sparse; isidia short to long, occurring on at least young soralia or isidialia, but easily abraded and lost; on both hardwoods and conifers, to at least 1100 m elevation; BC to n Cal, w Cas.. *U. pacificana*

36b Medulla P- (squamatic acid). Thallus shrubby to subpendent; base conspicuously blackened, longitudinal cracks rare; cortex 9-15% of diam; medulla usually dense, 8-19% of diam; axis thick, 40-54% of diam (photo p. 442); white medullary rings occasional; mature soralia or isidialia often rounded, borne on tubercles, varying from minute to enlarged, raised to slightly concave; fibrils often abundant near the base; branches sometimes sparsely foveolate; on conifers and hardwoods, rarely on rock; widespread; common at low to mid elevations in and w Cas, rare e Cas ....*U. subfloridana* (squamatic acid chemotype)

35b Medulla K+Y, O, or R (reaction is often slow), P+ or P-, weakly UV+ (medulla often appearing pale in contrast to the cortex which appears dark because of strong UV absorbance by usnic acid)

37a Medulla and soralia K+ deep Y, P+ deep Y or O (thamnolic acid); isidalia or soralia often becoming broad. Cortex with few to scattered annular cracks; thallus rarely becoming pendent; medulla lacking salacinic acid, low elevations w Cas ....*U. subfloridana* (thamnolic acid chemotype)

37b Medulla K+ Y to O or R, P+ Y or O (salacinic acid)

38a Cortex thin, usually < 12% of diam (photo this page). Thallus tufted to frequently subpendent; branches ± irregularly swollen, slender, sometimes with foveoles and depressions; branch tips often sinuose; base pale or blackened; medulla with salacinic ± barbatic acid; papillae present, short to cylindric; fibrils variable; medulla lax to dense, variable in thickness; on bark and wood; widespread, occasional, on both sides of the Cas but more common w Cas .... *U. diplotypus* (Small individuals of *U. scabrata* may key here; these tend to have sparse fibrils; long, tapered branch tips; and irregular flat spots and ridges.)

38b Cortex relatively thick, usually > 12% of diam (photo p. 437); papillae usually warty; terminal branches tapering

39a Cortex with many annular cracks, especially at the base; main branches broadly divergent .... ***U. silesiaca*** (Lead 16a)

39b Cortex with few annular cracks unless in poor condition; main branches broadly or narrowly divergent .... *U. dasopoga* (Lead 18a)

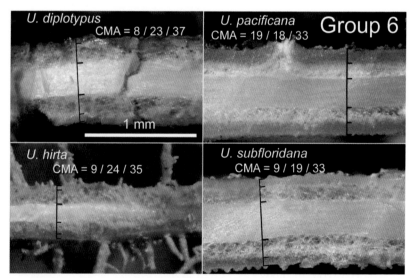

Longitudinal sections of main branches in Usnea. "CMA" refers to relative thicknesses of the cortex, medulla, and axis, expressed as percentages of the branch diameter (see details under genus description). The CMA values shown are examples only; they will vary among branches and individuals.

Usnea cavernosa

Description: Thallus pendulous, to 50(100) cm or more long, soft, draped over branches, the base seldom apparent; branching sparse, irregularly dichotomous with few or no fibrils; main branches to 0.8(2.0) mm diam, becoming strongly pitted and ridged, bluntly or sharply angular in cross section but the finest branches cylindrical; papillae lacking; central strand and terminal branches ± sinuous; cortex with sparse to abundant annular cracks, especially on main branches; cortex thin (4-10% of branch diam); medulla compact, 18-25% of diam; axis 30-45% of diam; isidia and soredia lacking; apothecia rare; medulla K+Y to O and P+O (salacinic acid), rarely K-, P- (usnic acid only).

Range: Widespread, boreal, s in mts to Ariz, Cal, and Mex; uncommon but locally abundant in the PNW, both e and w Cas.

Substrate: Bark and wood of conifers, hardwoods, and shrubs.

Habitat: Most common in old northern boreal and montane forests but also in oak woodlands and savannas.

Notes: This species is usually easily recognized by its dangling habit with irregularly pitted and ridged branches, and absence of papillae, soredia, and isidia. At a distance it has the growth form of *Alectoria sarmentosa*, with few perpendicular side branches or fibrils. Sometimes the much more common *U. scabrata* approaches *U. cavernosa* in morphology, but *U. scabrata* usually has papillae, isidia, or both. The patchy local abundance of *U. cavernosa* suggests dispersal limitation, which could be expected because it lacks specialized asexual or sexual propagules.

Usnea hirta

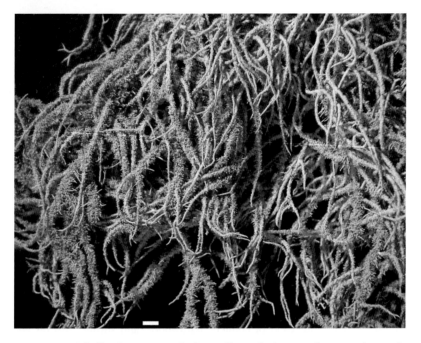

Description: Thallus fruticose, tufted, usually < 5(10) cm; pale greenish or yellowish tinged; main branches with numerous fibrils; main branches with an uneven, irregular surface, lacking papillae; base not blackened; cortex thin (5-10% of diam); medulla lax and thick (20-32% of diam); axis thin or thick (20-40% of diam, photo p. 365); isidia present; apothecia not seen; medulla K-, P- (usnic acid ± fatty acids) or K+Y to R, P+O (norstictic, usnic, and fatty acids).

Range: Widespread, boreal, temperate, and montane; in our area more common e Cont Div but also rare in dry forests w Cont Div.

Substrate: Bark and wood.

Habitat: Usually in dry, open forests or savanna; often the predominant *Usnea* in continental climates in the RM but scarce in oceanic and suboceanic climates.

Notes: The densely fibrillose main branches that lack papillae are diagnostic. In addition, the branches are usually weakly to distinctly uneven with ridges and shallow depressions. The species frequently co-occurs with *U. lapponica* and both have a densely tufted growth form. In the field, check for papillae and soredia with a hand lens. Presence of either of those suggests *U. lapponica* rather than *U. hirta*.

Usnea lambii

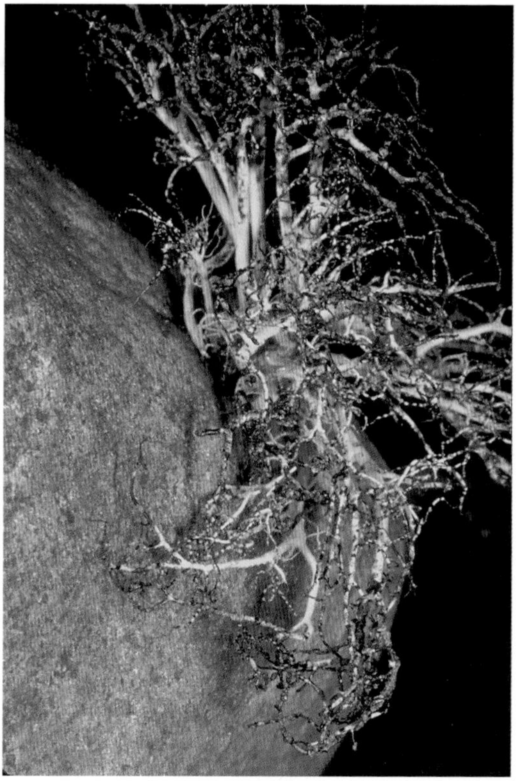

Description: Thallus fru
ticose, tufted, to 3(5) cm
long, pale yellowish green
at the base, black toward
the tips, often with black
spotting or banding;
soredia present; isidia
lacking; apothecia not
seen; medulla K-, P- or
P+ deep Y (hypostrep-
silic ± psoromic acid or
usnic acid only).

Range: Antarctic n to
disjuncts on high vol-
canoes in Wash and Ore
Cas.

Substrate: Rock
(noncalcareous).

Habitat: Alpine to sub-
alpine windswept areas,
especially steep rock surfaces that are free of snow
for considerable periods in the winter.

Notes: This species is distinct from all other local *Usnea* species in terms of habitat, substrate, and coloration. It and its close relatives have an unusual world distribution, being known from both the Antarctic and N Am, with a sprinkling of alpine sites n to Wash Cas. For *U. lambii*, the closest known sites to the Ore and Wash populations are in Mexico. Arctic material belongs to the closely related *U. sphacelata*. Finding the rare *U. lambii* is one of the rewards of climbing the shoulders of the huge volcanoes in the Cascade Range. You are most likely to find it on the sides of steep windswept outcrops and promontories on spur ridges well below the highest summits.

Usnea lapponica

Description: Thallus fruticose, tufted, mostly < 8 cm long, pale greenish or yellowish tinged; main branches with few to numerous fibrils; main branches roundish in cross section, densely papillose; medulla thick and loose; soralia  becoming strongly concave, often exposing the central axis, the edges of the ruptured cortex flexed outward, eventually the soralia wrapping around the branches; isidia lacking; apothecia not seen; medulla spot tests various, often negative or K+Y to O (most commonly usnic acid only or usnic and salacinic acids; occasionally with accessory barbatic, protocetraric, or psoromic acids or terpenoids; never with the stictic acid group).

Range: Widespread, throughout the PNW but rare w Cas, abundant e Cas; also present in outlying mountains e Cont Div.

Substrate: Bark and wood of hardwoods and conifers, also on shrubs in steppe.

Habitat: A wide variety of habitats, lowlands to subalpine, forests, savanna, and shrub steppe.

Notes: This is easily the most common *Usnea* e Cas. It is readily recognized by the soralia that become broadly flared at the edges, sometimes exposing the central cord, in combination with the tufted habit and papillose main branches. Another species with flaring soralia and papillae is *U. fulvoreagens*, much more common than *U. lapponica* w Cas. Numerous species are similar in general appearance to *U. lapponica* but bear isidia as well as soredia, most commonly *U. diplotypus, U. pacificana,* and *U. subfloridana. Usnea lapponica* differs chemically from *U. fulvoreagens* in lacking the stictic/norstictic acid group.

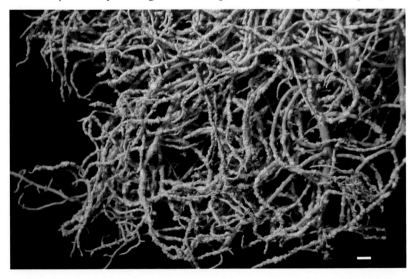

Usnea longissima

Description: Thallus fruticose, pendulous, to 2 m or more long, pale greenish or yellowish tinged; main branches with numerous fibrils; cortex of main branches soon scaling off, leaving a rough, dull surface; papillae lacking; central cord very thick (usually > 0.6× the diam of the branch, photo p. 437); isidia lacking; soredia rarely present, emerging from raised pale spots; apothecia rare; medulla I+B; other spot tests negative or rarely K+O, P+O. The most common chemotype in the PNW has barbatic and norbarbatic acids; next most common contains diffractaic acid, then salacinic acid. The evernic acid chemotype reported by Halonen et al. (1998) from BC has not been found in Oregon.

Range: Alas to Cal, w Cas.

Substrate: Bark and wood.

Habitat: In Ore and Wash usually on riparian trees, both conifers and hardwoods, at low elevations, but seldom in broad open valleys; most abundant in a narrow band transitional between the mountains and foothills.

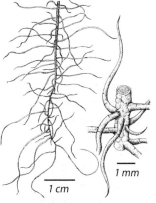

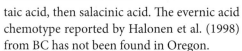

Notes: Unlike the other pendulous *Usnea* species, *U. longissima* has an I+ violet or dark bluish central cord (expose with razor blade). *Usnea longissima* is threatened or extirpated throughout most of its world range and is in danger of being overharvested by the local "moss harvest" trade. Its highly local distributions suggest dispersal limitations that will impede its recovery from disturbances to its habitat. On the other hand, the species shows fast growth rates in well-lit sites that help to counter its poor dispersal abilities.

Usnea rubicunda

Description: Thallus fruticose, tufted to drooping, to 15 cm long but usually much smaller, reddish brown on at least the lower 2/3 of the thallus when fresh, the reddish hue produced by coalescing reddish spots in the cortex; main branches with few to numerous fibrils; central cord rather thick (photo p. 363); isidia usually numerous and conspicuous; soredia usually present, punctiform, often on top of warts (tubercles), sometimes coalescing; apothecia not seen; medulla K+Y to R, P+Y to R (salacinic, norstictic, and miscellaneous other compounds in the stictic acid group).

Range: Widespread in tropics and subtropics, n into oceanic temperate areas, n on the Pacific coast to s BC.

Substrate: Bark and wood.

Habitat: Coastal forests and shrublands, including headlands, estuaries, and deflation plains.

Notes: The general reddish color of this species is distinctive, making it easy to spot at a distance. The chemistry is also unusual in containing both salacinic and norstictic acids. Occurrences of *U. rubicunda* are quite spotty in the PNW; if the species was not so distinctively colored it would probably be overlooked. Some individuals almost completely lack the reddish pigment and might be confused with the chemically similar *U. fulvoreagens*. That species, however, lacks isidia and tends to form large, splayed-open soralia.

Usnea scabrata

Description: Thallus fruticose, pendulous, to 70 cm long, pale greenish or yellowish tinged; main branches with few to numerous fibrils; cortex of main branches slightly to densely papillate near the base, with sparse to numerous raised wrinkles, with intervening flat or depressed spots, roundish in cross section but often uneven in thickness; branch tips often sinuose; cortex thin (2-8% of diam); isidia emerging from raised pale spots; apothecia rare; medulla K+Y to O (salacinic acid) or K- (usnic acid only).

Range: Widespread, boreal and montane, s to Cal and NM.

Substrate: Bark and wood.

Habitat: Mesic forests at low to mid elevations, often a dominant species in mountain forests between about 300 and 1000 m in w Ore.

Notes: The polymorphic *Usnea scabrata* is variable in production of papillae, the degree of ridging of the branches, and the density of fibrils. In the PNW it can be very similar to *U. dasopoga*, formerly *U. filipendula*, which is distinguished by dense papillae and fibrils and, most importantly, a thicker cortex and absence of ridging on the branches. *Usnea dasopoga* is most common in the valleys and foothills, while *U. scabrata* is most prominent in the mountains. The usnic-only chemotype is more common inland (this has been called *U. scabrata* subsp. *nylanderiana* Motyka) while w Cas the salacinic chemotype is most common.

1 cm

Usnea silesiaca

Description: Thallus initially shrubby, forming tufts about 6-12 cm broad; becoming subpendent to pendent, to 25 cm long, base blackened; branching mainly radiating from near the center, with 5-10 main arms, but with few main branches beyond that, other than the abundant side branches (fibrils); papillae usually abundant and distinct, especially near the base, but sometimes indistinct; cortex with abundant annular cracks, especially at the base; cortex thick (10-20% of radius); medulla compact, thin (7-15% of radius); axis thick, about ½ the diameter of the branch; soralia typically larger than ½ of the branch diameter, isidiate at least when young; soralia often numerous, distinctly raised (tuberculate) to slightly excavate, minute to enlarged, sometimes flaring at the edges as in *U. lapponica*, usually abundant but occasionally absent or sparse; medulla K+Y or R, P+Y or O, with salacinic acid ± protocetraric acid, rarely with accessory barbatic acid.

Range: Common in oceanic areas from Cal to BC, usually coastal, sometimes in the Coast Ranges, sporadic in the Willamette-Puget trough.

Substrate: Bark and wood of conifers, hardwoods, and shrubs.

Habitat: Moist forests, woodlands, and semi-open to open areas such as roadsides, residential areas, and riparian edges.

Notes: This species can be recognized in the field from the general appearance, with radiating main arms like a brittle seastar, each main arm with dense, short perpendicular branches (fibrils). Its identity can be confirmed by slicing a main branch lengthwise; look for the combination of a thick cortex, thick axis, and a very thin medulla, which is unique in our *Usnea* species. Some authors have used the name *U. madeirensis* for this species, but this is no longer accepted.

Usnea subgracilis

Description: Thallus fruticose, pendulous, to 50 cm long or more, resembling *Alectoria* at a distance, pale greenish or yellowish tinged; main branches usually with few fibrils, but occasionally fibrils numerous; base not blackened; papillae absent; annular cracks common on main branches; medulla as thin or thinner than cortex; cortex relatively thick (8-14% of diam); medulla thin and compact (6-16% of diam); axis thick (43-64% of diam); isidia absent to sparse, soon abraded; apothecia not seen; medulla K-, P+R (protocetraric ± lobaric acid).

Range: BC to Mex, usually near the ocean.

Substrate: Bark and wood.

Habitat: Usually on conifers near the coast.

Notes: Usnea chaetophora and *U. subgracilis* are similar in their annular cracking and relatively sparse or absent fibrils, the latter character giving an *Alectoria*-like appearance in the field. *Usnea subgracilis* has papillae weak or lacking while *U. chaetophora* has more distinct papillae. *Usnea subgracilis* is also distinguished by the presence of protocetraric acid and absence of salacinic acid.

Xanthomendoza

Thallus foliose, small (generally < 1 cm), appressed or ascending; lobes < 2 mm broad; upper surface yellow or orange, occasionally greenish or tan in shaded sites; lower surface white, yellow, or orange; rhizines present; soredia or blastidia (partly corticate granular soredia) present or not; apothecia lecanorine, the disk orange or yellow; spores 1-septate, polarilocular; pycnidia generally immersed in thalloid warts, producing rod-shaped spermatia; upper cortex K+ purple or R and lw UV+O or R (anthraquinones); on substrates. See key to species and comparison with *Xanthoria* under *Xanthoria*.

Sources include: Lindblom (1997, 2004, 2006), Poelt (1969), Søchting et al. (2002).

ID tips: *Xanthomendoza* species can be challenging to separate, so we recommend using the distributional differences among species (**Table 9**) to help narrow down your choices. For example, *X. hasseana* and *X. montana* are nearly identical in external appearance, but if you are west of the Cascade crest, *X. hasseana* is much more likely. Although experts use a compound microscope to separate these two species based on their spores, you might not have access to that.

Table 9. Distributional summary of primarily corticolous *Xanthoria* and *Xanthomendoza* species in western North America.

	w Cas	Cal	n RM	c RM	GB	Bor
Sorediate						
Xanthoria candelaria	C	C	C			R
Xanthomendoza fallax	C	C	O	C	C	C
Xanthomendoza fulva	C	O	C	O	O	
Xanthomendoza galericulata	O	O	C	C	C	
Xanthomendoza mendozae	O	O	O	O	O	
Xanthomendoza oregana	C	C	O			
Xanthomendoza ulophyllodes	R	R		R		R
Esorediate						
Xanthomendoza hasseana	C	C	O			C
Xanthomendoza montana	O		O	C	C	
Xanthoria parietina	U	U				
Xanthoria polycarpa	C	C	C			U
Xanthoria tenax		C				

C = common, O = occasional, U = uncommon, R = rare;
w Cas = west of Cascade crest; Cal = mediterranean climate in California; n RM = northern Rocky Mountains; c RM = central and southern Rocky Mountains; GB = Great Basin; Bor = boreal Canada and U.S.

Xanthomendoza fallax

Description: Thallus foliose, yellow orange, small, mostly < 2 cm but coalescing to cover larger areas; marginal lobes ± appressed, the lobe margins wavy, often both ± downcurved and reflexed, giving a ruffled appearance, thickish, to 7 × 1.5 mm; soredia present, mainly on lobe margins and undersides of reflexed lobe tips; apothecia uncommon, margins of older apothecia occasionally breaking open into soralia; spores (11)12.5-15.5(17) × 5.5-7(9) μm, septum (2)2.3-3.6(5) μm wide; upper cortex K+ purple or R.

Range: Widespread, common in drier, more continental climates e Cas.

Substrate: Bark (especially *Populus* and other broad-leaved spp), occasionally on rock.

Habitat: Mostly open to somewhat sheltered, often nutrient-enriched sites.

Notes: The name *Xanthomendoza* (*Xanthoria*) *fallax* has been broadly applied in w N Am to what we now recognize as *Xanthomendoza galericulata, X. mendozae, X. oregana, X. ulophyllodes,* and *X. fulva. Xanthomendoza fallax* is, however, usually a distinctive species. Small individuals in compact growth forms can resemble *X. fulva*, but *X. fallax* has wider lobes than *X. fulva*. In addition, *X. fallax* produces soredia from crescent-shaped lobe margins, while *X. fulva* produces soredia from the lower surface.

Xanthomendoza fulva

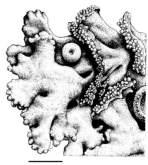

Description: Thallus to 1 cm diam or coalescing over much larger areas, yellow orange to deep red orange, forming small rosettes or colonies of small ascending lobes; lobes 0.2-0.6(1.0) mm wide, flat or more often ascending; lobe tips often fan shaped and upturned to expose the soredia; soredia present, mainly on the lower surface of the lobe tips; apothecia fairly common, the disk and margin similar in color to the thallus; spores 12-18 × 5.5-7.5 μm; pycnidia sparse but almost always present, orange to red; rhizines usually sparse, sometimes abundant; upper cortex K+ purple or R.

1 mm

Range: Common e Cas in the coastal states and provinces as well as the RM (and eastward), but rare w Cas and absent on the immediate coast.

Substrate: On bark, wood, and rock (both calcareous and noncalcareous).

Habitat: In both semi-open and shaded sites, commonly in quite dry low-elevation sites, including steppe and riparian habitats.

Notes: Xanthomendoza fulva can resemble *X. galericulata* when fan-shaped lobe tips are present. In *X. fulva* the tips vary from slightly hooded to flat or splayed back, while distinctly hooded lobe tips predominate in *X. galericulata*. Furthermore, pycnidia are common in *X. fulva*. Lindblom (2004, 2006) resolved confusion between her earlier concepts of *X. fulva* and *X. oregana* by describing *X. galericulata*. That species can be similar to *X. oregana* but with helmet-shaped lobe tips with soredia beneath, and to *X. fulva* but with broader lobes and tips more distinctly helmet shaped.

Xanthomendoza hasseana

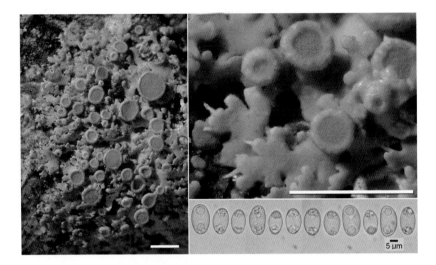

Description: Thallus small to medium, to 3 cm diam, forming rosettes but often coalescing over larger areas, yellow orange to deep orange; lobes mostly 0.1-1.0 mm broad; rhizines frequent, often projecting beyond the lobe margins; apothecia almost always present; spores 15.5-20 × 7.5-9.5 μm; pycnidia common, darker than the surrounding thallus; cortex K+ purple or R.

Range: Common in the coastal states and provinces, inland to Mont and across the boreal forest to ne US, but apparently absent from the Great Basin and c and s RM.

Substrate: On bark (esp. *Populus, Quercus*, and other hardwoods), occasionally on wood or rock.

Habitat: Semi-open to open, nutrient-rich habitats.

Notes: Xanthomendoza hasseana is very similar to *X. montana*, but *X. hasseana* has distinctly larger spores. The two species overlap little in range, so that w Cas and in n Cal one expects *X. hasseana*, while *X. montana* is abundant in more continental climates of Id, Mont, Wyo, Colo, e Wash, and e Ore.

Xanthomendoza montana

Description: Thallus small to medium, to 3 cm diam, forming rosettes that often coalesce over larger areas, yellow orange to deep red orange; lobes mostly 0.2-0.5 mm wide; rhizines frequent; apothecia almost always present; spores 13-15.5 × 5-7.5 μm; pycnidia fairly common, darker than the surrounding surface; cortex K+ purple or R.

Range: Widespread e Cas, only occasionally w Cas.

Substrate: On bark (both hardwoods and conifers), occasionally on wood.

Habitat: Usually in fairly open, dry habitats.

Notes: Examining the ascospores is the only reliable way to distinguish *X. montana* from *X. hasseana*. As a shortcut, the two differ markedly in range (see under *X. hasseana*). Like most *Xanthoria* species, the thallus color varies from deep reddish orange in full sun to yellowish in partial shade and yellowish green in shade.

5 μm

Xanthoparmelia

Thallus foliose, yellowish green or pale greenish, to 20 cm diam or more, closely or loosely appressed or unattached; lobes mostly < 5 mm wide; pseudocyphellae lacking; lower surface cream, brown, or black; rhizines present in most species, simple, sparse to abundant; soredia or isidia present in some spp; apothecia lecanorine, the disk brown; pycnidia laminal, immersed; spot tests various but cortex KC+Y (usnic acid); on rock, moss over rock, and soil; rarely on wood in dry habitats.

Sources include: Blanco et al. (2004), Esslinger (1977, 1978a), Hale (1990), Hodkinson and Lendemer (2011), Leavitt et al. (2011), McCune and Rosentreter (2007), Rosentreter (1993).

ID tips: Unfortunately, chromatography is necessary to identify some *Xanthoparmelia* specimens to species. In the PNW, however, our *Xanthoparmelia* diversity is much lower than in Cal and sw US. And the diversity is even lower in Ore and Wash w Cas than in the more continental climates e Cas. In more oceanic areas of the PNW one species predominates, *X. cumberlandia*, with much smaller amounts of *X. mougeotii. Xanthoparmelia mougeotii* is readily separated from all others by its small size, narrow lobes, and soredia.

A perennial sticking point in identifying *Xanthoparmelia* is the color of the lower surface. While some cases are clearly brown or tan, the full range from tan to black is apparent, sometimes varying considerably within a population or even an individual. The late expert of this group, Mason Hale, recognized this issue and would respond to questions about it by saying things like, "If you say it is brown then it is *X. cumberlandia*, but if you argue for black then it is *X. angustiphylla*" (two species with identical chemical content but supposed to differ in color of the lower surface). We would hope that molecular studies would resolve problems like this, but so far they have revealed a bewildering complexity that seems to defy a simple, clear taxonomy, unless it is backed off to a very high level. Leavitt et al. (2011) concluded that "the present study suggests that species diversity has been overestimated in the species-rich genus *Xanthoparmelia*."

Introductory Key

1a Thallus generally free on soil or plant detritus or attached to soil; isidia and soredia lacking..**Group 1** (Lead 3)
1b Thallus attached to rock, moss, or plant detritus; isidia or soredia present or not
  2a Soredia or isidia present ....**Group 2** (Lead 6)
  2b Soredia and isidia lacking ....**Group 3** (Lead 11)

Group 1

Pale greenish; on soil

3a Lobes narrow (generally < 2 mm wide); see photos of spp in this group in McCune and Rosentreter (2007)
  4a Upper surface faintly white spotted. Arctic to BC and Colo; uncommon on soil in steppe and high plains..................................*X. camtschadalis*
  4b Upper surface not white spotted. Steppe to alpine, on soil and gravel, occasional e Cas ....*X. wyomingica*
3b Lobes broad (generally > 1.5 mm wide)
  5a Medulla with salacinic acid (TLC required). Occasional on soil in steppe e Cas, more common e Cont Div..........................*X. chlorochroa*
  5b Medulla with norstictic or stictic acid. Semiarid calcareous rangelands in Mont, Id, Wyo, and se Ore ....*X. neochlorochroa*

Group 2

Thallus pale greenish; on rock; sorediate or isidiate

6a Soredia present, isidia lacking. Lobes narrow, to 0.5 mm wide; soralia roundish; lower surface black; Cal to Wash, mainly w Cas, rarely inland to Colo and Mont.. *X. mougeotii*
6b Soredia absent, isidia present
  7a Medulla K-
    8a Medulla P+Y (psoromic acid). Isidia short, broad, inflated appearing; S Dak, Wyo to s Cal and Ariz .... [*X. lavicola*]
    8b Medulla P+R (fumarprotocetraric acid and other substances). Widespread, e N Am and Cal.................................. [*X. subramigera*]
  7b Medulla K+Y to O, or R
    9a Lower surface black. Medulla containing norstictic and stictic acid group; common in e N Am, rare in PNW, more common from Colo to Cal and s ....*X. conspersa*
    9b Lower surface light brown
      10a Medulla with salacinic acid (TLC required). Common e Cas, widespread .... *X. mexicana*
      10b Medulla with stictic and norstictic acids ....***X. plittii***

Group 3

Thallus pale greenish; on rock; no soredia or isidia

11a Medulla K-, P- (fatty acids)
  12a Thallus lacking lichesterinic acid, containing several unknown fatty acids. Lower surface light brown; widespread, uncommon in PNW, more often e Cont Div..*X. subdecipiens*
  12b Thallus containing a fatty acid near lichesterinic acid. Lower surface light brown; widely scattered, rare, known from Mont (near Kootenai Falls), Utah, and Ariz...*X. montanensis*
  (Cal records of *X. montanensis* belong to *X. oleosa*, according to Nash et al. 2004)

11b Medulla K+Y or R, P+O or R

    13a Thallus loosely attached to rock, generally easily removed with fingers; very common at mid to high elevations........................ **X. coloradoensis**

    13b Thallus generally closely attached to rock, occasionally overgrowing adjacent mosses on rock and then easily removed with your fingers; mostly at lower elevations

        14a Lower surface tan to brown

            15a Medulla with salacinic acid (TLC required). Occasional e Cas; low to mid elevations..*X. lineola*

            15b Medulla with stictic and norstictic acids.............**X. cumberlandia**

        14b Lower surface black

            16a Medulla with stictic, constictic, and norstictic acids (TLC required). Widespread, e N Am, Cal; uncommon in PNW ....*X. angustiphylla*

            16b Medulla with salacinic acid as major substance (with norstictic and protocetraric acids as minor or trace). Widespread, uncommon, more frequent s to Cal...........................*X. hypofusca*

Xanthoparmelia mougeotii

Xanthoparmelia coloradoensis

Description: Thallus foliose, yellowish green or pale greenish, mostly < 12 cm diam, loosely appressed; lobes mostly < 4 mm wide; lower surface tan to brown; soredia and isidia lacking; apothecia lecanorine, common, the disk brown; pycnidia laminal, immersed; cortex KC+Y (usnic acid); medulla K+Y to O, P+O (salacinic acid), C-, KC-.

Range: Widespread.

Substrate: Rock, moss over rock, and soil.

Habitat: Very common at mid to high elevations, usually in exposed to somewhat sheltered sites.

Notes: The relatively loose attachment to rocks (often you can separate it from the rock fairly easily with your fingers) is unusual among the local *Xanthoparmelia*. Other species, however, such as *X. cumberlandia*, can grow loosely enough to lift with your fingers, but only when overgrowing bryophytes or other organic matter over rock. *Xanthoparmelia coloradoensis* intergrades in appearance with *X. wyomingica* and *X. lineola*, differing only in the degree of attachment to the substrate. *X. wyomingica* is loose on soil or gravel while *X. lineola* is tightly appressed to rock.

Xanthoparmelia cumberlandia

Description: Thallus foliose, yellowish green or pale greenish, mostly < 20 cm diam, appressed; lobes mostly < 5 mm wide; lower surface tan to brown; soredia and isidia lacking; apothecia lecanorine, common, the disk brown; pycnidia laminal, immersed; cortex KC+Y (usnic acid); medulla K+Y to O, P+O (stictic and norstictic acids), C-, KC-.

Range: Widespread, common throughout the PNW.

Substrate: Rock, sometimes overgrowing moss or litter over rock.

Habitat: Exposed to somewhat sheltered rocks, mostly at low to mid elevations, in both forested and steppe regions.

Notes: This is easily the most common *Xanthoparmelia* w Cas and it is abundant e Cas. You can find it making large circular individuals on abandoned pavement, dry roadside outcrops, and exposed cliffs high above the ocean. *X. lineola* is essentially identical but has salacinic instead of stictic and norstictic acids (sorry, TLC is required to make this distinction). The color of the lower surface seems to intergrade with *X. angustiphylla*, which has a very dark brown or black lower surface.

Xanthoria

Thallus foliose, small (generally < 10 cm diam), appressed, ascending, or minutely fruticose; lobes < 2(4) mm broad; upper surface some shade of yellow or orange, occasionally greenish or tan in shaded sites; lower surface white, yellow, or orange; hapters or very short rhizines present, elongated rhizines lacking; soredia or blastidia (partly corticate granular soredia) present or not; apothecia lecanorine, the disk orange or yellow; spores hyaline, 1-septate, polarilocular (with a narrow isthmus connecting the two cells); pycnidia generally immersed in thalloid warts, producing ellipsoidal spermatia; upper cortex K+ purple or R; on a wide range of substrates, especially in nutrient-enriched environments.

Sources include: Lindblom (1997, 2004, 2006), Poelt (1969), Søchting et al. (2002).

ID tips: The genus *Xanthomendoza* is separated from *Xanthoria* by pycnidial characters, rhizines, and molecular data. *Xanthomendoza* has rod-shaped spermatia and usually well-developed rhizines, while *Xanthoria* has ellipsoidal spermatia and only short rhizines or hapters, lacking long rhizines. Apart from the rhizines, which often require manipulation under a dissecting scope to detect, *Xanthoria* and *Xanthomendoza* cannot be separated by an easy macroscopic character, so one simply needs to learn which species belongs to which genus. However, our most common epiphytes belong to *Xanthomendoza*, except for *Xanthoria polycarpa*, *X. candelaria*, and *X. parietina*. Further splits have moved all or our species except *X. parietina* into other genera (*Polycauliona* and *Rusavskia*), but we have taken a more conservative view until the taxonomy of the Teloschistaceae stabilizes.

In a few species of *Xanthoria* and *Xanthomendoza*, observing spore size and the thickness of the spore septum facilitate identification. To see this, examine a thin section of an apothecium under a compound microscope (for example, see inset photo for *Xanthomendoza hasseana*, p. 457). On fresh material the septum may be obscured by large oil drops in the cells, but this is less often a problem with herbarium specimens. To better resolve the septum with fresh material, gently heat the slide over a lighter, stopping short of the boiling point. If bubbles start to form under the cover slip, immediately remove from heat. Right away you should see more clearly the septum and the narrow channel that penetrates the septum.

Species of *Xanthomendoza* are keyed below with *Xanthoria*.

Introductory Key

1a Soredia present
  2a Soredia laminal, isidia-like or breaking down into soredia; usually on rock. Thallus appressed, foliose, to 4 cm broad; lower surface whitish, with sparse, coarse rhizines; apothecia generally lacking; usually on exposed nutrient-enriched calcareous rock, occasionally on shaded vertical rock faces, also on antlers, rarely on bark; widespread; occasional throughout w and boreal N Am; most common in the PNW at high elevations.. *Xanthoria sorediata*
  2b Soredia usually terminal, marginal, or below the lobe tips, fine or granular; on bark or rock
    3a Thallus becoming dwarf fruticose, steeply ascending to erect; spermatia ellipsoidal, about 2.3-3 × 1-1.3 μm (almost always present, section thallus-colored or darker warts on upper surface ....***Xanthoria candelaria***
    3b Thallus always dorsiventral, not erect fruticose, appressed at least near the base; spermatia ± rod-shaped or both rod-shaped and elliptical within the same pycnidium, the rod-shaped spermatia about 2.5 × 0.5 μm (thin section or squash mount of small reddish warts) ....**Group 1** (Lead 4)
1b Soredia lacking ....**Group 2** (Lead 12)

Group 1

Soredia present but not laminal; not fruticose

4a Substrate rock (rarely soil); lobes turned upward and ascending from the substrate, but with the lobe tips rolled under or downturned; soredia produced from the lower surface; thallus ± pruinose, often forming stands of erect or suberect lobes rather than discrete rosette-forming thalli
  5a Soredia similar in color to the upper surface, ± corticate (blastidia), irregular in shape; pycnidia immersed, forming reddish spots; lower surface white and ± shiny. On both calcareous and noncalcareous rock, on the ground among bryophytes, and on antlers and bones, often where nutrient-enriched; so far known in N Am only from the Arctic .... [*Xanthomendoza borealis*]
  5b Soredia usually paler than the upper surface (yellowish to greenish yellow), ecorticate, spherical; pycnidia immersed to protruding but similar in color to the upper surface; lower surface grayish, matte. Apothecia unknown; spermatia bacilliform; on noncalcareous or calcareous rock, usually steep faces where shaded, occasionally on bryophytes over rock; widespread in w N Am, including both continental and suboceanic climates, often at high elevations, but so far not known from the Coast Range or the immediate coast ....*Xanthomendoza mendozae*

4b Substrate rock, bark, or wood; lobes semi-
    erect to horizontal, the lobe tips various;
    soredia marginal, submarginal, or from the
    lower surface; thallus not pruinose, often
    forming discrete rosettes

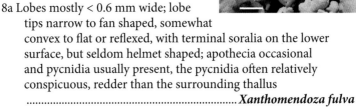

X. mendozae

6a Lobes narrow, usually < 0.6 mm wide
    7a Lobes ± parallel to the substrate; soralia
      marginal ...... ***Xanthomendoza oregana***
      (Narrow-lobed but otherwise typical
      forms of *Xanthomendoza oregana* will
      key here.)
    7b Lobes soon ascending to erect; soralia
      apical and on the lower side of the
      lobes
      8a Lobes mostly < 0.6 mm wide; lobe
        tips narrow to fan shaped, somewhat
        convex to flat or reflexed, with terminal soralia on the lower
        surface, but seldom helmet shaped; apothecia occasional
        and pycnidia usually present, the pycnidia often relatively
        conspicuous, redder than the surrounding thallus
        ....***Xanthomendoza fulva***
      8b Lobes typically > 0.6 mm wide, but sometimes less; lobe tips
        becoming helmet-shaped with soredia on the lower surface;
        apothecia and pycnidia unknown
        .... *Xanthomendoza galericulata* (Lead 11b)
6b Lobes wider, typically 0.8-1.4 mm wide
    9a Soredia produced in marginal crescent-shaped slits that are bordered
      with the remaining upper and lower cortex, the soralia often
      broadening to half moons or almost circular ("bird nests"); soredia
      often paler (yellowish or greenish yellow) than the upper surface;
      spermatia bacilliform.....................................***Xanthomendoza fallax***
    9b Soredia marginal or produced below the lobe tips, soredia similar in
      color to the upper surface; spermatia bacilliform or variable
      10a Rhizines frequently visible from above; spermatia bacilliform.
        Thallus often forming largish rosettes (to over 3 cm); soralia
        marginal to submarginal, often laminal on well-developed thalli;
        usually on bark, occasionally on wood or rock; abundant in GL
        region but rare and widely scattered in w N Am
        .... *Xanthomendoza ulophyllodes*
      10b Rhizines rarely visible from above; spermatia ellipsoid to
        bacilliform or irregular, often variable within a single pycnidium
        11a Soralia marginal and submarginal
        ....***Xanthomendoza oregana***
        11b Soralia initially on the underside and margin of the lobe tips,
          eventually lining distinctly hood-shaped lobe tips, similar to

Physcia adscendens. Individual thalli small, but coalescing to cover large areas with ascending or suberect lobes, yellow orange to deep orange; apothecia and pycnidia unknown; mainly in arid to semiarid areas with a continental to suboceanic climate .... *Xanthomendoza galericulata*

Group 2

Soredia lacking

12a Lower cortex absent
  13a On bark and wood; apparently restricted to California. Thallus and lobes flat, lobate-crustose; lobes appressed; lower cortex absent except under the lobe tips. Usually on twigs, also on trunks and wood, generally on angiosperms in open to semi-open habitats, oak savanna and coastal scrub; Sacramento Valley, foothills and Coast Ranges of Cal, s to Baja; not yet known from PNW but to be sought in the dry valleys of sw Ore.. [*Xanthoria tenax*]
  13b On rock; widespread .... *Caloplaca*
12b Lower cortex present
  14a Thallus medium sized, often > 3 cm diam; lobes usually > 1 mm broad; on bark, wood, or rock
    15a Lobes concave to plane, mostly 1-3 mm broad; on rock, bark, or wood .... ***Xanthoria parietina***
    15b Lobes convex, mostly 0.5-1.5 mm broad; on rock
      ....***Xanthoria elegans***
  14b Thallus small, usually < 2 cm diam; lobes < 1 mm broad; usually on bark or wood
    16a Rhizines absent, the thallus a dense knobby cluster or tiny rosette, attached with short, obscure hapters............... ***Xanthoria polycarpa***
    16b Rhizines present, easily found
      17a Spores ellipsoidal (LM), 15-20 × 7.5-9.5 μm; septum wide (5.5-7.5 μm)...***Xanthomendoza hasseana***
      17b Spores cylindrical to ellipsoidal (LM), 13-15.5 × 5-7.5 μm; septum narrow (1.5-3 μm)...................***Xanthomendoza montana***

Xanthoria candelaria

Description: Thallus lobate to minutely fruticose, yellow or orange, < 1 cm but coalescing to cover larger areas; lobes < 1 mm broad, initially ± dorsiventral but becoming radially structured with proliferations on all sides; blastidia present; apothecia uncommon, thalline margin of apothecia often with lobules and blastidia; spores mostly 12-15 × 5.5-8 μm; spermatia ellipsoidal, 2.3-3 × 1-1.3 μm (almost always present, section thallus-colored or darker warts on upper surface); upper cortex K+ purple or R.

Range: Widespread, throughout the PNW.

Substrate: Rock, also occasionally on bark.

Habitat: Sheltered rock faces in a wide range of macroclimates and elevations; on bark in both open and sheltered sites; also on seashore rock.

Notes: Older reports in the literature that use the name *X. candelaria* are often based on other species. With a hand lens the distinguishing characteristic for *X. candelaria* is the minutely fruticose proliferations of the thallus. Small narrow-lobed forms of other species are commonly misidentified as *X. candelaria,* but the others always have lobes with a distinct upper and lower surface.

Xanthoria elegans

Description: Thallus foliose, small, mostly < 5 cm wide; lobes mostly 0.5-1.5(2) mm broad, appressed to loosely appressed; upper surface some shade of orange; lower surface white or pale yellowish, corticate (photo inset), with short, sparse hapters; soredia and isidia absent; apothecia common; spores (11)12-16(17.5) × (5.5)6-8(8.5) μm; septum 1.5-4.5 μm wide; spermatia ellipsoidal; upper cortex K+ purple or R.

Range: Widespread; throughout the PNW.

Substrate: Rock, both calcareous and noncalcareous, occasionally overgrowing moss or litter over rock.

Habitat: Exposed to somewhat sheltered sites, often where nutrient enriched by bird or small mammal droppings.

Notes: Xanthoria elegans forms bright orange rosettes on rock and moss over rock. Small forms are very similar to *Caloplaca saxicola*, which is very tightly appressed and lacks a lower cortex (and therefore is technically a crustose lichen). The lower surface of *X. elegans* has a smooth, pale lower cortex, except at the scattered points of attachment (hapters).

Xanthoria parietina

Description: Thallus foliose, mostly < 8(10) cm wide; lobes 1-4(7) mm broad, appressed; upper surface some shade of yellow, orange, or greenish yellow; lower surface white, corticate, with sparse pale rhizines or hapters; soredia and isidia absent; apothecia usually present; spores (12)13-16(17.5) × (5 6-9(10) μm; septum 4-8 μm wide; spermatia ellipsoidal; cortex K+ purple or R.

Range: BC to Baja; w Cas; so far known in the PNW only from the Willamette-Puget trough and a few coastal sites near Tillamook. The range and abundance appear to be rapidly expanding w Cas. The species is much more abundant in the Willamette Valley than it was 30 years ago.

Substrate: Bark and wood, rarely on rock (e.g., gravestones), roofing, and fiberglass.

Habitat: Hardwood forests in low-elevation broad valleys; scattered on *Populus* and other hardwoods in riparian areas in agricultural and populated areas.

Notes: This is the broadest-lobed epiphytic *Xanthoria* in the PNW. It is usually abundantly fertile and is never sorediate. The species has been postulated to have been inadvertently introduced by humans into Australia. The same is apparently true here, judging from its local center of distribution in the heavily populated Willamette Valley and its rapid recent increases in abundance.

Xanthoria polycarpa

Description: Thallus foliose, tiny, mostly < 1(2) cm wide; lobes < 1 mm broad, appressed or more often forming a knobby cluster or cushion on small twigs; upper surface some shade of yellow, orange, or greenish yellow; soredia and isidia absent; apothecia usually abundant, crowded; spores (10)11-15(16.5) × (4)5.5-8(9) µm; septum (2)3-6(7) µm; spermatia ellipsoidal; cortex K+ purple or R.

Range: Common in the coastal states and provinces inland to w Mont and YNP, but not known from the Great Basin, c or s RM.

Substrate: Bark and wood, rarely rock; mostly on hardwoods but occasionally on conifers, often where nutrient enriched.

Habitat: In a wide range of habitats at low to middle elevations, but most common on hardwood twigs; also on sagebrush in steppe and in other dry, open habitats.

Notes: W Cas *X. polycarpa* often cohabits twigs with the larger *Xanthomendoza hasseana*. When they grow together, *Xanthomendoza hasseana* has distinctly broader lobes and the rhizines peek out from beneath the lobes. Moving e Cas, *Xanthomendoza montana* largely replaces *X. polycarpa* and *Xanthomendoza hasseana*. *Xanthoria polycarpa* is one of the few lichenized fungi with a known mating system. Honegger et al. (2004) demonstrated that it is an outcrosser (heterothallic), unlike its common associate *Xanthoria parietina*, which is self fertile (homothallic).

SCIENTIFIC NAME	SYNONYM	COMMON NAME
Candelariella Müll. Arg. *(species not treated here)*		**yolk lichen**
Cavernularia Degel. included in *Hypogymnia*		
Cetraria Ach.		**ruffle lichen or Iceland lichen**
Cetraria aculeata (Schreber) Fr.	*Coelocaulon aculeatum* (Schreber) Link	
	Cornicularia aculeata (Schreber) Ach.	
Cetraria californica Tuck.	*Cornicularia californica* (Tuck.) Du Rietz	
	Kaernefeltia californica (Tuck.) A. Thell & Goward	
	Nephromopsis tuckermanii Divakar, Crespo & Lumbsch	
Cetraria canadensis (Räsänen) Räsänen	*Tuckermannopsis canadensis* (Räsänen) Hale	
	Vulpicida canadensis (Räsänen) J.-E. Mattsson & M.J. Lai	
Cetraria chlorophylla (Willd.) Vainio	*Tuckermannopsis chlorophylla* (Willd.) Hale	
	Nephromopsis chlorophylla (Willd.) Divakar, Crespo & Lumbsch	
Cetraria commixta (Nyl.) Th. Fr.	*Melanelia commixta* (Nyl.) A. Thell	
	Cetrariella commixta (Nyl.) A. Thell & Kärnefelt	
	Cetraria fahlunensis (L.) Schreber	
Cetraria coralligera (W. A. Weber) Hale	*Tuckermannopsis coralligera* (W. A. Weber) W. A. Weber	
	Nephromopsis coralligera (W. A. Weber) Divakar, Crespo & Lumbsch	
	Tuckermanella coralligera (W. A. Weber) Essl.	
Cetraria cucullata (Bellardi) Ach.	*Flavocetraria cucullata* (Bellardi) Kärnefelt & A. Thell	
	Nephromopsis cucullata (Bellardi) Divakar, Crespo & Lumbsch	
Cetraria ericetorum Opiz		
Cetraria fendleri (Nyl.) Tuck.	*Nephromopsis fendleri* (Nyl.) Divakar, Crespo & Lumbsch	
	Tuckermannopsis fendleri (Nyl.) Hale	
	Tuckermanella fendleri (Nyl.) Essl.	
Cetraria halei W. L. Culb. & C. F. Culb.	*Cetraria ciliaris var. halei* (W. L. Culb. & C. F. Culb.) Ahti	
	Tuckermannopsis americana (Sprengel) Hale	
	Nephromopsis americana (Sprengel) Divakar, Crespo & Lumbsch	
	Tuckermannopsis halei (W. L. Culb. & C. F. Culb.) M. J. Lai	
Cetraria islandica (L.) Ach.		
Cetraria juniperina (L.) Ach.	*Cetraria tilesii* Ach.	
	Vulpicida tilesii (Ach.) J.-E. Mattsson & M. J. Lai	
Cetraria laevigata Rass.		
Cetraria madreporiformis (Ach.) Müll. Arg.	*Allocetraria madreporiformis* (Ach.) Kärnefelt & A. Thell	
	Dactylina madreporiformis (Ach.) Tuck.	
	Dufourea madreporiformis (Ach.) Ach.	
Cetraria merrillii Du Rietz	*Nephromopsis merrillii* (Du Rietz) Divakar, Crespo & Lumbsch	
	Tuckermannopsis merrillii (Du Rietz) Hale	
Cetraria muricata (Ach.) Eckfeldt	*Coelocaulon muricatum* (Ach.) J. R. Laundon	
	Cornicularia muricata (Ach.) Ach.	
Cetraria nivalis (L.) Ach.	*Flavocetraria nivalis* (L.) Kärnefelt & A. Thell	
	Nephromopsis nivalis (L.) Divakar, Crespo & Lumbsch	
Cetraria orbata (Nyl.) Fink	*Nephromopsis orbata* (Nyl.) Divakar, Crespo & Lumbsch	
	Tuckermannopsis orbata (Nyl.) M. J. Lai	
Cetraria pallidula Tuck. ex Riddle	*Ahtiana pallidula* (Tuck. ex Riddle) Goward & A. Thell	
	Tuckermannopsis pallidula (Tuck. ex Riddle) Hale	
Cetraria pinastri (Scop.) Gray	*Tuckermannopsis pinastri* (Scop.) Hale	
	Vulpicida pinastri (Scop.) J.-E. Mattson & M. J. Lai	
Cetraria platyphylla Tuck.	*Tuckermannopsis platyphylla* (Tuck.) Hale	
Cetraria sepincola (Ehrh.) Ach.	*Tuckermannopsis sepincola* (Ehrh.) Hale	

SCIENTIFIC NAME	SYNONYM	COMMON NAME
Cetraria sphaerosporella (Müll. Arg.) McCune	*Ahtiana sphaerosporella* (Müll. Arg.) Goward	
	Nephromopsis sphaerosporella (Müll. Arg.) Divakar, Crespo & Lumbsch	
	Parmelia sphaerosporella Müll. Arg.	
Cetraria subalpina Imshaug	*Nephromopsis subalpina* (Imshaug) Divakar, Crespo & Lumbsch	
	Tuckermannopsis subalpina (Imshaug) Kärnefelt	
Cetrelia W. L. Culb. & C. F. Culb.		**curly shield**
Cetrelia cetrarioides (Duby) W. L. Culb. & C. F. Culb.		
Cladina Nyl. = Cladonia subg. *Cladina*		**reindeer lichen**
Cladonia P. Browne		**cladonia**
Cladonia albonigra Brodo & Ahti		
Cladonia andereggii S. Hammer		
Cladonia arbuscula (Wallr.) Flotow	*Cladina arbuscula* (Wallr.) Hale & W. L. Culb.	
Cladonia artuata S. Hammer		
Cladonia asahinae J. W. Thomson		
Cladonia bacillaris (Ach.) Genth	*Cladonia macilenta* var. *bacillaris* (Ach.) Schaerer	
Cladonia bellidiflora (Ach.) Schaerer		
Cladonia borealis S. Stenroos		
Cladonia botrytes (K. G. Hagen) Willd.		
Cladonia cariosa (Ach.) Sprengel		
Cladonia carneola (Fr.) Fr.		
Cladonia cenotea (Ach.) Schaerer		
Cladonia chlorophaea (Flörke ex Sommerf.) Sprengel		
Cladonia ciliata Stirton	*Cladina ciliata* (Stirton) Trass	
Cladonia ciliata var. *tenuis* (Flörke) Ahti		
Cladonia coccifera (L.) Willd.		
Cladonia concinna Ahti & Goward		
Cladonia coniocraea (Flörke) Sprengel	*Cladonia ochrochlora* Flörke	
Cladonia conista (Nyl.) Robbins		
Cladonia cornuta (L.) Hoffm.		
Cladonia crispata (Ach.) Flotow		
Cladonia crispata var. *cetrariiformis* (Delise) Vainio		
Cladonia cryptochlorophaea Asahina		
Cladonia dahliana Kristinsson		
Cladonia deformis (L.) Hoffm.		
Cladonia digitata (L.) Hoffm.		
Cladonia dimorpha S. Hammer		
Cladonia ecmocyna Leighton		
Cladonia fimbriata (L.) Fr.	*Cladonia major* (K. G. Hagen) Sandst.	
Cladonia furcata (Hudson) Schrader	*Cladonia herrei* Fink ex J. Hedrick	
Cladonia gracilis (L.) Willd.		
Cladonia grayi G. Merr. ex Sandst.		
Cladonia homosekikaica Nuno		
Cladonia humilis (With.) J. R. Laundon	*Cladonia conistea* auct.	
Cladonia japonica Vainio		
Cladonia luteoalba Wheldon & A. Wilson		
Cladonia macilenta Hoffm.		
Cladonia macrophyllodes Nyl.		
Cladonia mitis Sandst.	*Cladina mitis* (Sandst.) Hustich	

SCIENTIFIC NAME	SYNONYM	COMMON NAME
Cladonia multiformis G. Merr.		
Cladonia norvegica Tønsberg & Holien		
Cladonia novochlorophaea (Sipman) Ahti & Brodo		
Cladonia phyllophora Hoffm.		
Cladonia pleurota (Flörke) Schaerer		
Cladonia pocillum (Ach.) O. J. Rich.		
Cladonia poroscypha S. Hammer		
Cladonia portentosa (Dufour) Coem.	*Cladina portentosa* (Dufour) Follmann	
Cladonia portentosa subsp. *pacifica* (Ahti) Ahti	*Cladina portentosa* subsp. *pacifica* (Ahti) Ahti	
Cladonia prolifica Ahti & S. Hammer		
Cladonia pseudostellata Asahina		
Cladonia pyxidata (L.) Hoffm.		
Cladonia rangiferina (L.) F. H. Wigg.	*Cladina rangiferina* (L.) Nyl.	
Cladonia rei Schaerer	*Cladonia nemoxyna* (Ach.) Arnold	
Cladonia scabriuscula (Delise) Nyl.		
Cladonia schofieldii Ahti & Brodo		
Cladonia singularis S. Hammer		
Cladonia squamosa Hoffm.		
Cladonia squamosa var. *subsquamosa* (Nyl. ex Leighton) Vainio	*Cladonia subsquamosa* (Nyl. ex Leighton) Crombie nom. illeg.	
Cladonia stellaris (Opiz) Pouzar & Vezda	*Cladina stellaris* (Opiz) Brodo	
Cladonia straminea (Sommerf.) Flörke	*Cladonia metacorallifera* Asahina	
Cladonia stricta (Nyl.) Nyl.		
Cladonia stygia (Fr.) Ruoss	*Cladina stygia* (Fr.) Ahti	
Cladonia subulata (L.) Weber ex F. H. Wigg.		
Cladonia sulphurina (Michaux) Fr.	*Cladonia gonecha* (Ach.) Asahina	
Cladonia symphycarpa (Flörke) Fr.		
Cladonia transcendens (Vainio) Vainio		
Cladonia umbricola Tønsberg & Ahti		
Cladonia uncialis (L.) F. H. Wigg.		
Cladonia verruculosa (Vainio) Ahti		
Cladonia verticillata (Hoffm.) Schaerer	*Cladonia cervicornis* subsp. *verticillata* (Hoffm.) Ahti	

Coelocaulon Link included in *Cetraria*

Collema F. H. Wigg.   **jelly lichen**
(see also Blennothallia, Enchylium, Lathagrium, Rostania)
Collema coniophilum Goward
Collema curtisporum Degel.
Collema furfuraceum (Arnold) Du Rietz

Collema furfuraceum var. *luzonense* (Räsänen) Degel.	*Collema subfurfuraceum* Degel.	

Collema glebulentum (Nyl. ex Crombie) Degel.
Collema nigrescens (Hudson) DC.
Collema subflaccidum Degel.

Cornicularia (Schreber) Hoffm.   **wire lichen**
Cornicularia normoerica (Gunnerus) Du Rietz

Cystocoleus Thwaites   **black velvet lichen**
Cystocoleus ebeneus (Dillwyn) Thwaites

SCIENTIFIC NAME	SYNONYM	COMMON NAME
Dactylina Nyl.		**finger lichen**
(see also Cetraria)		
Dactylina arctica (Hook. f.) Nyl.		
Dactylina ramulosa (Hook. f.) Tuck.		
Dendriscosticta B. Moncada & Lücking		
Dendriscosticta oroborealis (Goward & Tønsberg) B. Moncada & Lücking	*Sticta oroborealis* Goward & Tønsberg	
Dermatocarpon Eschw.		**leather lichen or stippleback**
Dermatocarpon atrogranulosum Breuss		
Dermatocarpon deminuens Vainio	*Dermatocarpon bachmannii* Anders	
	Dermatocarpon meiophyllizum Vainio (misapplied in w N Am)	
Dermatocarpon intestiniforme (Körber) Hasse		
Dermatocarpon leptophyllodes (Nyl.) Zahlbr.	*Dermatocarpon lorenzianum* Anders	
Dermatocarpon "micropapillatum" Mc-Cune ined. (unpubl.)		
Dermatocarpon miniatum (L.) W. Mann		
Dermatocarpon moulinsii (Mont.) Zahlbr.		
Dermatocarpon "oleosum" McCune ined. (unpubl.)		
Dermatocarpon polyphyllizum (Nyl.) Blomb. & Forssell		
Dermatocarpon reticulatum H. Magn.		
Dermatocarpon rivulorum (Arnold) Dalla Torre & Sarnth.		
Dermatocarpon taminium Heiðmarsson		
Dermatocarpon vellereum Zschacke		
Enchylium (Ach.) Gray		
Enchylium bachmanianum (Fink) Otalora et al.	*Collema bachmanianum* (Fink) Degel.	
Enchylium coccophorum (Tuck.) Otalora et al.	*Collema coccophorum* Tuck.	
Enchylium polycarpon (Hoffm.) Otalora et al.	*Collema polycarpon* Hoffm.	
Enchylium tenax (Sw.) Gray	*Collema tenax* (Sw.) Ach.	
Ephebe Fr.		**thread lichen**
See McCune (2017)		
Epilichen Clem.		
Epilichen scabrosus (Ach.) Clem.	*Buellia scabrosa* (Ach.) A. Massal.	
Erioderma Fée		
Erioderma pedicellatum (Hue) P. M. Jørg.		
Erioderma sorediatum D. J. Galloway & P. M. Jørg.		
Esslingeriana Hale & M. J. Lai		**Esslinger's lichen**
Esslingeriana idahoensis (Essl.) Hale & M. J. Lai	*Cetraria idahoensis* Essl.	
Evernia Ach.		**oakmoss lichen or antler lichen**
Evernia divaricata (L.) Ach.		
Evernia mesomorpha Nyl.		
Evernia prunastri (L.) Ach.		
Flavoparmelia Hale		**yellow shield**
Flavoparmelia caperata (L.) Hale	*Parmelia caperata* (L.) Ach.	
	Pseudoparmelia caperata (L.) Hale	

SCIENTIFIC NAME	SYNONYM	COMMON NAME
Flavopunctelia (Krog) Hale		**speckled yellow shield**
Flavopunctelia flaventior (Stirton) Hale	Parmelia flaventior Stirton	
	Punctelia flaventior (Stirton) Krog	
Flavopunctelia soredica (Nyl.) Hale	Parmelia soredica Nyl.	
	Parmelia ulophyllodes (Vainio) Savicz	
	Punctelia soredica (Nyl.) Krog	
Fuscopannaria P. M. Jørg.		
Fuscopannaria ahlneri (P. M. Jørg.) P. M. Jørg.		
Fuscopannaria alaskana P. M. Jørg. & Tønsberg		
Fuscopannaria aurita P. M. Jørg.		
Fuscopannaria cheiroloba (Müll. Arg.) P. M. Jørg.		
Fuscopannaria convexa P. M. Jørg.		
Fuscopannaria coralloidea P. M. Jørg.		
Fuscopannaria crustacea P. M. Jørg.		
Fuscopannaria cyanolepra (Tuck.) P. M. Jørg.	Pannaria cyanolepra Tuck.	
Fuscopannaria hookerioides P. M. Jørg.		
Fuscopannaria incisa (Mull. Arg.) P. M. Jørg.		
Fuscopannaria leprosa P. M. Jørg. & Tønsberg		
Fuscopannaria leucostictoides (Ohlsson) P. M. Jørg.	Pannaria leucostictoides Ohlsson	
Fuscopannaria maritima (P. M. Jørg.) P. M. Jørg.	Pannaria maritima P. M. Jørg.	
Fuscopannaria mediterranea (Tav.) P. M. Jørg.	Pannaria mediterranea Tav.	
Fuscopannaria pacifica P. M. Jørg.	Pannaria saubinetii (Mont.) Nyl. (misidentifications in PNW)	
Fuscopannaria praetermissa (Nyl.) P. M. Jørg.	Pannaria praetermissa Nyl.	
Fuscopannaria pulveracea (P. M. Jørg. & Henssen) P. M. Jørg.		
Fuscopannaria ramulina P. M. Jørg. & Tønsberg	Fuscopannaria confusa (P. M. Jørg.) P. M. Jørg. (misapplied in N Am)	
Fuscopannaria thiersii P. M. Jørg.		
Gabura Adans.		
Gabura insignis (P. M. Jørg. & Tønsberg) Magain & Sérus.	Leptogium brebissonii Mont. (misapplied)	
	Leptogium insigne P. M. Jørg. & Tønsberg	
	Arctomia insignis (P. M. Jørg. & Tønsberg) Ertz	
Glypholecia Nyl.		
Glypholecia scabra (Pers.) Mull. Arg.	Acarospora scabra (Pers.) Th. Fr.	
Heterodermia Trevis.		**fringed lichen**
Heterodermia erinacea (Ach.) W. A. Weber	Anaptychia erinacea (Ach.) Trevis.	
Heterodermia japonica (M. Sato) Swinsc. & Krog		
Heterodermia leucomela (L.) Poelt	Anaptychia "leucomelaena" auct.	
	Heterodermia leucomelaena (L.) Poelt	
Heterodermia sitchensis Goward & W. J. Noble		
Heterodermia speciosa (Wulfen) Trevis.	Anaptychia pseudospeciosa var. tremulans (Müll. Arg.) Kurok.	
	Anaptychia speciosa (Wulfen) A. Massal.	
	Heterodermia tremulans (Müll. Arg.) W. L. Culb.	

SCIENTIFIC NAME	SYNONYM	COMMON NAME
Heteroplacidium Breuss		
Heteroplacidium compuctum (A. Massal.) Gueidan & Cl. Roux)		
Hydrothyria J. L. Russell included in *Peltigera*	*Verrucaria compacta* (A. Massal.) Jatta	
Hypogymnia (Nyl.) Nyl.		**tube lichen**
Hypogymnia apinnata Goward & McCune		
Hypogymnia austerodes (Nyl.) Räsänen	*Hypogymnia dichroma* Goward	
	Hypogymnia salsa Goward	
	Hypogymnia verruculosa Goward	
Hypogymnia bitteri (Lynge) Ahti	*Hypogymnia protea* Goward, T. Sprib. & Ahti	
Hypogymnia canadensis Goward & McCune		
Hypogymnia duplicata (Ach.) Rass.		
Hypogymnia enteromorpha (Ach.) Nyl.		
Hypogymnia farinacea Zopf	*Hypogymnia bitteriana* (Zahlbr.) Räsänen	
Hypogymnia heterophylla L. H. Pike		
Hypogymnia hultenii (Degel.) Krog	*Cavernularia hultenii* Degel.	
Hypogymnia imshaugii Krog	*Hypogymnia amplexa* Goward, Björk & Wheeler	
Hypogymnia inactiva (Krog) Ohlsson		
Hypogymnia lophyrea (Ach.) Krog	*Cavernularia lophyrea* (Ach.) Degel.	
Hypogymnia occidentalis L. H. Pike		
Hypogymnia oceanica Goward		
Hypogymnia physodes (L.) Nyl.		
Hypogymnia pulverata (Nyl.) Elix		
Hypogymnia rugosa (G. Merr.) L. H. Pike		
Hypogymnia subobscura (Vainio) Poelt		
Hypogymnia subphysodes (Kremp.) Filson		
Hypogymnia tubulosa (Schaerer) Hav.		
Hypogymnia tuckerae McCune & S. Conway		
Hypogymnia vittata (Ach.) Parrique		
Hypogymnia wilfiana Goward, T. Sprib. & Ahti	*Hypogymnia metaphysodes* (Asahina) Rass. misapplied in N Am	
	Hypogymnia recurva Goward, C. R. Björk & Hollinger	
Hypotrachyna (Vainio) Hale		**forked lichen**
Hypotrachyna afrorevoluta (Krog & Swinscow) Krog & Swinscow	*Hypotrachyna revoluta* (Flörke) Hale (misapplied)	
Hypotrachyna riparia McCune		
Hypotrachyna sinuosa (Sm.) Hale	*Parmelia sinuosa* (Sm.) Ach.	
Imshaugia S. L. F. Mey.		**Imshaug's lichen**
Imshaugia aleurites (Ach.) S. L. F. Mey.	*Parmeliopsis aleurites* (Ach.) Nyl.	
Imshaugia placorodia (Ach.) S. L. F. Mey.	*Parmeliopsis placorodia* (Ach.) Nyl.	
Koerberia A. Massal.		
(see also *Tingiopsidium*)		
Koerberia biformis A. Massal.		
Kohlmeyera S. Schatz		**seaweed lichen**
Kohlmeyera complicatula (Nyl.) S. Schatz	*Turgidosculum complicatulum* (Nyl.) Kohlm. & E. Kohlm.	
Lasallia Mérat included in **Umbilicaria**		
Lathagrium (Ach.) Gray		
Lathagrium cristatum (L.) Otálora, P. M. Jørg. & Wedin	*Collema cristatum* (L.) Weber ex F. H. Wigg.	

SCIENTIFIC NAME	SYNONYM	COMMON NAME
Lathagrium fuscovirens (With.) Otálora, P. M. Jørg. & Wedin	*Collema fuscovirens* (With.) J. R. Laundon	
	Collema tuniforme	
Leciophysma Th. Fr.		
See McCune (2017)		
Leioderma *Nyl.*		**mouse ears**
Leioderma sorediatum D. J. Galloway & P. M. Jørg.		
Lempholemma Körber		**simple spored jelly lichen**
Lempholemma chalazanum (Ach.) B. de Lesd.		
Lempholemma cladodes (Tuck.) Zahlbr.		
Lempholemma intricatum (Arnold) Zahlbr.	*Lempholemma fennicum* (Räsänen) Degel.	
Lempholemma polyanthes (Bernh.) Malme	*Lempholemma myriococcum* (Ach.) Th. Fr.	
Lempholemma radiatum (Sommerf.) Henssen		
Leprocaulon Nyl. ex Lamy		**mealy lichen**
Leprocaulon adhaerens (K. Knudsen, Elix & Lendemer) Lendemer & Hodkinson	*Lepraria adhaerens* K. Knudsen, Elix & Lendemer	
Leprocaulon albicans (Th. Fr.) Nyl.	*Stereocaulon albicans* Th. Fr. ex Hue	
Leprocaulon gracilescens (Nyl.) I. M. Lamb & A. M. Ward		
Leprocaulon microscopicum (Vill.) Gams ex D. Hawksw.	*Stereocaulon microscopicum* (Vill.) Frey	
Leprocaulon santamonicae (K. Knudsen & Elix) Lendemer & Hodkinson	*Lepraria santamonicae* K. Knudsen & Elix	
Leprocaulon subalbicans (I. M. Lamb) I. M. Lamb & A. M. Ward	*Stereocaulon subalbicans* I. M. Lamb	
Leptochidium M. Choisy		**hairy jelly lichen**
Leptochidium albociliatum (Desm.) M. Choisy	*Leptogium albociliatum* Desm.	
	Polychidium albociliatum (Desm.) Zahlbr.	
Leptogidium Nyl.		
Leptogidium contortum (Henssen) T. Sprib. & Muggia	*Polychidium contortum* Henssen	
Leptogidium dendriscum (Nyl.) Nyl.	*Polychidium dendriscum* (Nyl.) Henssen	
Leptogium (Ach.) Gray		**wrinkled jelly lichen**
Leptogium arsenei Sierk		
Leptogium compactum D. F. Stone, F. L. Anderson & J. W. Hinds		
Leptogium cyanescens (Rabenh.) Körber		
Leptogium pseudofurfuraceum P. M. Jørg. & A. K. Wallace	*Leptogium furfuraceum* (Harm.) Sierk (misapplied)	
Leptogium saturninum (Dickson) Nyl.		
Leptogium stancookii D. F. Stone & Lendemer	*Leptogium cookii* D. F. Stone & Lendemer	
Leptogium umpquaense D. F. Stone & McCune		
Letharia (Th. Fr.) Zahlbr.		**wolf lichen**
Letharia 'barbata' (unpublished)		
Letharia columbiana (Nutt.) J. W. Thomson	*Letharia californica* (Lév. ex Nyl.) Hue	
Letharia gracilis Kroken in McCune & Altermann		
Letharia 'lucida' (unpublished)		

SCIENTIFIC NAME	SYNONYM	COMMON NAME
Letharia lupina Altermann, Leavitt & Goward		
Letharia 'rugosa' (unpublished)		
Letharia vulpina (L.) Hue	*Evernia vulpina* (L.) Ach.	
Lichenomphalia Redhead et al.		**mushroom lichen**
Lichinella Nyl.		
Lichinella nigritella (Lettau) P. P. Moreno & Egea	*Gonohymenia nigritella* (Lettau) Henssen	
Lichinodium Nyl.		
Lichinodium canadense Henssen		
Lobaria (Schreber) Hoffm.		**lung lichen or lungwort**
Lobaria anomala (Brodo & Ahti) T. Sprib. & McCune	*Pseudocyphellaria anomala* Brodo & Ahti	
Lobaria anthraspis (Ach.) T. Sprib. & McCune	*Pseudocyphellaria anthraspis* (Ach.) H. Magn.	
Lobaria hallii (Tuck.) Zahlbr.		
Lobaria linita (Ach.) Rabenh.		
Lobaria oregana (Tuck.) Müll. Arg.	*Lobaria silvae-veteris* (Goward & Goffinet) Goward & Goffinet	
	Nephroma silvae-veteris Goward & Goffinet	
Lobaria pulmonaria (L.) Hoffm.		
Lobaria retigera (Bory) Trevisan		
Lobaria scrobiculata (Scop.) DC.	*Lobarina scrobiculata* (Scop.) Nyl.	
Loxosporopsis Henssen		**tiny tree coral**
Loxosporopsis corallifera Brodo, Henssen & Imshaug		
Massalongia Körber		**rockmoss rosette**
Massalongia carnosa (Dickson) Körber		
Massalongia microphylliza (Nyl. ex Hasse) Henssen		
Melanelia Essl.		**camouflage lichen**
Melanelia agnata (Nyl.) A. Thell	*Cetraria agnata* (Nyl.) Kristinsson	
Melanelia hepatizon (Ach.) A. Thell	*Cetraria hepatizon* (Ach.) Vainio	
Melanelia stygia (L.) Essl.	*Parmelia stygia* (L.) Ach.	
Melanelixia Blanco et al.		
Melanelixia californica A. Crespo & Divakar	*Melanelixia glabra* (Schaerer) Blanco et al.	
	Melanelia glabra (Schaerer) Essl.	
	Parmelia glabra (Schaerer) Nyl.	
Melanelixia glabratula (Lamy) Sandler & Arup	*Melanelixia fuliginosa* (Fr. ex Duby) Blanco et al.	
	Melanelia fuliginosa (Fr. ex Duby) Essl.	
	Melanelia glabratula (Lamy) Essl.	
	Parmelia glabratula (Lamy) Nyl.	
Melanelixia glabroides (Essl.) Blanco et al.	*Melanelia glabroides* (Essl.) Essl.	
	Parmelia glabroides Essl.	
Melanelixia subargentifera (Nyl.) Blanco et al.	*Melanelia subargentifera* (Nyl.) Essl.	
	Parmelia subargentifera Nyl.	
Melanelixia subaurifera (Nyl.) Blanco et al.	*Melanelia subaurifera* (Nyl.) Essl.	
	Parmelia subaurifera Nyl.	
Melanohalea Blanco et al.		
Melanohalea elegantula (Zahlbr.) Blanco et al.	*Melanelia elegantula* (Zahlbr.) Essl.	
	Melanelia incolorata (Parrique) Essl.	

SCIENTIFIC NAME	SYNONYM	COMMON NAME
Melanohalea exasperatula (Nyl.) Blanco et al.	*Melanelia exasperatula* (Nyl.) Essl.	
Melanohalea infumata (Nyl.) Blanco et al.	*Melanelia infumata* (Nyl.) Essl.	
Melanohalea multispora (A. Schneider) Blanco et al.	*Melanelia multispora* (A. Schneider) Essl.	
Melanohalea septentrionalis (Lynge) O. Blanco et al.		
Melanohalea subelegantula (Essl.) Blanco et al.	*Melanelia subelegantula* (Essl.) Essl.	
Melanohalea subolivacea (Nyl. ex Hasse) Blanco et al.	*Melanelia subolivacea* (Nyl. ex Hasse) Essl.	
Melanohalea trabeculata (Ahti) O. Blanco et al.		
Menegazzia A. Massal.		**tree flute**
Menegazzia subsimilis (H. Magn.) R. Sant.		
Menegazzia terebrata (Hoffm.) A. Massal.		
Moelleropsis Gyeln.		
Moelleropsis nebulosa (Hoffm.) Gyeln.		
Montanelia Divakar et al.		
Montanelia disjuncta (Erichsen) Divakar et al.	*Melanelia disjuncta* (Erichsen) Essl.	
	Melanelia granulosa (Lynge) Essl.	
Montanelia panniformis (Nyl.) Divakar et al.	*Melanelia panniformis* (Nyl.) Essl.	
Montanelia sorediata (Ach.) Divakar et al.	*Melanelia sorediata* (Ach.) Goward & Ahti	
	Melanelia sorediosa (Almb.) Essl.	
Montanelia tominii (Oxner) Divakar	*Melanelia tominii* (Oxner) Essl.	
	Melanelia substygia (Räsänen) Essl.	
Multiclavula R. H. Petersen		**club lichen**
Neofuscelia Essl.		
Neofuscelia loxodes (Nyl.) Essl.	*Xanthoparmelia loxodes* (Nyl.) O. Blanco et al.	
Neofuscelia subhosseana (Essl.) Essl.	*Xanthoparmelia subhosseana* (Essl.) O. Blanco et al.	
Neofuscelia verruculifera (Nyl.) Essl.	*Xanthoparmelia verruculifera* (Nyl.) O. Blanco et al.	
Nephroma Ach.		**kidney lichen**
Nephroma arcticum (L.) Torss.		
Nephroma bellum (Sprengel) Tuck.		
Nephroma expallidum (Nyl.) Nyl.		
Nephroma isidiosum (Nyl.) Gyeln.		
Nephroma laevigatum Ach.		
Nephroma occultum Wetmore		
Nephroma orvoi Timdal et al.		
Nephroma parile (Ach.) Ach.		
Nephroma resupinatum (L.) Ach.		
Nephroma tropicum (Müll. Arg.) Zahlbr.	*Nephroma helveticum* Ach. (misapplied)	
Niebla Rundel & Bowler		**fog lichen**
Niebla cephalota (Tuck.) Rundel & Bowler	*Desmazieria cephalota* (Tuck.) Follmann & Huneck	
	Ramalina cephalota Tuck.	
Nodobryoria Common & Brodo		**chestnut beard**
Nodobryoria abbreviata (Müll. Arg.) Common & Brodo	*Alectoria abbreviata (Müll. Arg.) R. Howe*	
	Bryoria abbreviata (Müll. Arg.) Brodo & D. Hawksw.	
Nodobryoria oregana (Tuck.) Common & Brodo	*Alectoria oregana Tuck.*	
	Bryoria oregana (Tuck.) Brodo & D. Hawksw.	

SCIENTIFIC NAME	SYNONYM	COMMON NAME
*Nodobryoria subdivergens (*E. Dahl) Common & Brodo	*Alectoria subdivergens* E. Dahl	
	Bryoria subdivergens (E. Dahl) Brodo & D. Hawksw.	
Omphalina Quélet see *Lichenomphalia*		
Pannaria Delise		**shingled lichen**
Pannaria conoplea (Ach.) Bory		
Pannaria hookeri (Borrer) Nyl.		
Pannaria oregonensis McCune & M. Schultz	*Pannaria malmei* C. W. Dodge (misapplied)	
	Pannaria rubiginella P. M. Jørg. & Sipman (misapplied)	
	Pannaria rubiginosa (Thunb.) Delise (misapplied)	
Pannaria tavaresii P. M. Jørg.		
Parmelia Ach.		**shield lichen**
Parmelia barrenoae Divakar et al.		
Parmelia fraudans (Nyl.) Nyl.		
Parmelia hygrophila Goward & Ahti		
Parmelia imbricaria Goward et al.		
Parmelia omphalodes (L.) Ach.		
Parmelia pseudosulcata Gyeln.		
Parmelia saxatilis (L.) Ach.		
Parmelia skultii Hale		
Parmelia squarrosa Hale		
Parmelia sulcata Taylor		
Parmelia sulymae Goward et al.		
Parmeliella Müll. Arg.		**shingled lichen**
Parmeliella parvula P. M. Jørg.		
Parmeliella triptophylla (Ach.) Müll. Arg.		
Parmelina Hale		
Parmelina coleae Argüello & Crespo	*Parmelina quercina* (Willd.) Hale	
Parmeliopsis Nyl.		**starburst**
Parmeliopsis ambigua (Wulfen) Nyl.		
Parmeliopsis hyperopta (Ach.) Arnold		
Parmotrema A. Massal.		**eyelash lichen**
Parmotrema arnoldii (Du Rietz) Hale	*Parmelia arnoldii* Du Rietz	
Parmotrema chinense (Osbeck) Hale & Ahti	*Parmelia perlata* (Hudson) Ach.	
	Parmotrema perlatum (Hudson) Choisy	
Parmotrema crinitum (Ach.) M. Choisy	*Parmelia crinita* Ach.	
Parmotrema herrei (Zahlbr.) Spielmann & Marcelli		
Parmotrema stuppeum (Taylor) Hale	*Parmelia stuppea* Taylor	
Peltigera Willd.		**veined lichen**
Peltigera aphthosa (L.) Willd.		
Peltigera britannica (Gyeln.) Holt.-Hartw. & Tønsberg		
Peltigera canina (L.) Willd.		
Peltigera castanea Goward, Goffinet & Miadl.		
Peltigera chionophila Goward & Goffinet		
Peltigera cinnamomea Goward		
Peltigera collina (Ach.) Schrader	*Peltigera scutata* (Dickson) Duby	
Peltigera degenii Gyeln.		
Peltigera didactyla (With.) J. R. Laundon	*Peltigera canina* var. *spuria* (Ach.) Schaerer	
	Peltigera spuria (Ach.) DC.	

SCIENTIFIC NAME	SYNONYM	COMMON NAME
Platismatia wheeleri Goward, Altermann & Björk		
Polychidium (Ach.) Gray		**intricate lichen**
(see also *Leptogidium*)		
Polychidium muscicola (Sw.) Gray		
Protopannaria (Gyeln.) P. M. Jørg. & S. Ekman		
Protopannaria pezizoides (Weber) P. M. Jørg.	*Pannaria pezizoides* (Weber) Trevis.	
Pseudephebe M. Choisy		**rock wool**
Pseudephebe minuscula (Nyl. ex Arnold) Brodo & D. Hawksw.	*Alectoria minuscula* (Nyl. ex Arnold) Degel.	
	Pseudephebe pubescens (L.) M. Choisy (misapplied in N Am)	
Pseudocyphellaria Vainio		**specklebelly**
(see also *Lobaria*)		
Pseudocyphellaria citrina (Gyeln.) Lücking, B. Moncada & S. Stenroos	*Pseudocyphellaria crocata* (L.) Vainio (misapplied in N Am)	
Pseudocyphellaria hawaiiensis H. Magn.	*Pseudocyphellaria perpetua* McCune & Miadl.	
Pseudocyphellaria mallota (Tuck.) H. Magn.		
Pseudocyphellaria rainierensis Imshaug		
Psoroma Michaux		**beaded moss lover**
Psoroma hypnorum (Vahl) Gray	*Pannaria hypnorum* (Vahl) Körber	
Punctelia Krog		**speckled shield**
Punctelia borreri (Sm.) Krog		
Punctelia caseana Lendemer & Hodkinson		
Punctelia jeckeri (Roum.) Kalb	*Punctelia perreticulata* (Räsänen) G. Wilh. & Ladd (misapplied)	
	Punctelia subrudecta (Nyl.) Krog (misapplied)	
Punctelia rudecta (Ach.) Krog		
Punctelia stictica (Duby) Krog	*Parmelia stictica* (Duby) Nyl.	
Punctelia sp. (undescribed)		
Racodium Pers.: Fr.		
Racodium rupestre Pers.		
Ramalina Ach.		**ramalina**
Ramalina americana Hale		
Ramalina dilacerata (Hoffm.) Hoffm.	*Fistulariella minuscula* (Nyl.) Bowler & Rundel	
	Ramalina minuscula (Nyl.) Nyl.	
Ramalina farinacea (L.) Ach.		
Ramalina intermedia (Delise ex Nyl.) Nyl.		
Ramalina labiosorediata Gasparyan, Sipman & Lücking	*Ramalina pollinaria* (Westr.) Ach. (possibly misapplied)	
Ramalina menziesii Taylor	*Ramalina reticulata* (Nohden) Kremp.	
Ramalina obtusata (Arnold) Bitter		
Ramalina roesleri (Hochst. ex Schaerer) Hue	*Fistulariella roesleri* (Hochst. ex Schaerer) Bowler & Rundel	
Ramalina sinensis Jatta		
Ramalina subleptocarpha Rundel & Bowler		
Ramalina thrausta (Ach.) Nyl.	*Alectoria thrausta* Ach.	
Rhizoplaca Zopf		**rimmed navel lichen or rock brooch**
Rhizoplaca chrysoleuca (Sm.) Zopf	*Lecanora chrysoleuca* (Sm.) Ach.	
	Lecanora rubina (Vill.) Ach.	
Rhizoplaca haydenii (Tuck.) W. A. Weber		
Rhizoplaca melanophthalma (DC.) Leuck.	*Lecanora melanophthalma* (DC.) Ramond	

SCIENTIFIC NAME	SYNONYM	COMMON NAME
Rhizoplaca peltata (Ramond) Leuck. & Poelt		
Rhizoplaca subdiscrepans (Nyl.) R. Sant.		
Ricasolia De Not.		
Ricasolia amplissima (Scop.) De Not.	*Lobaria amplissima* (Scop.) Forssell	
Ricasolia amplissima subsp. *sheiyi* Derr & Dillman		
Rostania Trevis.		
Rostania ceranisca (Nyl.) Otálora, P. M. Jørg. & Wedin	*Collema ceraniscum* Nyl.	
Rostania occultata (Bagl.) Otálora, P. M. Jørg. & Wedin	*Collema occultatum* Bagl.	
Scytinium (Ach.) Gray		
Scytinium aquale (Arnold) Otálora et al.	*Leptogium aquale* (Arnold) P. M. Jørg.	
Scytinium californicum (Tuck.) Otálora et al.	*Leptogium californicum* Tuck.	
Scytinium callopismum (A. Massal.) Otálora et al.	*Collema callopismum* A. Massal.	
Scytinium cellulosum (P. M. Jørg. & Tønsberg) Otálora et al.	*Leptogium cellulosum* P. M. Jørg. & Tønsberg	
Scytinium intermedium (Arnold) Otálora et al.	*Leptogium intermedium* (Arnold) Arnold	
	Leptogium minutissimum (Flörke) Fr.	
Scytinium nanum (Herre) D. F. Stone & McCune	*Leptogium nanum* Herre	
Scytinium palmatum (Hudson) Gray	*Leptogium palmatum* (Hudson) Mont.	
	Leptogium corniculatum (Hoffm.) Minks	
Scytinium platynum (Tuck.) Otálora et al.	*Leptogium platynum* (Tuck.) Herre	
Scytinium plicatile (Ach.) Otálora et al.	*Leptogium plicatile* (Ach.) Leighton	
Scytinium polycarpum (P. M. Jørg. & Goward) Otálora et al.	*Leptogium polycarpum* P. M. Jørg. & Goward	
Scytinium pulvinatum (Hoffm.) Otálora et al.		
Scytinium quadrifidum (D. F. Stone & McCune) A. Košuth. & Wedin	*Collema quadrifidum* D. F. Stone & McCune	
	Rostania quadrifida (D. F. Stone & McCune) McCune	
Scytinium rivale (Tuck.) Otálora et al.	*Leptogium rivale* Tuck.	
Scytinium schraderi (Bernh.) Otálora et al.	*Leptogium schraderi* (Bernh.) Nyl.	
Scytinium singulare T. Carlberg & P. M. Jørg.		
Scytinium siskiyouense (D. F. Stone & Ruchty) Otálora et al.	*Leptogium siskiyouensis* D. Stone & Ruchty	
Scytinium subaridum (P. M. Jørg. & Goward) Otálora et al.	*Leptogium subaridum* P. M. Jørg. & Goward	
Scytinium subtile (Schrad.) Otálora et al.	*Leptogium subtile* (Schrader) Torss.	
Scytinium tacomae (P. M. Jørg. & Tønsberg) McCune	*Leptogium tacomae* P. M. Jørg. & Tønsberg	
Scytinium tenuissimum (Dickson) Otálora et al.	*Leptogium tenuissimum* (Dickson) Körber	
Scytinium teretiusculum (Wallr.) Otálora et al.	*Leptogium teretiusculum* (Wallr.) Arnold	
Scytinium turgidum (Ach.) Otálora et al.	*Leptogium turgidum* (Ach.) Crombie	
Seirophora Poelt		
Seirophora contortuplicata (Ach.) Frödén	*Teloschistes contortuplicatus* (Ach.) Clauzade & Rondon ex Vezda	
	Xanthaptychia contortuplicata (Ach.) S. Y. Kondr. & Ravera	

SCIENTIFIC NAME	SYNONYM	COMMON NAME
Siphula Fr.		
Siphula ceratites (Wahlenb.) Fr.		
Solorina Ach.		chocolate chip lichen
Solorina bispora Nyl.		
Solorina crocea (L.) Ach.		
Solorina monospora Gyeln.		
Solorina octospora (Arnold) Arnold		
Solorina saccata (L.) Ach.		
Solorina spongiosa (Ach.) Anzi		
Sphaerophorus Pers.		coral lichen
Sphaerophorus fragilis (L.) Pers.		
Sphaerophorus globosus (Hudson) Vainio		
Sphaerophorus tuckermanii Räsänen		
Sphaerophorus venerabilis Wedin, Högnabba & Goward		
Steineropsis T. Sprib. & Muggia		
Steineropsis laceratula (Hue) T. Sprib. & S. Ekman	*Fuscopannaria laceratula* (Hue) P. M. Jørg.	
Stereocaulon Hoffm.		snow lichen
Stereocaulon alpinum Laurer ex Funck		
Stereocaulon alpinum var. *erectum* Frey		
Stereocaulon botryosum Ach.		
Stereocaulon condensatum Hoffm.		
Stereocaulon coniophyllum I. M. Lamb		
Stereocaulon depressum (Frey) I. M. Lamb		
Stereocaulon glareosum (Savicz) H. Magn.		
Stereocaulon groenlandicum (E. Dahl) I. M. Lamb		
Stereocaulon incrustatum Flörke		
Stereocaulon intermedium (Savicz) H. Magn.		
Stereocaulon nivale (Follman) Fryday		
Stereocaulon oregonense McCune, E. Di Meglio & Tønsberg		
Stereocaulon paschale (L.) Hoffm.		
Stereocaulon pileatum Ach.		
Stereocaulon rivulorum H. Magn.		
Stereocaulon spathuliferum Vainio		
Stereocaulon sterile (Savicz) Lamb ex Krog		
Stereocaulon subcoralloides (Nyl.) Nyl.		
Stereocaulon symphycheilum I. M. Lamb		
Stereocaulon tomentosoides (I. M. Lamb) McCune	*Stereocaulon sasakii* var. *tomentosoides* I. M. Lamb *Stereocaulon sasakii* var. *simplex* I. M. Lamb	
Stereocaulon tomentosum Fr.		
Stereocaulon vesuvianum Pers.		
Sticta (Schreber) Ach.		moon lichen
Sticta arctica Degel.		
Sticta fuliginosa (Hoffm.) Ach. (broad sense)		
Sticta limbata (Sm.) Ach.		
Sticta rhizinata B. Moncada & Lücking	*Sticta beauvoisii* Delise (misapplied) *Sticta weigelii* (Ach.) Vainio (misapplied)	

SCIENTIFIC NAME	SYNONYM	COMMON NAME
Sticta torii Ant. Simon & Goward		
Sulcaria Bystrek		**grooved beard lichen**
Sulcaria badia Brodo & D. Hawksw.		
Sulcaria spiralifera (Brodo & D. Hawksw.) Myllys, Velmala & Goward	*Bryoria spiralifera* Brodo & D. Hawksw.	
Sulcaria spiralifera var. *pseudocapillaris* (Brodo & D. Hawksw.) McCune	*Bryoria pseudocapillaris* Brodo & D. Hawksw.	
Teloschistes Norman		**orange bush lichen**
Teloschistes flavicans (Sw.) Norman		
Thamnolia Ach. ex Schaerer		**whiteworm lichen**
Thamnolia subuliformis (Ehrh.) W. L. Culb.	*Thamnolia vermicularis* (Sw.) Ach. ex Schaerer (misapplied to thamnolic chemotype in our area)	
Tholurna Norman		**urn lichen**
Tholurna dissimilis (Norman) Norman		
Tingiopsidium Werner		
Tingiopsidium elaeinum (Wahlenb. ex Ach.) Hafellner & T. Sprib.	*Vestergrenopsis elaeina* (Wahlenb.) Gyeln.	
Tingiopsidium isidiatum (Degel.) Hafellner & T. Sprib.	*Vestergrenopsis isidiata* (Degel.) E. Dahl	
Tingiopsidium sonomense (Tuck.) Hafellner & T. Sprib.	*Vestergrenopsis sonomensis* (Tuck.) T. Sprib. & Muggia	
	Koerberia sonomensis (Tuck.) Henssen	
Turgidosculum Kohlm. & E. Kohlm.		
Turgidosculum ulvae (M. Reed) Kohlm. & E. Kohlm.		
Umbilicaria Hoffm.		**navel lichen or rocktripe**
Umbilicaria americana Poelt & T. H. Nash		
Umbilicaria angulata Tuck.		
Umbilicaria arctica (Ach.) Nyl.		
Umbilicaria caroliniana Tuck.	*Lasallia caroliniana* (Tuck.) Davydov et al.	
Umbilicaria cinereorufescens (Schaerer) Frey		
Umbilicaria cylindrica (L.) Delise ex Duby		
Umbilicaria decussata (Vill.) Zahlbr.	*Omphalodiscus decussatus* (Vill.) Schol.	
Umbilicaria deusta (L.) Baumg.		
Umbilicaria dura McCune		
Umbilicaria havaasii Llano		
Umbilicaria herrei Frey	*Umbilicaria hyperborea* var. *radicicula* (J. E. Zetterst.) Hasselrot	
Umbilicaria hirsuta (Sw. ex Westr.) Hoffm.		
Umbilicaria hyperborea (Ach.) Hoffm.		
Umbilicaria lambii Imshaug		
Umbilicaria lyngei Schol.	*Agyrophora lyngei* (Schol.) Llano	
Umbilicaria multistrata McCune		
Umbilicaria nodulospora McCune, Di Meglio & M. J. Curtis		
Umbilicaria nylanderiana (Zahlbr.) H. Magn.		
Umbilicaria papulosa (Ach.) Nyl.	*Lasallia papulosa* (Ach.) Llano	
Umbilicaria pensylvanica Hoffm.	*Lasallia pensylvanica* (Hoffm.) Llano	
Umbilicaria phaea Tuck.		
Umbilicaria phaea var. *coccinea* Llano		
Umbilicaria polaris (Schol.) Zahlbr.	*Umbilicaria krascheninnikovii* (Savicz) Zahlbr.	

SCIENTIFIC NAME	SYNONYM	COMMON NAME
	Omphalodiscus krascheninnikovii (Savicz) Schol.	
Umbilicaria polyphylla (L.) Baumg.		
Umbilicaria polyrhiza (L.) Fr.	*Actinogyra polyrhiza* (L.) Schol.	
Umbilicaria proboscidea (L.) Schrader		
Umbilicaria rigida (Du Rietz) Frey	*Agyrophora rigida* (Du Rietz) Llano	
Umbilicaria scholanderi (Llano) Krog	*Agyrophora scholanderi* Llano	
Umbilicaria semitensis Tuck.		
Umbilicaria torrefacta (Lightf.) Schrader		
Umbilicaria vellea (L.) Hoffm.		
Umbilicaria virginis Schaerer	*Omphalodiscus virginis* (Schaerer) Schol.	
Usnea Dill. ex Adans.		**beard lichen**
Usnea brasiliensis (Zahlbr.) Motyka		
Usnea cavernosa Tuck.		
Usnea ceratina Ach.	*Usnea californica* Herre	
Usnea chaetophora Stirton		
Usnea cornuta Körber	*Usnea inflata* (Duby) Motyka	
	Usnea occidentalis Motyka	
Usnea dasopoga (Ach.) Nyl.	*Usnea filipendula* Stirton	
	Usnea plicata (L.) Weber	
Usnea diplotypus Vainio		
Usnea esperantiana P. Clerc		
Usnea flavocardia Räsänen	*Usnea wirthii* P. Clerc	
Usnea florida (L.) F. H. Wigg.		
Usnea fragilescens Hav. ex Lynge		
Usnea fragilescens var. *mollis* (Vainio) P. Clerc		
Usnea fulvoreagens (Räsänen) Räsänen	*Usnea glabrescens* (Vainio) Vainio var. *fulvoreagens* Räsänen	
Usnea glabrata (Ach.) Vainio	*Usnea sorediifera* (Arnold) Lynge	
	Usnea kujalae Räsänen	
Usnea hirta (L.) F. H. Wigg.		
Usnea intermedia (A. Massal.) Jatta	*Usnea arizonica* Motyka	
Usnea lambii (Imshaug) Wirtz & Lumbsch	*Usnea sphacelata* R. Br. (not in PNW)	
	Neuropogon lambii Imshaug	
Usnea lapponica Vainio	*Usnea laricina* Vainio ex Räsänen	
	Usnea sorediifera sensu Motyka	
	Usnea substerilis Motyka	
Usnea longissima Ach.	*Dolichousnea longissima* (Ach.) Articus	
Usnea nidulans Motyka		
Usnea pacificana Halonen		
Usnea quasirigida Lendemer & I. I.Tav.	*Usnea rigida* Motyka non *U. rigida* Vainio	
Usnea rubicunda Stirton		
Usnea scabrata Nyl.	*Usnea rugulosa* Vainio	
Usnea scabrata Nyl. sens. lat.	*Usnea alpina* Motyka	
	Usnea barbata (L.) F. H. Wigg.	
	Usnea catenulata Motyka	
	Usnea graciosa Motyka	
	Usnea scabiosa Motyka	
Usnea silesiaca Motyka	*Usnea madeirensis* Motyka	
Usnea subfloridana Stirton		
Usnea subgracilis Vainio	*Usnea hesperina* Motyka	
	Usnea schadenbergiana Göpp. & Stein	
Usnea trichodea Ach.		
Usnea wasmuthii Räsänen		

SCIENTIFIC NAME	SYNONYM	COMMON NAME
Vahliella P. M. Jørg.		
Vahliella californica (Tuck.) P. M. Jørg.	*Fuscopannaria californica* (Tuck.) P. M. Jørg.	
Verrucaria Schrad.		
Verrucaria sphaerospora Anzi		
Vestergrenopsis Gyeln. (see *Tingiopsidium*)		
Vulpicida J.-E. Mattsson & M. J. Lai (see *Cetraria*)		
Xanthomendoza S. Y. Kondr. & Kärnefelt		**orange lichen or sunburst**
Xanthomendoza borealis (R. Sant. & Poelt) Søchting et al.	*Xanthoria borealis* R. Sant. & Poelt	
Xanthomendoza fallax (Hepp) Søchting et al.	*Xanthoria fallax* (Hepp) Arnold	
Xanthomendoza fulva (Hoffm.) Søchting et al.	*Xanthoria fulva* (Hoffm.) Poelt. & Petutschnig	
	Xanthoria subramulosa Räsänen	
Xanthomendoza galericulata L. Lindblom		
Xanthomendoza hasseana (Räsänen) Søchting et al.	*Xanthoria hasseana* Räsänen	
Xanthomendoza mendozae (Räsänen) S. Y. Kondr. & Kärnefelt	*Xanthoria mendozae* Räsänen	
Xanthomendoza montana (L. Lindblom) Søchting et al.	*Xanthoria montana* Lindblom	
Xanthomendoza oregana (Gyeln.) Søchting et al.	*Xanthoria oregana* Gyeln.	
Xanthomendoza poeltii (S. Y. Kondr. & Kärnefelt) Søchting et al.		
Xanthomendoza ulophyllodes (Räsänen) Søchting et al.	*Xanthoria ulophyllodes* Räsänen	
Xanthoparmelia (Vainio) Hale		**yellow rock shield or rock frog**
Xanthoparmelia camtschadalis (Ach.) Hale		
Xanthoparmelia chlorochroa (Tuck.) Hale	*Parmelia chlorochroa* Tuck.	
Xanthoparmelia coloradoensis (Gyeln.) Hale		
Xanthoparmelia conspersa (Ehrh. ex Ach.) Hale	*Parmelia conspersa* (Ehrh. ex Ach.) Ach.	
Xanthoparmelia cumberlandia (Gyeln.) Hale	*Parmelia cumberlandia* (Gyeln.) Hale	
Xanthoparmelia hypofusca (Gyeln.) B. P. Hodk. & Lendemer	*Xanthoparmelia tasmanica* (Hook. f. & Taylor) Hale	
Xanthoparmelia lavicola (Gyeln.) Hale	*Xanthoparmelia kurokawae* (Hale) Hale	
Xanthoparmelia lineola (E. C. Berry) Hale	*Parmelia lineola* E. C. Berry	
Xanthoparmelia mexicana (Gyeln.) Hale	*Parmelia mexicana* Gyeln.	
Xanthoparmelia montanensis Hale		
Xanthoparmelia mougeotii (Schaerer) Hale	*Parmelia mougeotii* Schaerer	
Xanthoparmelia neochlorochroa Hale		
Xanthoparmelia planilobata (Gyeln.) Hale	*Xanthoparmelia angustiphylla* (Gyeln.) Hale	
	Xanthoparmelia hypopsila (Müll. Arg.) Hale	
Xanthoparmelia plittii (Gyeln.) Hale	*Parmelia plittii* Gyeln.	
Xanthoparmelia subdecipiens (Vainio) Hale	*Parmelia subdecipiens* Vainio	
Xanthoparmelia subramigera (Gyeln.) Hale	*Parmelia subramigera* Gyeln.	
Xanthoparmelia wyomingica (Gyeln.) Hale	*Parmelia wyomingica* (Gyeln.) Hale	

Beck, J. N., and G. J. Ramelow. 1990. Use of lichen biomass to monitor dissolved metals in natural waters. *Bulletin of Environmental Contamination and Toxicology* 44: 302–308.

Belnap, J. 2001. Microbes and microfauna associated with biological soil crusts. Pp. 167–174 in: J. Belnap and O. Lange (eds.), *Biological Soil Crusts: Structure, Function, and Management.* Ecological Studies, Springer-Verlag, Berlin, Heidelberg.

Belnap, J., and O. L. Lange (eds.). 2001. *Biological Soil Crusts: Structure, Function, and Management.* Springer-Verlag, Berlin.

Bird, C. D. 1974. Studies on the lichen genus *Evernia* in North America. *Canadian Journal of Botany* 52: 2427–2434.

Bjerke, J. W. 2003. *Menegazzia subsimilis,* a widespread sorediate lichen. *Lichenologist* 35: 393–396.

Björk, C. R. 2009. Additions to the lichen flora of Washington state, United States. I. *Evansia* 27: 18–20.

Blanco, O., A. Crespo, P. K. Divakar, T. L. Esslinger, D. L. Hawksworth, and H. T. Lumbsch. 2004. *Melanelixia* and *Melanohalea,* two new genera segregated from *Melanelia* (Parmeliaceae) based on molecular and morphological data. *Mycological Research* 108: 873–884.

Boluda, C. G., D. L. Hawksworth, P. K. Divakar, A. Crespo, V. J. Rico. 2016. Microchemical and molecular investigations reveal *Pseudephebe* species as cryptic with an environmentally modified morphology. *Lichenologist* 48: 527–544.

Boluda, C. G., and 8 authors. 2019. Evaluating methodologies for species delimitation: The mismatch between phenotypes and genotypes in lichenized fungi (*Bryoria* sect. Implexae, Parmeliaceae). *Persoonia* 42: 75–100.

Bowler, P. A. 1977. *Ramalina thrausta* in North America. *Bryologist* 80: 529–532.

Bowler, P. A., and J. E. Marsh. 2004. *Niebla.* Pp. 368–380 in: T. H. Nash, III, B. D. Ryan, P. Diederich, C. Gries, and F. Bungartz (eds.), *Lichen Flora of the Greater Sonoran Desert Region. Vol. II.* Lichens Unlimited, Arizona State University, Tempe.

Bowler, P. A., and P. W. Rundel. 1977. Synopsis of a new lichen genus, *Fistulariella* Bowler & Rundel (Ramalinaceae). *Mycotaxon* 6: 195–202.

Bowler, P. A., and P. W. Rundel. 1978. The *Ramalina farinacea* complex in North America: Chemical, ecological and morphological variation. *Bryologist* 81: 386–403.

Bowler, P. A., R. E. Riefner Jr., P. W. Rundel, J. Marsh and T. H. Nash, III. 1994. New species of *Niebla* (Ramalinaceae) from western North America. *Phytologia* 77: 23–37.

Brahney, J., M. Hallerud, E. Heim, M. Hahnenberger, and S. Sukumaran. 2020. Plastic rain in protected areas of the United States. *Science* 368: 1257–1260.

Brahney, J., N. Mahowald, M. Prank, G. Cornwell, Z. Klimont, H. Matsui, K. A. Prather. 2021. Constraining the atmospheric limb of the plastic cycle. *Proceedings of the National Academy of Sciences* 118(16): e2020719118.

Breuss, O. 2003. Eine neue *Dermatocarpon*-Art (lichenisierte Ascomyceten, Verrucariales) aus Kanada. *Bibliotheca Lichenologica* 86: 99–102.

A mycophagist in southwestern Oregon. *Canadian Journal of Zoology* 64: 2086–2089.

Maser, Z., C. Maser, and J. M. Trappe. 1985. Food habits of the northern flying squirrel (*Glaucomys sabrinus*) in Oregon. *Canadian Journal of Zoology* 63: 1085–1088.

Mattson, J.-E. 1993. A monograph of the genus *Vulpicida* (Parmeliaceae, Ascomycetes). *Opera Botanica* 119: 1–61.

McCune, B. 1984. Lichens with oceanic affinities in the Bitterroot Mountains of Montana and Idaho. *Bryologist* 87: 44–50.

McCune, B. 1987. Distribution of chemotypes of *Rhizoplaca* in North America. *Bryologist* 90: 6–14.

McCune, B. 1993. Gradients in epiphyte biomass in three *Pseudotsuga-Tsuga* forests of different ages in western Oregon and Washington. *Bryologist* 96: 405–411.

McCune, B. 1998. *Hypotrachyna riparia*, a new lichen from western North America. *Bryologist* 101: 21–31.

McCune, B. 2000. Lichen communities as indicators of forest health. *Bryologist* 103: 353–356.

McCune, B. 2007. *Parmelia skultii* new to the lower 48 states. *Evansia* 24: 76.

McCune, B. 2017. *Microlichens of the Pacific Northwest.* Vols. 1 & 2. Wild Blueberry Media, Corvallis.

McCune, B. 2018. Two new species in the *Umbilicaria torrefacta* group from Alaska and the Pacific Northwest of North America. *Graphis Scripta* 30: 65–77.

McCune, B. 2019a. *Lichen Spot Test Reactions.* 17 x 22-inch poster. Wild Blueberry Media, Corvallis. wildblueberrymedia.net.

McCune, B. 2019b. *Lichen Color Chart.* 10 x 29-inch poster. Wild Blueberry Media, Corvallis. wildblueberrymedia.net.

McCune, B., and S. Altermann. 2009. *Letharia gracilis* (Parmeliaceae), a new species from California and Oregon. *Bryologist* 112: 375–378.

McCune, B., and S. N. Conway. 2022. Two new species, *Hypogymnia tuckerae* and *H. discopruina* (Parmeliaceae), from North America and China. *Bryologist* 125: 406–419.

McCune, B., and M. J. Curtis. 2012. *Umbilicaria semitensis* (lichenized fungi: Umbilicariaceae) resurrected. *Bryologist* 115: 255–264.

McCune, B., and J. Di Meglio. 2021. Revision of the *Aspicilia reptans* group in western North America, an important component of soil biocrusts. *Monographs in North American Lichenology* 5: 1–92.

McCune, B., and T. Goward. 1995. *Macrolichens of the Northern Rocky Mountains.* Mad River Press, Eureka, California.

McCune, B., and R. Rosentreter. 1997. *Hypogymnia subphysodes* new to North America. *Evansia* 14: 106.

McCune, B., and R. Rosentreter. 2007. Biotic Soil Crust Lichens of the Columbia Basin. *Monographs in North American Lichenology* 1: 1–105.

McCune, B., and D. F. Stone. 2022. Eight new combinations of North American macrolichens. *Evansia* 39(3): 123–128.

McCune, B., and S. Yang. 2021. *Common Macrolichens of the Pacific Northwest.* Online interactive key with photos. https://lichens.twinferntech.net/pnw/index.shtml.

McCune, B., R. Rosentreter, and A. DeBolt. 1997. Biogeography of rare lichens from the coast of Oregon. Pp. 234–241 in: T. N. Kaye et al. (eds.), *Conservation and Management of Native Flora and Fungi.* Native Plant Society of Oregon.

McCune, B., K. Glew, P. Nelson, and J. Villella. 2007. Lichens from the Matterhorn and Ice Lake, northeastern Oregon. *Evansia* 24: 72–75.

McCune, B., C. Schoch, H. T. Root, S. A. Kageyama, and J. Miadlikowska. 2011. Geographic, climatic, and chemical differentiation in the *Hypogymnia imshaugii* species complex (Parmeliaceae). *Bryologist* 114: 526–544.

McCune, B., J. Di Meglio, and M. J. Curtis. 2014a. An unusual ascospore shape and a new species, *Umbilicaria nodulospora* (Umbilicariaceae), from California and Oregon. *Bryologist* 117: 170–178.

McCune, B., R. Rosentreter, T. Spribille, O. Breuss, and T. B. Wheeler. 2014b. Montana Lichens: An Annotated List. *Monographs in North American Lichenology* 2: 1–183.

McCune, B., E. Di Meglio, T. Tønsberg, and R. Yahr. 2019. Five new crustose *Stereocaulon* species in western North America. *Bryologist* 122: 197–218.

McCune, B., M. Schultz, T. Fennell, A. Passo, and J. M. Rodriguez. 2022. A new endemic, *Pannaria oregonensis*, replaces two misapplied names in the Pacific Northwest of North America. *Bryologist* 125: 170–185.

McDonald, T., J. Miadlikowska, and F. Lutzoni. 2003. The lichen genus *Sticta* in the Great Smoky Mountains: A phylogenetic study of morphological, chemical, and molecular data. *Bryologist* 106: 61–79.

McHenry, G., and T. Tønsberg. 2002. *Heterodermia sitchensis* found in Oregon, U.S.A. *Evansia* 19: 158–160.

Meyer, S. L. F. 1982. Segregation of the new lichen genus *Foraminella* from *Parmeliopsis*. *Mycologia* 74: 592–598.

Meyer, S. L. F. 1985. The new lichen genus *Imshaugia* (Ascomycotina, Parmeliaceae). *Mycologia* 77: 336–338.

Miadlikowska, J., and F. Lutzoni. 2000. Phylogenetic revision of the genus *Peltigera* (lichen-forming Ascomycota) based on morphological, chemical and large subunit nuclear ribosomal DNA data. *International Journal of Plant Sciences* 161(6): 925–958.

Miadlikowska, J., and F. Lutzoni. 2004. Phylogenetic classification of peltigeralean fungi (Peltigerales, Ascomycota) based on ribosomal RNA small and large subunits. *American Journal of Botany* 91: 449–464.

Miadlikowska, J., B. McCune, and F. Lutzoni. 2002. *Pseudocyphellaria perpetua*, a new lichen from western North America. *Bryologist* 105: 1–10.

Miadlikowska, J., F. Lutzoni, T. Goward, S. Zoller, and D. Posada. 2003. New approach to an old problem: Incorporating signal from gap-rich regions of ITS and nrDNA large subunit into phylogenetic analyses to resolve the *Peltigera canina* species complex. *Mycologia* 95: 1181–1203.

Miadlikowska, J., C. L. Schoch, S. A. Kageyama, K. Molnar, F. Lutzoni, and B. McCune. 2011. *Hypogymnia* phylogeny, including *Cavernularia*, reveals biogeographic structure. *Bryologist* 114: 392–400.

Miadlikowska, J., and 8 authors. 2014. Phylogenetic placement, species delimitation, and cyanobiont identity of endangered aquatic *Peltigera* species (lichen-forming *Ascomycota, Lecanoromycetes*). *American Journal of Botany* 101: 1141–1156.

Miller, J. E. D. 2011. The *Usnea rigida* group in California and the Pacific Northwest. *Bulletin of the California Lichen Society* 18: 1–4.

Moberg, R. 1977. The lichen genus *Physcia* and allied genera in Fennoscandia. *Symbolae Botanicae Upsalienses* 22: 1–108.

Moberg, R. 1997. The lichen genus *Physcia* in the Sonoran Desert and adjacent areas. *Symbolae Botanicae Upsalienses* 31: 163–186.

Moberg, R. 2011. The lichen genus *Heterodermia* (Physciaceae) in South America — a contribution including five new species. *Nordic Journal of Botany* 29: 129–147.

Molina, M. C., P. K. Divakar, T. Goward, A. M. Millanes, H. T. Lumbsch, and A. Crespo. 2017. Neogene diversification in the temperate lichen-forming fungal genus *Parmelia* (Parmeliaceae, Ascomycota). *Systematics and Biodiversity* 15: 166–181.

Molnár, K., and E. Farkas. 2010. Current results on biological activities of lichen secondary metabolites: A review. *Zeitschrift für Naturforschung C* 65: 157–173.

Moncada, B., and R. Lücking. 2012. Ten new species of *Sticta* and counting: Colombia as a hot spot for unrecognized diversification in a conspicuous macrolichen genus. *Phytotaxa* 74: 1–29.

Moncada, B., R. Lücking, and L. Betancourt-Macuase. 2013. Phylogeny of the *Lobariaceae* (lichenized Ascomycota: Peltigerales) with a reappraisal of the genus *Lobariella*. *Lichenologist* 45: 203–263.

Moncada, B., B. Reidy, and R. Lücking. 2014. A phylogenetic revision of Hawaiian *Pseudocyphellaria* sensu lato (lichenized Ascomycota: Lobariaceae) reveals eight new species and a high degree of inferred endemism. *Bryologist* 117: 119–160.

Moncada, B., C. W. Smith, and R. Lücking. 2021. A taxonomic reassessment of the genus *Sticta* (lichenized Ascomycota: *Peltigeraceae*) in the Hawaiian archipelago. *Lichenologist* 53: 117–133.

Morris, C., and D. Stone. 2022. *Nephroma orvoi* in western North America. *Evansia* 39: 82–86.

Muggia, L., P. Nelson, T. Wheeler, L. S. Yakovchenko, T. Tønsberg, and T. Spribille. 2011. Convergent evolution of a symbiotic duet: The case of the lichen genus *Polychidium* (Peltigerales, Ascomycota). *American Journal of Botany* 98: 1647–1656.

Muir, P. S. 1991. Fog water chemistry in a wood-burning community, western Oregon. *Journal of the Air and Waste Management Association* 41: 32–38.

Myllys, L., P. Halonen, and S. Velmala. 2006. Notes on some rare species of *Bryoria* from Finland. *Graphis Scripta* 18: 23–26.

Myllys, L., S. Velmala, H. Holien, P. Halonen, L.-S. Wang, and T. Goward. 2011. Phylogeny of the genus *Bryoria*. *Lichenologist* 43: 617–638.

Myllys, L., S. Velmala, H. Lindgren, D. Glavich, T. Carlberg, L. S. Wang and T. Goward. 2014. Taxonomic delimitation of the genera *Bryoria* and *Sulcaria*, with a new combination *Sulcaria spiralifera* introduced. *Lichenologist* 46: 737–752.

Myllys, L., S. Velmala, R. Pino-Bodas, and T. Goward. 2016a. New species in *Bryoria* (Parmeliaceae, Lecanoromycetes) from north-west North America. *Lichenologist* 48: 355–366.

Myllys, L., H. Lindgren, S. Aikio, L. Häkkinen, and F. Högnabba. 2016b. Chemical diversity and ecology of the genus *Bryoria* section *Implexae* (Parmeliaceae) in Finland. *Bryologist* 119: 29–38.

Nash, T. H., III. 2002. Lichens as bioindicators of sulfur dioxide. *Symbiosis* 33:1–21.

Nash, T. H., III. 2008. *Lichen Biology*. 2nd edition. Cambridge University Press, Great Britain.

Nash, T. H., III, B. D. Ryan, C. Gries, and F. Bungartz. 2002. *Lichen Flora of the Greater Sonoran Desert Region. Vol. I*. Lichens Unlimited, Arizona State University, Tempe.

Nash, T. H., III, B. D. Ryan, P. Diederich, C. Gries, and F. Bungartz. 2004. *Lichen Flora of the Greater Sonoran Desert Region. Vol. II*. Lichens Unlimited, Arizona State University, Tempe.

Nash, T. H., III, C. Gries, and F. Bungartz. 2008. *Lichen Flora of the Greater Sonoran Desert Region. Vol. III*. Lichens Unlimited, Arizona State University, Tempe.

Neitlich, P., and B. McCune. 1997. Hotspots of epiphytic lichen diversity in two young managed stands. *Conservation Biology* 11: 172–182.

Nelsen, M. P., and A. Gargas. 2009. Assessing clonality and chemotype monophyly in *Thamnolia* (Pertusariales: Icmadophilaceae). *Bryologist* 112: 42–53.

Nelsen, M. P., and 7 authors. 2011. The cetrarioid core group revisited (Lecanorales: Parmeliaceae). *Lichenologist* 43: 537–551.

Nelson, P., J. Walton, and C. Roland. 2009. *Erioderma pedicellatum* (Hue) P. M. Jørg, new to the United States and western North America, discovered in Denali National Park and Preserve and Denali State Park, Alaska. *Evansia* 26: 19–21.

Nelson, P., J. Walton, H. Root, and T. Spribille. 2011. *Hypogymnia pulverata* (Parmeliaceae) and *Collema leptaleum* (Collemataceae), two macrolichens new to Alaska. *North American Fungi* 6(7): 1–8.

Nitare, J., and M. Norén. 1992. Woodland key-habitats of rare and endangered species will be mapped in a new project of the Swedish National Board of Forestry. *Svensk Botanisk Tidskrift* 86: 219–226.

Noble, W. J. 1982. *The Lichens of the Coastal Douglas-Fir Dry Subzone*. Part II from PhD thesis. University of British Columbia, Vancouver, B.C. Reprinted in 2017 with nomenclatural updates. *Monographs in North American Lichenology* 3: 1–260.

Ohlsson, K. 1973a. New and interesting macrolichens of British Columbia. *Bryologist* 76: 366–387.

Ohlsson, K. 1973b. *A Revision of the Lichen Genus Sphaerophorus*. PhD thesis. University Microfilms, Ann Arbor, Michigan.

Onut-Brännström, I., L. Tibell, and H. Johannesson. 2017. A worldwide phyloge-
ography of the whiteworm lichens *Thamnolia* reveals three lineages with distinct
habitats and evolutionary histories. *Ecology and Evolution* 7: 3602–3615.

Onut-Brännström, I., H. Johannesson, and L. Tibell. 2018. *Thamnolia tundrae* sp.
nov., a cryptic species and putative glacial relict. *Lichenologist* 50: 59–76.

Otálora, M. A. G., I. Martinez, M. Carmen Molina, G. Aragon, and F. Lutzoni.
2008. Phylogenetic relationships and taxonomy of the *Leptogium lichenoides*
group (Collemataceae, Ascomycota) in Europe. *Taxon* 57: 907–921.

Otálora, M. A. G., P. M. Jørgensen, and M. Wedin. 2014. A revised generic classification of the jelly lichens, Collemataceae. *Fungal Diversity* 64: 275–293.

Ott, S. 1987. Reproductive strategies in lichens. *Bibliotheca Lichenologica* 25:
81–93.

Otto, G. F. 1983. *Tholurna dissimilis* well established in western North America.
Bryologist 86: 263–265.

Pardo, L. H., and 10 authors. 2011. Effects of nitrogen deposition and empirical nitrogen critical loads for ecoregions of the United States. *Ecological Applications*
21: 3049–3082.

Pearson, L. D., and D. B. Lawrence. 1965. Lichens as microclimate indicators in
northwestern Minnesota. *American Midland Naturalist* 74: 257–268.

Perkins, D. F. 1992. Relationship between fluoride contents and loss of lichens
near an aluminum works. *Water, Air, and Soil Pollution* 64: 503–510.

Peterson, E. B., and B. McCune. 2003. The importance of hotspots for lichen diversity in forests of western Oregon. *Bryologist* 106: 246–256.

Peterson, E. B., D. M. Greene, B. McCune, E. T. Peterson, M. A. Hutten, P.
Weisberg, and R. Rosentreter. 1998. *Sulcaria badia*, a rare lichen in western
North America. *Bryologist* 101: 112–115.

Pike, L. H. 1978. The importance of epiphytic lichens in mineral cycling. *Bryologist*
81: 247–257.

Pike, L. H., and M. E. Hale Jr. 1982. Three new species of *Hypogymnia* from western North America (Lichenes: Hypogymniaceae). *Mycotaxon* 16: 157–161.

Pino-Bodas, R., A. R. Burgaz, M. P. Martin, and H. T. Lumbsch. 2011. Phenotypical
plasticity and homoplasy complicate species delimitation in the *Cladonia gracilis* group (Cladoniaceae, Ascomycota). *Organisms Diversity and Evolution* 11:
343–355.

Pino-Bodas, R., A. R. Burgaz, M. P. Martin, and H. T. Lumbsch. 2012. Species
delimitation in the *Cladonia cariosa* group (Cladoniaceae, Ascomycota).
Lichenologist 44: 121–135.

Pino-Bodas, R., T. Ahti, S. Stenroos, M. P. Martin, and A. R. Burgaz. 2013.
Multilocus approach to species recognition in the *Cladonia humilis* complex
(Cladoniaceae, Ascomycota). *American Journal of Botany* 100: 664–678.

Pino-Bodas, R., A. R. Burgaz, M. P. Martín, T. Ahti, S. Stenroos, M. Wedin, and
H. T. Lumbsch. 2015. The phenotypic features used for distinguishing species
within the *Cladonia furcata* complex are highly homoplasious. *Lichenologist* 47:
287–303.

Poelt, J. 1969. *Bestimmungsschlüssel europäischer Flechten.* J. Cramer, Lehre.

Poelt, J., and T. H. Nash, III. 1993. Studies in the *Umbilicaria vellea* group (Umbilicariaceae) in North America. *Bryologist* 96: 422–430.

Printzen, C., S. Domaschke, F. Fernández-Mendoza, and S. Pérez-Ortega. 2013. Biogeography and ecology of *Cetraria aculeata*, a widely distributed lichen with a bipolar distribution. *MycoKeys* 6: 33–53.

Purvis, W. 2000. *Lichens*. Smithsonian Institution Press, Washington DC, and The Natural History Museum, London.

Pypker, T. G., B. J. Bond, T. E. Link, D. Marks, and M. H. Unsworth. 2005. The importance of canopy structure in controlling the interception loss: Examples from a young and old-growth Douglas-fir forests. *Agricultural and Forest Meteorology* 130: 113–129.

Pypker, T. G., M. H. Unsworth, and B. J. Bond. 2006a. The role of epiphytes in rainfall interception by forests in the Pacific Northwest. I. Laboratory measurements of water storage. *Canadian Journal of Forest Research* 36: 808–818.

Pypker, T. G., M. H. Unsworth, and B. J. Bond. 2006b. The role of epiphytes in rainfall interception by forests in the Pacific Northwest. II. Field measurements at the branch and canopy scale. *Canadian Journal of Forest Research* 36: 819–832.

Pyrczak-Felczykowska, A., and 10 authors. 2019. Synthesis of usnic acid derivatives and evaluation of their antiproliferative activity against cancer cells. *Journal of Natural Products* 82: 1768–1778.

Ra, H.-S.Y., L. Rubin, and R. F. E. Crang. 2004. Structural impacts on thallus and algal cell components of two lichen species in response to low-level air pollution in Pacific Northwest forests. *Microscopy and Microanalysis* 10: 270–279.

Ra, H. S. Y., L. H. Geiser, and R. F. E. Crang. 2005. Effects of season and low-level air pollution on physiology and element content of lichens from the U.S. Pacific Northwest. *Science of the Total Environment* 343: 155–167.

Ranković, B. 2019. *Lichen Secondary Metabolites*. Springer Nature Switzerland AG.

Richardson, D. H. S., and C. M. Young. 1977. Lichens and vertebrates. Pp. 121–144 in: M. R. D. Seaward (ed.), *Lichen Ecology*. Academic Press, London.

Riddell, J., T. H. Nash, III, and P. Padgett. 2008. The effect of HNO_3 gas on the lichen *Ramalina menziesii*. *Flora* 203: 47–54.

Riley, J., B. McCune, and P. Neitlich. 1995. Range extensions of *Usnea sphacelata* from Oregon and Washington. *Evansia* 12: 24–26.

Roblin, B., and J. Aherne. 2020. Moss as a biomonitor for the atmospheric deposition of anthropogenic microfibres. *Science of the Total Environment* 715: 136973.

Rochelle, J. A. 1980. *Mature Forests, Litterfall and Patterns of Forage Quality as Factors in the Nutrition of Black-tailed Deer on Northern Vancouver Island*. PhD thesis. Univ. of British Columbia.

Rodriguez, C. M., J. P. Bennett, and C. J. Johnson. 2012. Lichens, unexpected antiprion agents? *Prion* 6: 1: 11–16.

Root, H. T., and 9 authors. 2013. A simple tool for estimating throughfall nitrogen deposition in forests of western North America using lichens. *Forest Ecology and Management* 306: 1–8.

Rose, F. 1976. Lichenological indicators of age and environmental continuity in woodlands. Pp. 297–307 in: D. H. Brown, D. I.. Hawksworth, and R. H. Bailey (eds.), *Lichen Ecology: Progress and Problems*. Academic Press, London.

Rosentreter, R. 1993. Vagrant lichens in North America. *Bryologist* 96: 333–338.

Rosentreter, R., and B. McCune. 1992. Vagrant *Dermatocarpon* in western North America. *Bryologist* 95: 15–19.

Rosentreter, R., and B. McCune. 1996. Distribution and ecology of *Teloschistes contortuplicatus* in North America. *Evansia* 13: 10–13.

Rundel, P. W., and P. A. Bowler. 1976. *Ramalina leptocarpha* and *Ramalina subleptocarpha*: A fertile-sorediate species pair. *Bryologist* 79: 364–369.

Rundel, P. W., and P. A. Bowler. 1978. *Niebla*, a new generic name for the genus *Desmazieria* (Ramalinaceae). *Mycotaxon* 6: 497–499.

Saag, L., K. Mark, A. Saag, and T. Randland. 2014. Species delimitation in the lichenized fungal genus *Vulpicida* (Parmeliaceae, Ascomycota) using gene concatenation and coalescent-based species tree approaches. *American Journal of Botany* 101: 2169–2182.

Sanders, W. B., and H. Masumoto. 2021. Lichen algae: The photosynthetic partners in lichen symbioses. *Lichenologist* 53: 347–393.

Santesson, R. 1943. The South American Menegazziae. *Arkiv für Botanik* 30A: 1–35.

Sato, M. M. 1962. Mixture ratio of the lichen genus *Thamnolia*. *Nova Hedwigia* 5: 149–155.

Schultz, M. 2004. *Lempholemma*. Pp. 320–322 in: Nash, T. H., III, B. D. Ryan, P. Diederich, C. Gries, and F. Bungartz (eds.), *Lichen Flora of the Greater Sonoran Desert Region. Vol. II*. Lichens Unlimited, Arizona State University, Tempe.

Schultz, M., and B. Büdel. 2002. Key to the genera of the Lichinaceae. *Lichenologist* 34: 39–62.

Sheard, J. W. 1977. Paleogeography, chemistry and taxonomy of the lichenized ascomycetes *Dimelaena* and *Thamnolia*. *Bryologist* 80: 100–118.

Sierk, H. A. 1964. The genus *Leptogium* in North America north of Mexico. *Bryologist* 67: 245–317.

Sillett, S. C., B. McCune, J. E. Peck, T. R. Rambo, and A. Ruchty. 2000. Dispersal limitations of epiphytic lichens result in species dependent on old-growth forests. *Ecological Applications* 10: 789–799.

Simon, A., T. Goward, J. Di Meglio, K. Dillman, T. Spribille, and B. Goffinet. 2018. *Sticta torii* sp. nov., a remarkable lichen of high conservation priority from northwestern North America. *Graphis Scripta* 30: 105–114.

Singh, B. N., D. K. Upreti, V. K. Gupta, X. F. Dai, and Y. Jiang. 2017. Endolichenic fungi: A hidden reservoir of next generation biopharmaceuticals. *Trends in Biotechnology* 35: 808–813.

Skult, H. 1987. The *Parmelia omphalodes* complex in the northern hemisphere. Chemical and morphological aspects. *Annales Botanici Fennici* 24: 371–383.

Smith, C. W., A. Aptroot, B. J. Coppins, A. Fletcher, O. L. Gilbert, P. W. James, and P. A. Wolseley. 2009. *The Lichens of Great Britain and Ireland*. British Lichen Society, London.

Smith, R. J., and 22 authors. 2012. Rare inland reindeer lichens at Mima Mounds in southwest Washington State. *North American Fungi* 7(3): 1–25.

Smith, R. J., S. Jovan, and B. McCune. 2017. *Lichen communities as climate indicators in the U.S. Pacific States.* Gen. Tech. Rep. PNW-GTR-952. USDA Forest Service, Pacific Northwest Research Station, Portland, Oregon.

Smith, R. J., S. Jovan, and B. McCune. 2020. Climatic niche limits and community-level vulnerability of obligate symbioses. *Journal of Biogeography* 47: 382–395.

Søchting, U., I. Kärnefelt, and S. Kondratyuk. 2002. Revision of *Xanthomendoza* (Teloschistaceae, Lecanorales) based on morphology, anatomy, secondary metabolites and molecular data. *Mitteilungen aus dem Institut für allgemeine Botanik in Hamburg* 30-32: 225–240.

Spies, T. A., and 10 authors. 2019. Twenty-five years of the Northwest Forest Plan: What have we learned? *Frontiers in Ecology and the Environment* 17: 511–520.

Spribille, T., and L. Muggia. 2013. Expanded taxon sampling disentangles evolutionary relationships and reveals a new family in Peltigerales (Lecanoromycetidae, Ascomycota). *Fungal Diversity* 58: 171–184.

Spribille, T., and 7 authors. 2009. Contributions to an epiphytic lichen flora of northwest North America: I. Eight new species from British Columbia inland rain forests. *Bryologist* 112: 109–137.

Spribille T., S. Pérez-Ortega, T. Tønsberg, and D. Schirokauer. 2010. Lichens and lichenicolous fungi of the Klondike Gold Rush National Historic Park, Alaska, in a global biodiversity context. *Bryologist* 113: 439–515.

Spribille, T., et al. 2016. Basidiomycete yeasts in the cortex of ascomycete macrolichens. *Science* 353: 488–492.

Spribille T., and 12 authors. 2020. Lichens and associated fungi from Glacier Bay National Park, Alaska. *Lichenologist* 52: 61–181.

Spribille, T., P. Resl, D. E. Stanton, and G. Tagirdzhanova. 2022. Evolutionary biology of lichen symbioses. *New Phytologist* 234: 1566–1582.

Stenroos, S. 1989a. Taxonomy of the *Cladonia coccifera* group. 1. *Annales Botanici Fennici* 26: 157–168.

Stenroos, S. 1989b. Taxonomy of the *Cladonia coccifera* group. 2. *Annales Botanici Fennici* 26: 307–317.

Stenroos, S. 1990. *Cladonia luteoalba*—an enigmatic *Cladonia*. *Karstenia* 30: 27–32.

Stenroos, S., J. Hyvönen, L. Myllys, A. Thell, and T. Ahti. 2002. Phylogeny of the genus *Cladonia* s.lat. (Cladoniaceae, Ascomycetes) inferred from molecular, morphological, and chemical data. *Cladistics* 18: 237–278.

Stenroos, S., R. Pino-Bodas, J. Hyvönen, H. T. Lumbsch, and T. Ahti. 2019. Phylogeny of the family Cladoniaceae based on sequences of multiple loci. *Cladistics* 35: 351–384.

Stevenson, S. K., and J. A. Rochelle. 1984. Lichen litterfall—its availability and utilization by black-tailed deer. Pp. 391–396 in: W. R. Meehan et al. (eds.), *Proceedings: Symposium on Fish and Wildlife Relationships in Old-Growth Forests.*

Stocker-Wörgötter, E., L. M. C. Cordeiro, M. Iacomini. 2013. Accumulation of potential pharmaceutically relevant lichen metabolites in lichens and cultured lichen symbionts. Pp. 337–380 in: Atta-ur-Rahman (ed.), *Studies in Natural Products Chemistry. Vol. 39.* Elsevier.

Stone, D. F. 2018. *Usnea* in the Pacific Northwest, Aide Mémoire. Northwest Lichenologists, Corvallis, Oregon, U.S.A.

Stone, D., and A. Ruchty. 2008. *Leptogium siskiyouensis,* a new epiphytic lichen species from the Pacific Northwest of the United States. *North American Fungi* 3(2): 1–7.

Stone, D. F., and B. McCune. 2010. *Collema quadrifidum,* a new epiphytic lichen species from the Pacific Northwest of the United States. *North American Fungi* 5(2): 1–6.

Stone, D. F., and B. McCune. 2022. Two new hairy *Leptogium* (Collemataceae) species from western North America. *Bryologist* 125: 157–169.

Stone, D. F., J. W. Hinds, F. L. Anderson, and J. C. Lendemer. 2016. A revision of the *Leptogium saturninum* group in North America. *Lichenologist* 48: 387–422.

Stone, D., M. Gordon, and B. McCune. 2020. *Pseudocyphellaria holarctica* (Lobariaceae) specimens from Oregon are referable to *P. hawaiiensis. Bryologist* 123: 260–264.

Storeheier P. V., S. Mathiesen, N. Tyler, and M. Olsen. 2002. Nutritive value of terricolous lichens for reindeer in winter. *Lichenologist* 34: 247–257.

Tavares, I. I. 1997. A preliminary key to *Usnea* in California. *Bulletin of the California Lichen Society* 4: 19–23.

Taylor, R. J., and M. A. Bell. 1983. Effects of SO_2 on the lichen flora in an industrial area: Northwest Whatcom County, Washington. *Northwest Science* 57: 157–166.

Tehler, A. 1996. Systematics, phylogeny and classification. Pp. 217–239 in: T .H. Nash III (ed.), *Lichen Biology.* Cambridge University Press.

Thell, A. 1995. A new position of the *Cetraria commixta* group in *Melanelia* (Ascomycotina, Parmeliaceae). *Nova Hedwigia* 60: 407–422.

Thell, A., and T. Goward. 1996. The new cetrarioid genus *Kaernefeltia* and related groups in the Parmeliaceae (lichenized Ascomycotina). *Bryologist* 99: 125–136.

Thell, A., T. Goward, T. Randlane, E. I. Kärnefelt, and A. Saag. 1995. A revision of the North American lichen genus *Ahtiana* (Parmeliaceae). *Bryologist* 98: 596–605.

Thell, A., S. Stenroos, T. Feuerer, I. Kärnefelt, L. Myllys, and J. Hyvönen. 2002. Phylogeny of cetrarioid lichens (Parmeliaceae) inferred from ITS and β-tubulin sequences, morphology, anatomy and secondary chemistry. *Mycological Progress* 1: 335–354.

Thell, A., and 10 authors. 2009. Phylogeny of the cetrarioid core (Parmeliaceae) based on five genetic markers. *Lichenologist* 41: 489–512.

Thell, A., I. Kärnefelt, and M. R. D. Seaward. 2018. Splitting or synonymizing – genus concept and taxonomy exemplified by the Parmeliaceae in the Nordic region. *Graphis Scripta* 30: 130–137.

Thomas, A., and R. Rosentreter. 1992. Utilization of lichens by pronghorn antelope in three valleys in east-central Idaho. *Idaho Bureau of Land Management Technical Bulletin* No. 92-93. 13 pp.

Thomson, J. W. 1950. The species of *Peltigera* of North America north of Mexico. *American Midland Naturalist* 44: 1–68.

Thomson, J. W. 1967a. The lichen genus *Baeomyces* in North America north of Mexico. *Bryologist* 70: 285–298.

Thomson, J. W. 1967b. *The lichen genus* Cladonia *in North America.* Univ. of Toronto Press, Toronto.

Thomson, J. W. 1969. *Letharia californica* is *Letharia columbiana* (Lichenes). *Taxon* 18: 535–537.

Thomson, J. W. 1984. *American Arctic Lichens I. The Macrolichens.* Columbia Univ. Press, New York.

Thomson, J. W. 1997. *American Arctic Lichens II. Crustose Lichens.* University of Wisconsin Press, Madison.

Thomson, J. W., and C. D. Bird. 1978. The lichen genus *Dactylina* in North America. *Canadian Journal of Botany* 56: 1602–1624.

Tibell, L. 1975. The Caliciales of boreal North America. *Symbolae Botanicae Upsalienses* 21: 1–128.

Tibell, L. 1992. Crustose lichens as indicators of forest continuity in boreal coniferous forests. *Nordic Journal of Botany* 12: 427–450.

Tibell, L. 1996. Caliciales. *Flora Neotropica* 69, New York Botanical Garden, New York.

Timdal, E., and T. Tønsberg. 2012. *Cladonia straminea*, the correct name for *C. metacorallifera. Graphis Scripta* 24: 33–35.

Timdal, E., and 7 authors. 2020. Integrative taxonomy reveals a new species, *Nephroma orvoi*, in the *N. parile* species complex (lichenized Ascomycota). *Graphis Scripta* 32: 70–85.

Timdal, E., T. H. Hofton, M. Westberg, and M. Bendiksby. 2021. The *Nephroma helveticum* complex (Peltigerales, lichenized Ascomycota) in the Nordic countries. *Graphis Scripta* 33: 86–110.

Tønsberg, T. 1999. *Pseudocyphellaria arvidssonii* new to Africa, and *P. mallota* new to North America. *Bryologist* 102: 128.

Tønsberg, T., and T. Ahti. 1980. *Cladonia umbricola*, a new lichen species from NW Europe and western North America. *Norwegian Journal of Botany* 27: 307–309.

Tønsberg, T., and T. Goward. 2001. *Sticta oroborealis* sp. nov. and other Pacific North American lichens forming dendriscocauloid cyanotypes. *Bryologist* 104: 12–23.

Truong, C., J. M. Rodriguez, and P. Clerc. 2013. Pendulous *Usnea* species (Parmeliaceae, lichenized Ascomycota) in tropical South America and the Galapagos. *Lichenologist* 45: 505–543.

Ure, D. C., and C. Maser. 1982. Mycophagy of red-backed voles in Oregon and Washington. *Canadian Journal of Zoology* 60: 3307–3315.

U.S. EPA (United States Environmental Protection Agency). 2020. 2017 National Emissions Inventory (NEI) Data. https://www.epa.gov/air-emissions-inventories/2017-national-emissions-inventory-nei-data.

U.S. EPA (United States Environmental Protection Agency). 2019. CMAQ Model Version 5.0.2 Output Data – 2010 CONUS_12km. UNC Dataverse, V1. https://doi.org/10.15139/S3/YIPTDY.

Velmala, S., L. Myllys, P. Halonen, T. Goward, and T. Ahti. 2009. Molecular data show that *Bryoria fremontii* and *B. tortuosa* (Parmeliaceae) are conspecific. *Lichenologist* 41: 231–242.

Velmala, S., L. Myllys, T. Goward, H. Holien, and P. Halonen. 2014. Taxonomy of *Bryoria* section Implexae (Parmeliaceae, Lecanoromycetes) in North America and Europe, based on chemical, morphological and molecular data. *Annales Botanici Fennici* 51: 345–371.

Vitikainen, O. 1985. Three new species of *Peltigera* (lichenized Ascomycetes). *Annales Botanici Fennici* 22: 291–298.

Vondrak, J., Z. Palice, J. Mareš, and J. Kocourková. 2013. Two superficially similar lichen crusts, *Gregorella humida* and *Moelleropsis nebulosa*, and a description of the new lichenicolous fungus *Llimoniella gregorellae*. *Herzogia* 26: 31–48.

Wang, T., A. Hamann, D. Spittlehouse, and C. Carroll. 2016. Locally downscaled and spatially customized climate data for historical and future periods for North America. PLoS ONE 11(6): e0156720. https://doi.org/10.1371/journal.pone.0156720.

Ward, R. L., and C. L. Marcum. 2005. Lichen litterfall consumption by wintering deer and elk in western Montana. *Journal of Wildlife Management* 69: 1081–1089.

Weber, W. A. 1965. The lichen flora of Colorado: II. Pannariaceae. *University of Colorado Studies, Series in Biology* 16: 1–10.

Wedin, M. 1993a. A phylogenetic analysis of the lichen family Sphaerophoraceae (Caliciales); a new generic classification and notes on character evolution. *Plant Systematics and Evolution* 187: 213–241.

Wedin, M. 1993b. The lichen family Sphaerophoraceae (Caliciales, Ascomycotina) in temperate areas of the southern hemisphere. *Symbolae Botanicae Upsaliensis* 31: 1–102.

Wedin, M. 1995. *Bunodophoron melanocarpum*, comb. nov. (Sphaerophoraceae, Caliciales s. lat.). *Mycotaxon* 55: 383–384.

Wedin, M., F. Högnabba, and T. Goward. 2009. A new species of *Sphaerophorus*, and a key to the family Sphaerophoraceae in western North America. *Bryologist* 112: 368–374.

Westberg, M., and U. Arup. 2011. *Candelaria pacifica* sp. nova (Ascomycota, Candelariales) and the identity of *Candelaria vulgaris*. *Bibliotheca Lichenologica* 106: 353–364.

Westberg, M., and T. H. Nash, III. 2002. *Candelaria*. Pp. 116–117 in: Nash, T. H., III, B. D. Ryan, C. Gries, and F. Bungartz (eds.), *Lichen Flora of the Greater Sonoran Desert Region. Vol. I*. Lichens Unlimited, Arizona State University, Tempe.

Wetmore, C. M. 1960. The lichen genus *Nephroma* in North and Middle America. *Publications of the Museum, Michigan State University, Biological Series* 1: 369–452.

Wetmore, C. M. 1970. The lichen family Heppiaceae in North America. *Annals of the Missouri Botanical Garden* 57: 158–209.

Wetmore, C. M. 1980. A new species of *Nephroma* from North America. *Bryologist* 83: 243–247.

White, F. J., and P. W. James. 1985. A new guide to microchemical techniques for the identification of lichen substances. *British Lichen Society Bulletin* 57 (supplement): 1–41.

White, F. J., and P. W. James. 1988. Studies on the genus *Nephroma*. II. The southern temperate species. *Lichenologist* 20: 103–166.

Will-Wolf, Susan. 2010. Analyzing lichen indicator data in the Forest Inventory and Analysis Program. U.S. Department of Agriculture, Forest Service, General Technical Report PNW-GTR-818.

Wirth, V., M. Hauck, and M. Schultz. 2013. *Die Flechten Deutschlands*. Ulmer, Stuttgart.

Wirtz, N., C. Printzen, L. G. Sancho, and H. T. Lumbsch. 2006. The phylogeny and classification of *Neuropogon* and *Usnea* (Parmeliaceae, Ascomycota) revisited. *Taxon* 55: 367–376.

Wirtz, N., C. Printzen, and H. T. Lumbsch. 2007. The delimitation of Antarctic and bipolar species of neuropogonoid *Usnea* (Ascomycota, Lecanorales): A cohesion approach of species recognition for the *Usnea perpusilla* complex. *Mycological Research* 112: 472–484.

Wright, D. 2001. Some species of the genus *Usnea* (lichenized Ascomycetes) in California. *Bulletin of the California Lichen Society* 8: 1–22.

Zhao, Y., M. Wang, and B. Xu. 2021. A comprehensive review on secondary metabolites and health-promoting effects of edible lichen. *Journal of Functional Foods* 80: 104283.

Illustrated Glossary

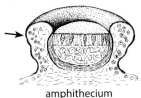
amphithecium

accessory: Applied to lichen substances that may or may not be present in a particular species, in addition to the usual substances detected by TLC.

amphitheci(um)(a) (= thalline margin): The rim surrounding the disk of an apothecium, usually containing algae.

anisotomic: Applied to dichotomous branching with unequal-sized branches.

annular cracks: Applied to circular cracks in the cortex of *Usnea*, each crack forming a ring around the branch.

apiculate: Abruptly narrowed to a short point.

apotheci(um)(a): A type of ascocarp; usually a disk- or cup-shaped structure, lined at maturity with an exposed spore-producing surface.

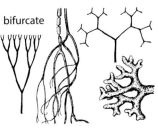
apothecia

appressed: Pressed closely against a surface.

arachnoid: Cobwebby.

areolate: Divided by cracks into small areas or **areoles**, often forming a mosaic.

areole: See areolate.

asci: See ascus.

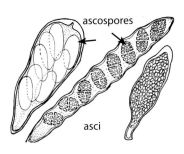
ascospores

asci

ascocarp (= ascoma): The sexual reproductive body of an ascomycete, usually consisting of covering or supporting layers and the spore-producing layer.

ascomycete: A fungus in the Class Ascomycetes (or Subdivision Ascomycotina) producing spores in sacs (the asci).

ascospore: A spore produced in an ascus.

asc(us)(i): A sac-like structure, usually opening at the tip, in which sexually produced spores are produced.

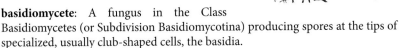
bifurcate

basidiomycete: A fungus in the Class Basidiomycetes (or Subdivision Basidiomycotina) producing spores at the tips of specialized, usually club-shaped cells, the basidia.

bifurcate: Dichotomously branched or forked.

blastidia: Asexual propagules containing both the photobiont and fungus, rather like soredia, but relatively coarse, partly corticate, and budding like yeast.

botryose: Like clusters of grapes.

calcareous: Applied to substrates rich in calcium carbonate (lime); usually referring to rock, especially limestone (fizzes in 1 M HCl) and dolomite (fizzes in 6 M HCl).

capitate: In the form of an expanded head- or cap-like structure, at the tip of a stalk or nearly sessile; usually applied to apothecia and soralia.

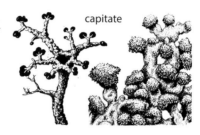

capitate

cephalodia: Cyanobacteria-containing structures in lichens which otherwise contain green algae; these structures usually appear as irregular warts or inclusions, or occasionally as squamules or minutely fruticose structures.

channeled: Applied to elongate lobes that are shallowly U-shaped in cross section.

cilia: Hair-like projections; usually applied to thread-like multicellular marginal hairs on a thallus or apothecium.

clavate: Club-shaped.

concolorous: Having the same color as another part.

confluent: Running together or merging.

coralloid: Coral-like, having dense, repeated branching of fine segments.

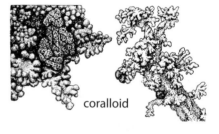

coralloid

cortex: The outer "skin" of lichens, generally ± smooth, often glossy, composed of closely packed fungal hyphae.

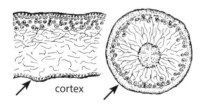

cortex

corticate: Having a cortex.

corticolous: Growing on bark, or more loosely applied to growing on bark or wood.

crenate: With a scalloped or round-toothed edge.

crustose: Forming a crust-like growth form that is closely applied to the substrate, like paint, generally attached by all of the lower surface and lacking a lower cortex and rhizines.

cyanobacteria: Prokaryotic photosynthetic organisms formerly known as "blue-green algae," generally having a blue-green tint and lacking chloroplasts. In the text and keys, we refer to photobionts as "blue green," meaning cyanobacteria.

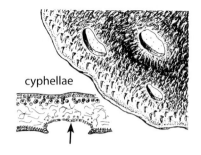

cyphellae

cyphellae: Crater-like pores, usually in the lower surface of lichens, that open into the medulla and are lined with differentiated cells; characteristic of the genus *Sticta*.

decorticate: Having lost a cortex through intrinsic developmental processes.

decumbent: Prostrate, but not closely appressed to a surface, often with the ends upturned.

dichotomously branched: Y-shaped branching.

disk: In surface view of an apothecium, the central part enclosed by, but not including, the exciple. Anatomically the disk is the upper surface of the hymenium in an apothecium.

dorsiventral: Having differentiated upper and lower surfaces.

ecorticate: Lacking a cortex throughout development.

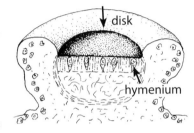

disk

hymenium

epinecral: A surface layer composed of dead cortical cells with indistinct lumina.

epispore: The outer spore wall.

epruinose: Lacking pruina.

exciple (= excipulum): The usually cup- or dish-shaped walls of an apothecium or flask-shaped walls of a perithecium that contain and support the hymenium and hypothecium. Viewed from above, the exciple is usually seen as a raised rim around the disk.

exciple

everted: Turned inside out.

fenestrate: Windowed, or with small openings or perforations.

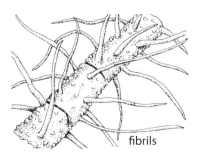

fibrils

fibril(s): A small fiber projecting from the thallus, roundish in cross section, usually produced at ± right angles to the direction of growth, usually applied in *Usnea*.

fibrillose: Bearing fibrils.

fimbriate: With an edge of narrow projections forming a fringe.

fissured: Having fissures or vertical cracks.

foliose: A lichen growth form with dorsiventral lobes, usually loosely to tightly appressed to the substrate, 2-dimensional or weakly 3-D, and usually with a cortex on upper and lower surfaces.

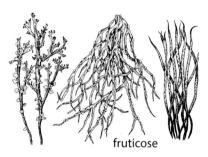

fruticose

foveolate: With small pits or shallow surface depressions.

fruticose: A 3-dimensional growth form of a lichen, not differentiated into upper and lower surfaces, and including pendulous and stringy, upright, or bushy thalli, or with a 2-part thallus of primary squamules and upright stalks.

glabrous: Smooth and not hairy.

hapters: Short, peg-like attachments of the lower surface of lichens to the substrate.

heteromerous: Applied to a thallus that is distinctly layered (stratified), usually with a white layer (the medulla), with the photobionts in a distinct green or blue-green band. Contrast with **homoiomerous**.

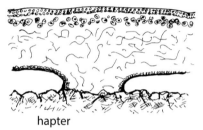

hapter

homoiomerous: Applied to a thallus where the fungus and photobiont are not separated into distinct bands (unstratified) but are evenly dispersed; most of these have a dark, gelatinous texture when moist.

hyaline: Transparent and colorless.

hymenium: The spore-bearing layer of fungal reproductive structures (see **disk**).

hypha(e): Fungal filaments.

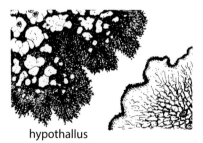

hypothallus

hypothallus (used here = **prothallus**): A thin, tightly appressed, differentiated fungal layer below the main portion of the thallus. A more restrictive use refers to a spongy tissue on the underside of certain lichens, while the prothallus is the initial hyphal mat from which a lichen develops and is often visible as an appressed fringe along the edges of the thallus or areoles.

hypothecium (= **medullary excipulum**): The fungal layer just below the hymenium.

intermontane: Of the region of valleys, plateaus, and hills occurring between the main masses of the Rocky Mountains and the Cascade Range.

isidali(um)(a): Discrete patches on the surface of a lichen that bear a cluster of isidia.

isidia: Asexual reproductive structures that are minute and finger-like or globular, branched or unbranched, covered with a cortex, and contain both the fungus and the photobiont.

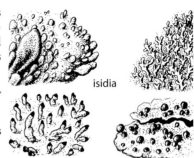

isidia

isodiametric: Having approximately equal diameters in all directions.

isotomic: Applied to dichotomous branching with equal-sized branches.

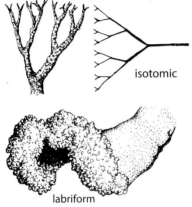

isotomic

labriform

labriform: Lip-shaped.

laminal: Occurring on the lobe surface, as opposed to the lobe margins.

lecanorine (= **thalline**): Applied to apothecial margins that are colored like the thallus and usually unlike the disk and usually contain the photobiont.

lecideine (= **proper**): Applied to apothecial margins that are colored like the disk and unlike the thallus and usually lack the photobiont.

lichenicolous: Growing on lichens.

lichenized: Applied to a fungus that has an intimate mutualistic association with a photosynthetic green alga or cyanobacteria.

lobate: Having lobes.

lobe: A flattened branch or projection. (Measure lobe width away from branch points, and away from lobe tips.)

lobulate: Bearing lobules.

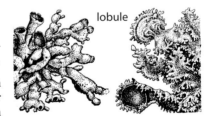

lobule

lobule: Tiny lobe-like asexual reproductive structures, usually dorsiventral.

maculate: Applied to a thallus with small, light-colored spots on the upper surface, often caused by differences in the thickness of the cortex or clumping of algae beneath the cortex.

margin: In reference to lichen apothecia, the upper edge of the exciple, surrounding the disk.

marginal: Situated on the margin of a lobe or apothecium or other structure.

marginate: Having a margin differentiated from the remaining structure.

mazaedium

mazaedium: A modified hymenium of ascomycetes in which the asci quickly disintegrate, producing a loose, powdery spore mass that is added to from below and sloughed off at the top.

medulla: The interior layer of most lichens, composed mainly of fungal hyphae, and therefore white (rarely orange or yellow).

melanized or **melanotic:** Turning brown or black, often applied to the basal portions of certain *Cladonia* species, where the exposed medulla darkens to brown or black, contrasting with paler patches of cortex.

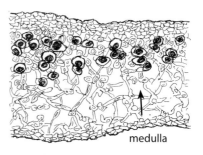

medulla

monophyllous: Applied to a thallus which is single-lobed, usually roundish in outline, the margin more or less entire or shallowly incised.

morph: A form without any implied taxonomic rank.

muriform: Applied to spores that have both internal crosswalls and longitudinal walls (i.e., dictyospores).

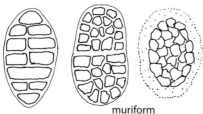

muriform

muscicolous: Growing on mosses.

mycobiont: The fungal partner of a lichen.

orbicular: Round.

ostiole: The pore or opening into a perithecium or pycnidium.

palmately branched: Having several to many branches or lobes radiating from a single point.

papilla(e): Minute, discrete, usually rounded bump(s).

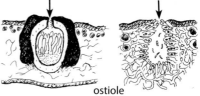

ostiole

paraphyses: Sterile filaments (simple or branched) in the hymenium and usually surrounding the asci.

paraplectenchyma: A tissue type with compactly massed hyphae having ± isodiametric lumina; essentially the fungal equivalent of parenchyma, but differing in development.

parenchyma: In plants a tissue composed of roundish to rectangular cells; parenchyma-like tissues in fungi are often called paraplectenchymatous.

peltate: Plate- or shield-shaped and generally raised above the surface by a central stalk, or at least with a narrow point of attachment.

peritheci(um)(a): A type of globose or flask-shaped ascocarp in which the hymenium is completely enclosed by protective sterile tissue, except for a small opening at the tip, the ostiole.

perithecium

photobiont: The photosynthetic partner in a lichen, consisting of either green algae or cyanobacteria or both.

photomorph: A form of a lichen that has one of two alternative photobionts. A blue-green photomorph has cyanobacteria as the primary photobiont, while a green photomorph has green algae as the primary photobiont. Because alternative photomorphs share the same fungus, these morphs cannot be separated taxonomically above the level of "form."

phyllocladia: photosynthetic units in the genus *Stereocaulon*, consisting of warty, branched lobate, coralloid, or squamulose protuberances from the nonphotosynthetic stalk.

placodioid

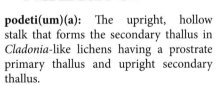

placodioid: Crustose but with a definite lobed margin, generally rosettiform.

plicate: Pleated or folded like a fan.

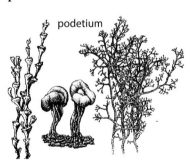

podetium

podeti(um)(a): The upright, hollow stalk that forms the secondary thallus in *Cladonia*-like lichens having a prostrate primary thallus and upright secondary thallus.

polarilocular: Applied to septate spores in which the two locules (cells) are connected by a channel through a thick septum.

polymorphic: An organism or part of an organism that takes more than one form.

polyphyllous: Applied to a thallus which is multiple-lobed, as opposed to monophyllous.

primary squamules: Squamules forming the primary thallus in lichens that differentiate into primary and secondary thalli.

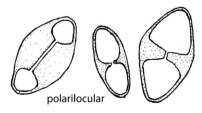

polarilocular

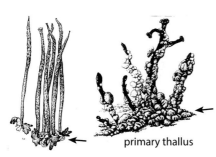

primary thallus

primary thallus: The horizontal, usually squamulose or crustose portion of a lichen that is differentiated into upright (secondary) and horizontal portions, as in *Cladonia* and *Pilophorus*.

proper margin (= proper exciple): The "excipulum proprium," a usually cup-shaped structure of exclusively fungal tissues that surrounds the hymenium in an apothecium.

prosoplectenchyma: A tissue type of compactly massed hyphae with elongate lumina.

pruina: Superficial chemical deposits, usually whitish and usually formed from calcium oxalates, causing a frost-covered or floury appearance.

pruinose: Having pruina.

pseudocyphellae: A broad term referring to any breaks in the cortex of lichens which are differentiated but lack specialized cells surrounding the opening; these may be round, irregular, angular, or minuscule pores.

pseudocyphellae

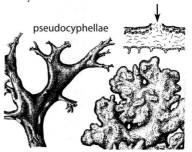

pseudopodeti(um)(a): An upright, generally solid stalk, similar in appearance to a podetium, but differing according to some authors in the fundamentals of development.

punctiform: Punctate; with small spots or hollows.

pustular: Usually applied to soredia that begin as a lump or small blister, then break down into (usually granular) soredia; also applied to blister-like swellings.

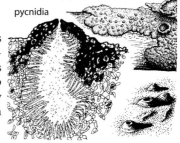

pustular

pycnidia

pycnidi(um)(a): A structure of the fungus that produces spores mitotically, usually imbedded in the thallus and visible externally as a black dot, occasionally in a projecting bump from the thallus (as in *Cetraria*); anatomically a generally flask-shaped structure lined with cells that produce pycnospores.

pycnospores (= **pycnoconidia, conidia,** or **spermatia**): Spores produced from pycnidia. When a sexual fertilization function is inferred, these are called "spermatia." When considered asexual propagules, these are called "conidia" or "pycnoconidia."

reticulate: Having a net-like pattern.

rhizine(s): Multicellular root-like structures, arising mostly from the lower surface.

rhizohyphae: Attachment structures more slender than rhizines, one cell thick in diameter, often occurring as part of a felty mass of hyphae.

rosettiform: Arranged in a rosette, a radially symmetric pattern of lobes.

saxicolous: Growing on rock.

scabrous (= **scabrid**): A surface having a roughened appearance.

rhizines

scale-like: Of small flakes or squamules, resembling fish scales.

scleroplectenchyma: A tissue type like paraplectenchyma with roundish but very thick-walled cells.

scurfy: Minutely scaly, having very small flakes.

secondary thallus: The upright portion of a lichen that is differentiated into upright and horizontal portions, as in *Cladonia* and *Pilophorus*.

septa

sessile: Attached to a surface without a stalk but not imbedded in the surface.

sept(um)(a): A partitioning cell wall.

siliceous: Applies to rocks in which the primary minerals are silicates rather than carbonates. In practice, applied to rocks that react HCl- (no fizzing) in 6 M HCl.

simple: Unbranched; of spores, lacking septa.

sorali(um) (a): Discrete areas on a lichen where soredia are produced.

soralia

soredi(um) (a): Asexual reproductive structures that are powdery to granular, not covered with a cortex, and contain both photobiont and fungus. Soredia are borne in discrete patches (soralia) or as a diffuse powder.

soredia

sorediose: Said of isidia that have an appearance of soredia, but they actually have at least a partially formed cortex and thus are intermediate between soredia and isidia.

spermati(um)(a): Spore-like gametes, the same structures referred to by various authors as "**conidia**" or "**pycnospores**," but implying a role in sexual recombination; (i.e., non-motile gametes.)

spinule: A spine-like asexual propagule or short, sharply pointed branch.

spinulose: Having small spines or spinules.

spot tests: Chemical tests for color reactions obtained by applying a liquid reagent to a lichen.

squamule(s): Small flakes or scales of a lichen, often rounded, ear-like, or lobed.

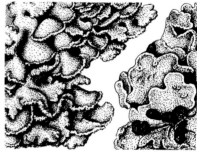

squamules

squamulose: Having squamules or applied to a growth form consisting of squamules.

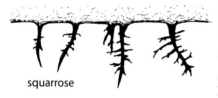

squarrose

squarrose: Branching by many short perpendicular branches from a single main axis.

stellate: Star-shaped.

steppe: Semiarid shrublands or grasslands, usually with cold winters, hot summers, and a continental climate.

stratified: Differentiated into layers; heteromerous.

subpendent: Somewhat pendent; intermediate between tufted and pendent.

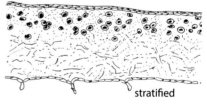

stratified

terete: Rounded in cross section; cylindrical to slightly tapering.

terricolous: Growing on the ground, including mineral soil, humus, litter, and organic sods.

thalline margin (= **thalline exciple** or **lecanorine margin**): An apothecial margin similar in color and structure to the thallus and usually containing the photobiont.

thalloconidia: Conidia arising from the surface of a thallus or prothallus; applied especially to the lower surface and rhizines of many *Umbilicaria* species that produce black, powdery conidia, giving a sooty black appearance under a hand lens or stereoscope.

thall(us) (i): The vegetative body of a lichen.

TLC: Thin-layer chromatography, used to identify secondary lichen substances.

tomentose: Having tomentum.

tomentum: Fine, short, superficial fuzz or minute hairs.

tuberculate: Applied to a surface having small, raised, wartlike projections (tubercles).

umbilicate: Applied to a thallus attached by a single holdfast, the umbilicus, as in *Dermatocarpon*, *Rhizoplaca*, and *Umbilicaria*.

umbo: A raised central bump; in *Umbilicaria* the raised portion above the umbilicus.

unstratified: Without separation of the photobiont and fungus into distinct layers; homoiomerous.

veins: Raised branching or anastomosing strands, usually on the lower surface of foliose lichens.

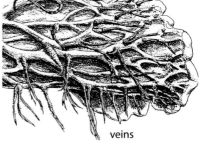

veins

vermiform: Worm-like.

verruca(e): Small bumps, warts, or projections.

verrucose: Having verrucae.

weft: A loose filamentous mat.

Acknowledgments

To bring a guide book of this scope to fruition requires the interest, support, contributions, and commitment of many people. The Bureau of Land Management, through the initiative of Jennie Sperling and Ron Exeter, supported publication of the second edition. Tom Booth, Mary Braun, Karyle Butcher, Jim Corbett, Ron Exeter, Patricia Hawk, and Kelli Van Norman facilitated the complicated process that led to the creation of contracts and partnership agreements between the photographers, illustrator, BLM, and OSU Press to produce this guide. For providing the personnel and resources to research and write the original edition and to revise sections of the second edition, we thank the U.S. Forest Service Air Program, especially Ken Snell and Jim Russell. For the first and second editions, Jo Alexander and Judy Radovsky did the technical editing and design with contributions from Amy Charron and David Drummond. Kim Hogeland and Micki Reaman provided editorial guidance and support for the third edition, and Micki Reaman created the layout. Technical editing by Jeanne Ponzetti greatly improved the readability, consistency, and accuracy of the third edition. Marty Brown orchestrated the cover design and Richard Droker graciously provided the cover photo.

About 45 biologists from the U.S. Forest Service Air and Forest Inventory and Analysis programs contributed many hours surveying and identifying lichens, contributing greatly to our understanding of the distribution and abundance of species and to the air scores described in the introduction. Especially important were Amanda Hardman, Jim Riley, Anne Ingersoll, Doug Glavich, Mark Boyll, Shanti Berryman, Elisa Di Meglio, Joe Di Meglio, Jenifer Ferriel, Daphne Stone, Riban Ulrich, Alexander Mikulin, Larissa Lasselle, Jill McMurray, Sarah Uebel, Elyse Woda, Maysa Miller, Linda Hasselbach, and Abbey Rosso. We thank Marie Antoine, Katie Glew, Rick Graw, Dave Kofranek, James Lendemer, Eric B. Peterson, Jeanne Ponzetti, Jim Riley, Abbey Rosso, Stephen Sillett, Daphne Stone, the many lichenology students at OSU, and botanists and ecologists of the U.S. Forest Service and the U.S. Bureau of Land Management who provided corrections and comments. We thank Jeremy Adams for distribution maps for range assessments. Robert J. Smith contributed numerous skillful analyses to the sections on air quality and climate change. Stuart Crawford provided helpful comments on ethnolichenology and graciously granted permission to reuse his map of *Bryoria* uses in the PNW.

We are also grateful to the lichenologists who assisted with specimens: Teuvo Ahti, Susanne Altermann, Andre Aptroot, Othmar Breuss, Irwin Brodo, Philippe Clerc, Evgeny Davydov, Jonathan Dey, Joe Di Meglio, Theodore Esslinger, Doug Glavich, Trevor Goward, Pekka Halonen, Starri Heidmarsson,

Per Magnus Jørgensen, Samuel Hammer, James Lendemer, Louise Lindblom, Jolanta Miadlikowska, Josef Poelt, Roger Rosentreter, Matthias Schultz, John Thomson, Tor Tønsberg, Orvo Vitikainen, and Mats Wedin. In addition, we appreciate the curators of herbaria at the Universities of Bergen, British Columbia, California at Berkeley, Colorado, Helsinki, Idaho, Uppsala, Washington, Western Washington, and Wisconsin, Arizona State University, Montana State University, Oregon State University, the New York Botanical Garden, the Smithsonian Institution, and the National Museum of Canada. McCune's lab helpers have provided a fountain of data that helped substantiate this work, with major contributions in recent years by Claudia Corvalán, Elisa Di Meglio, Joe Di Meglio, Eleonore Jacobson, Grace O'Brien, Bailey Rodgers, Zane Walker, and Claire Whittaker.

Finally, we recognize several excellent local photographers for generously donating photographs: Joe Di Meglio (*Ricasolia amplissima*), Richard Droker (*Alectoria imshaugii, Baeomyces rufus, Cladonia cariosa, Flavoparmelia caperata, Heterodermia sitchensis, Lichenomphalia, Lobaria pulmonaria, Pilophorus clavatus, Sphaerophorus fragilis*), Jim Riley (*Cetraria juniperina*), Martin Hutten (*Bunodophoron melanocarpum, Erioderma sorediatum, Pseudocyphellaria mallota, Umbilicaria havaasii*), Jason Hollinger (*Parmeliopsis ambigua* and *P. hyperopta*), and Peter Nelson (*Erioderma pedicellatum*). A full list of photographic credits is available at https://ir.library.oregonstate.edu/concern/defaults/gx41ms40q.

Index